DATE DUE

DEMCO 38-296

The Genus
Aeromonas

The Genus
Aeromonas

Edited by
B. AUSTIN
Heriot-Watt University, Edinburgh, Scotland

M. ALTWEGG
University of Zürich, Switzerland

P.J. GOSLING
Department of Health, London, England
and

S.W. JOSEPH
University of Maryland, College Park, USA

JOHN WILEY & SONS

Chichester · New York · Brisbane · Toronto · Singapore

Copyright © 1996 by John Wiley & Sons Ltd,

England

'77
779777

means,
guage

Other Wiley Editorial Offices

John Wiley & Sons, Inc., 605 Third Avenue,
New York, NY 10158-0012, USA

Jacaranda Wiley Ltd, 33 Park Road, Milton,
Queensland 4064, Australia

John Wiley & Sons (Canada) Ltd, 22 Worcester Road,
Rexdale, Ontario M9W 1L1, Canada

John Wiley & Sons (Asia) Pte Ltd, 2 Clementi Loop #02-01,
Jin Xing Distripark, Singapore 0512

Library of Congress Cataloging-in-Publication Data

The genus aeromonas / edited by B. Austin . . . [et al.].
    p. cm.
   Includes bibliographical references and index.
   ISBN 0-471-96741-6 (hardcover : alk. paper)
    1. Aeromonas.   I. Austin, B. (Brian). 1951–
OR82.P78G46   1996
   589.9'5--dc20   96-5722
                                  CIP

British Library Cataloguing in Publication Data

A catalogue record for this book is available from the British Library

ISBN 0 471 96741 6

Typeset in 10/12 pt Times by Mackreth Media Services, Hemel Hempstead.
Printed and bound in Great Britain by Biddles Ltd, Guildford.
This book is printed on acid-free paper responsibly manufactured from sustainable forestation,
for which at least two trees are planted for each one used for paper production.

Contents

List of Contributors

Sharon L. Abbott
Microbial Diseases Laboratory, Division of Communicable Disease Control, California Department of Health Services, Berkeley, CA 94704, USA

C. Adams
Department of Biological Sciences, Heriot-Watt University, Riccarton, Edinburgh EH14 4AS, Scotland

Martin Altwegg
Department of Medical Microbiology, University of Zürich, Gloriastrasse 30, Postfach, CH-8028, Zürich, Switzerland

Brian Austin
Department of Biological Sciences, Heriot-Watt University, Riccarton, Edinburgh EH14 4AS, Scotland

J. Thomas Buckley
Department of Biochemistry and Microbiology, University of Victoria, Victoria, British Columbia V8W 5C4, Canada

Amy M. Carnahan
Department of Medical and Research Technology, University of Maryland, Allied Health Building, Room 315C, 100 Penn Street, Baltimore, MD 21201-1082, USA

Peter J. Gosling
Department of Health, Skipton House, 80 London Road, London SE1 6LW, England

Philip Holmes
Severn Trent Water Limited, Quality and Environmental Services, Church Wiene Treatment Works, Draycott Road, Long Eaton, Nottingham NG10 3AZ, England

S. Peter Howard
Department of Biology, University of Regina, Regina, Saskatchewan S4S 0A2, Canada

J. Michael Janda
Microbial Diseases Laboratory, Division of Communicable Disease Control, California Department of Health Services, Berkeley, CA 94704, USA

S.W. Joseph
Department of Microbiology, University of Maryland, College Park, MD 20742, USA

Sheila MacIntyre
Department of Microbiology, University of Reading, London Road, Reading RG1 5AQ, Berkshire, England

Sally E. Millership
Department of Public Health and Clinical Policy, North Essex Health Authority, Collingwood Road, Witham, Essex CM8 2TT, England

Nelson P. Moyer
Hygienic Laboratory, Oakdale Hall, Iowa City, IA 52242, USA

Lynda M. Niccolls
Severn Trent Water Limited, Mythe Water Treatment Works, Ledbury Road, Tewkesbury GL20 8BR, England

Samuel A. Palumbo
Microbial Food Safety Research Unit, Eastern Regional Research Center, ARS/USDA, 600 E. Mermaid Lane, Wyndmoor, PA 19118, USA

David P. Sartory
Severn Trent Water Limited, Mythe Water Treatment Works, Ledbury Road, Tewkesbury GL20 8BR, England

Preface

Significant advances have been made during the past decade in the understanding of the biology of aeromonads. It is recognized that the organisms comprise the dominant component of the eutrophic freshwater aerobic bacterial population. Moreover, some of the species are associated with serious disease of humans and/or animals. The taxonomy of the group has changed radically since the publication of Volume 1 of *Bergey's Manual of Systematic Bacteriology* in 1984. In particular, *Aeromonas* has been reclassified in a new family, the Aeromonadaceae, and many new species have been described. With this level of interest, it is surprising that there has been a noticeable lack of a comprehensive text dealing with the biology of *Aeromonas*. Thus, during the organization of the 5th International *Aeromonas–Plesiomonas* Symposium, which was held in Edinburgh, Scotland in 1995, the current project was conceived. Potential contributors were identified, and willingly agreed to produce chapters within a remarkably short timescale. The resulting book has been targeted at research workers, including postgraduate students and diagnosticians. It is anticipated that the readership will include medical and veterinary microbiologists, public health scientists and microbial ecologists.

Foreword

The first monograph on *Aeromonas* was published in 1966. The knowledge of *Aeromonas* at this time was based on investigations carried out over a period of 13 years. Proof of the pathogenicity of aeromonads for humans was demonstrated in 1953, by fulfilling the postulate of Henle and Koch. During the following 30 years many important studies were made in this field. New species were found. The taxonomy of these microorganisms played a very important role and fundamental investigations were carried out. The results obtained may be the reason for the introduction of a new family – Aeromonadaceae. Furthermore, the better understanding of the biology of the aeromonads, the further progress in the detection of new toxins, their mechanisms of pathogenicity and the effect on humans and animals emphasized their importance for medical and veterinary microbiologists. The publication of this new monograph is therefore a welcome addition to the literature on this subject. It may serve as a key reference for questions on aeromonads.

F.-H. Caselitz
Hamburg, July 1995

1 Taxonomy

AMY M. CARNAHAN
University of Maryland, Baltimore, USA

MARTIN ALTWEGG
University of Zürich, Zürich, Switzerland

1.1 HISTORICAL ASPECTS

Aeromonads are ubiquitous, oxidase-positive, facultatively anaerobic, glucose-fermenting, Gram-negative bacilli that are autochthonous to aquatic environments. They have also been isolated from brackish, fresh, estuarine, marine, chlorinated and unchlorinated water supplies worldwide, with highest numbers achieved in the warmer months[1-6]. They have been isolated from diseased cold- and warm-blooded animals for over 100 years, and from humans since the early 1950s[7,8]. Thus, as both non-motile and motile aeromonads have caused serious fish disease for a number of decades, the motile aeromonads have also emerged as a serious microbial threat to human populations, both compromised and immunocompromised[9,10]. With the number of literature citations for the genus *Aeromonas* exceeding 2000, several excellent reviews can be found in recent literature[11-14].

As the history of these microorganisms has unfolded over the past 100 years, the taxonomy has followed a tortuous and often very confusing path. The information has become so interwoven that the best way to introduce a discussion of taxonomy is with a fairly strict chronological presentation of the literature. Perhaps in this way the reader can best decipher and collate all of the different lines of research that have led to our present understanding of the taxonomy of aeromonads. This discussion will focus on two major taxonomic groups of aeromonads, i.e. the motile, mesophilic aeromonads initially represented only as *Aeromonas hydrophila*, and the non-motile, psychrophilic aeromonads, initially represented as *A. salmonicida.*

The earliest reference to what we would consider today as a motile *Aeromonas* bacterium was Sanarelli's report in 1891 of *Bacillus hydrophilus fuscus* from the blood and lymph of an infected frog[7]. Sanarelli was able to prove the disease-causing potential of this bacillus by reinoculating it into cold- and warm-blooded animals and producing septicaemia. This invalid trinomial was employed by Chester in 1897[15], but then emended in 1901[16] to *Bacterium hydrophilum*, meaning a bacterium that was 'water loving'. While it is true that Zimmerman[17] isolated and described a bacterium isolated from

The Genus Aeromonas. Edited by B. Austin, M. Altwegg, P.J. Gosling and S. Joseph
© 1996 John Wiley & Sons Ltd

the drinking water in Chemnitz in 1890 as *Bacillus punctatus*, because of its dotted appearance on gelatin agar plates, it is a very scanty first description of a new genus and could easily have been a member of *Pseudomonas, Proteus* or *Serratia*. Likewise, *Bacillus rancida* described by Ernst in 1890[18] as a cause of 'red-leg' disease in frogs resembles upon close inspection a *Pseudomonas* species rather than an aeromonad because of the green pigmentation. Therefore, most taxonomists consider Sanarelli's work as the first valid description of the genus *Aeromonas*.

Over the next 60 years, this microorganism was isolated from a variety of cold- and warm-blooded animals, including fish, frogs, snakes, livestock, and birds[19–22]. During these years, it was classified as a member of many different genera, including *Aerobacter, Proteus, Pseudomonas, Escherichia, Achromobacter, Flavobacterium* and *Vibrio*, as bacteriologists struggled with the correct taxonomic position for this organism (Table 1.1). In 1936, Kluyver and van Niel[23] proposed the genus *Aeromonas*, meaning 'gas-producing unit', as a proper genus name which was then supported by Stanier's[24] taxonomic studies. This was eventually adopted in the Seventh Edition of *Bergey's Manual of Determinative Bacteriology* where Snieszko[25] proposed four species: three that were motile (*A. hydrophila, A. punctata, A. liquefaciens*) and one species that was non-motile (*A. salmonicida*) all residing within the family Pseudomonadaceae.

The historic path of the non-motile aeromonads began in 1894 with a

Table 1.1 Early designations recorded in literature for mesophilic members of the genus known currently as *Aeromonas*[a]

Name	Year of first description	Researcher
Bacillus punctatus[b]	1890	Zimmerman[17]
Bacillus ranicida[b]	1890	Ernst[18]
Bacillus hydrophilus fuscus	1891	Sanarelli[7]
Bacterium punctatum	1891	Lehmann and Neumann[27]
Aerobacter liquefaciens	1900	Beijerinck
Bacillus hydrophilus Sanarelli	1901	Chester[16]
Bacillus (Proteus, Pseudomonas, Escherichia) ichthyosmius	1917	Bergey *et al.* 1923, 1934
Achromobacter punctatum	1923	Bergey *et al.* 1934
Pseudomonas (Flavobacterium) fermentans	1930	Bergey *et al.* 1934
Pseudomonas punctata	1930	Breed *et al.* 1948
Proteus melanovogenes	1936	Miles and Halnan[37]
Pseudomonas caviae	1936	Scherago
Pseudomonas formicans	1954	Crawford
Vibrio jamaicensis	1955	Caselitz[38]

[a]Data derived from references 11, 13, 14, 39
[b]*B. punctatus* and *B. ranicida* were probably members of *Pseudomonas*

description by Emmerich and Weibel[26] of *Bacillus der Forellenseuche* as a cause of an epizoonosis in trout. Later work with similar bacteria that caused furunculosis in fish resulted in the name *Bacterium salmonicida*[15,27]. However, some taxonomists consider a description of *Bacillus devorans* by Zimmerman in 1890[28] as a synonym for *Bacterium salmonicida*[29]. Numerous studies ensued with this group of non-motile, pigmented psychrophilic (optimum growth at 10–15°C) microorganisms because of the economic implications for the fishing industry[30-33]. Smith[34] proposed in 1963 transferring the non-motile *A. salmonicida* to a new genus *Necromonas*, along with a new non-pigmented species, *N. achromogenes*. This proposal was not accepted because later DNA homology studies by MacInnes *et al.*[35] suggested a very close relationship to the motile aeromonads. These initial DNA–DNA hybridization experiments suggested that the genus *Aeromonas* consisted of two main evolutionary lines, i.e. a fairly homogenous group of non-motile aeromonads within the species *A. salmonicida* and a far more diverse group of motile aeromonads within which the species *A. hydrophila* resided[35]. Therefore, the Seventh Edition of *Bergey's Manual of Determinative Bacteriology* reported *A. salmonicida* and the non-motile species and *A. hydrophila*, *A. punctata*, and *A. liquefaciens* as the three motile species[25].

Sporadic reports had surfaced in the 1930s of a possible association between motile aeromonad species and human disease[36,37]. However, it is generally accepted that the report by Hill and colleagues[8] in 1954 of a 'new' bacterium from the autopsy material of a woman who died from a fulminant septicaemia with metastatic myositis was the first actual case report of human infection by an *Aeromonas* species. Although this organism was subsequently published in 1955 as *Vibrio jamaicensis* by Caselitz[38], it was later considered to be a motile *A. hydrophila* strain in the taxonomic studies of Ewing and co-workers at the Centers for Disease Control (CDC) in Atlanta, Georgia[39]. As this book unfolds, it will become apparent that several different aeromonad species are associated with a wide spectrum of disease states in humans as well as warm- and cold-blooded animals.

Several taxonomic studies based on phenetic similarity were conducted in the 1960s by Eddy and Carpenter[40,41], Ewing and colleagues[39,42], McCarthy[43], and Schubert[44-47]. Ewing and colleagues proposed three species to comprise the genus *Aeromonas*. These were the two species *A. hydrophila* and *A. shigelloides* for all motile, mesophilic aeromonads isolated from human sources, and *A. salmonicida*, for all non-motile, psychrophilic aeromonads which had only been isolated from diseased fish[39]. However, studies by Eddy and McCarthy[40,41,43] disagreed with this, and suggested that motile aeromonads should comprise two species, *A. punctata* and *A. caviae*, with *A. punctata* as the type species. Subsequent to these studies, Habs and Schubert proposed that the species *A. shigelloides* (formerly C27) be removed from the genus *Aeromonas* and reclassified as *Plesiomonas shigelloides*, which is where these organisms currently reside[48].

Schubert[49–51] proposed his own taxonomic classification of aeromonads to include three species and eight subspecies. These comprised the motile species *A. hydrophila* with subspecies *hydrophila, anaerogenes* and *proteolytica*, the motile species *A. punctata* with subspecies *punctata* and *caviae*; and the non-motile *A. salmonicida* with subspecies *A. salmonicida, achromogenes* and *masoucida*. This taxonomic scheme was published in the Eighth Edition of *Bergey's Manual of Determinative Bacteriology* with *Aeromonas* now considered a genus in the family Vibrionaceae[52].

Concurrent with this flurry of taxonomic work, a major paper was published in the *New England Journal of Medicine* by von Graevenitz and Mensch[53]. This study reported 27 cases of infection or colonization due to *A. hydrophila* in hospital patients in New Haven, Connecticut, together with a supplementary review of the literature. It further suggested a connection between diarrhoeal disease and the presence of large numbers of aeromonads in faecal cultures, although a small percentage of healthy persons harboured the organism as well.

1.2 NUMERICAL TAXONOMY

A landmark in the systematics of *Aeromonas* occurred in 1976 when Popoff and Véron[54] used numerical taxonomy to analyse 68 mesophilic aeromonads, mostly from environmental sources, for 203 morphological, biochemical and physiological characters at 30°C incubation temperature. Numerical taxonomy had just been defined by Sneath and Sokal as 'the grouping by numerical methods of taxonomic units into taxa on the basis of their character states[55]. First, the presence or absence of selected characters was determined in a group of microorganisms to be classified. This information was then converted into a binary form that was suitable for numerical analysis, and then compared using a computer. A minimum of 50 characters was recommended for analysis that would cover morphological, biochemical, and physiological properties[56].

Following character analysis, an 'association coefficient' was calculated for each pair of organisms that measured the agreement between characters possessed by two organisms. The most commonly used coefficient was the 'simple matching coefficient' (S_{SM}), which was the proportion of characters that matched regardless of whether the attribute was present or absent, i.e. total number of positive and negative matches divided by the total number of characters compared. Also useful was the calculation of the Jaccard coefficient (S_J), which ignored any characters that both organisms lacked, i.e. discounted any negative matches. Both coefficients increased linearly in value from 0.0 (no matches) to 1.0 (100% matches). These matching coefficients were arranged to form a similarity matrix where rows and columns represented the organisms as operational taxonomic units (OTUs) and each value in the matrix was an association coefficient measuring the similarity of two different organisms. Eventually each organism was compared with every other organism and those with the greatest similarity were grouped together as 'phenons' while dissimilar or 'atypical' organisms remained as separate smaller clusters or single isolated strains (Figure 1.1).

Another method of summarizing the numerical taxonomy results was in a treelike diagram known as a dendrogram (Figure 1.2). Here the diagram was placed on its side with the x-axis or abscissa denoted in units of similarity up to 1.0 (100%). Each branch point was located at that similarity value that related those two branch points, i.e. organisms below that branch point shared so many characteristics in common that they appeared as one, unless they were examined at an association coefficient greater than the magnitude of the branch point value. Phenons formed at 80–85% similarity were generally considered equivalent to bacterial species[56].

The study by Popoff and Véron of motile aeromonads used the Jaccard coefficient, and found that the 68 strains could be divided into two well-segregated classes on the basis of 50 variable characters, of which eight were of diagnostic value[54]. These were aesculin hydrolysis, fermentation of salicin, H_2S from GCF medium, growth in KCN medium, and growth on L-arabinose, salacin, L-arginine and L-histidine. They proposed that these two classes were actually two separate species. *A. hydrophila* (with biovars X_1 and X_2) and a new species, *A. sobria* (biovar Y).

Within months of this publication, Joseph and co-workers[57] at the Naval Medical Research Institute (NMRI) in Bethesda, Maryland reported on a primary wound infection sustained by a diver in polluted waters that revealed two different *Aeromonas* species. Both isolates were initially identified as *A. hydrophila* with an 18–24 h incubated API-20E identification strip (BioMérieux Vitek, Inc., Hazelwood, MO). Extended biotyping, using the scheme just published by Popoff, was then conducted on these isolates at NMRI by testing for aesculin hydrolysis, growth in KCN broth, elastase production, and gas from glucose. The last two tests had been published by Popoff and Véron[54] as useful for differentiating the two biovars of X_1. However, Joseph and colleagues chose an incubation temperature of 35–37°C instead of 30°C for their API-20E strips as well as the extended biochemical testing, since these were clinical isolates being examined by standard clinical microbiology laboratory procedures. One isolate was subsequently identified as *A. hydrophila* and the other as the newly proposed species, *A. sobria*[57].

Shaw and Hodder[58] used lipopolysaccharide analysis to examine a group of motile aeromonads as a means of taxonomic classification. The 12 aeromonads examined could be divided into three distinctly separate groups on the basis of the various combinations of hexose and heptose residues. The distribution of strains into these groups was substantially the same as the three groups outlined by Popoff in 1976[54].

Figure 1.1 Similarity matrix triangle of the 167 strains examined using the Simple Matching (S_{SM}) coefficient at a similarity level of S = 85%. AH denotes *A. hydrophila* (DNA Group 1), AV denotes *A. veronii* biovars sobria and veronii (DNA Groups 8/10), and AC denotes *A. caviae* (DNA Group 4). Reprinted with permission of Gustav Fischer Verlag from Reference 90

Popoff and co-workers followed their initial numerical taxonomy study with a genetic analysis of 55 motile *Aeromonas* strains using DNA–DNA hybridization by the S1 nuclease method[59]. They found three species: *A. hydrophila*, formerly biovar X_1 (type strain = ATCC 7966[T]), *A. caviae*, formerly biovar X_2 (type strain = ATCC 15468[T]), and *A. sobria*, formerly biovar Y (type strain = CIP 7433[T]). This new species, *A. caviae*, was the former 'anaerogenic' biovar of *A. hydrophila*, and could be differentiated from the 'aerogenic' *A. hydrophila* by the fact that *A. caviae* exhibited negative reactions for elastase production and the Voges–Proskauer reaction, no production of gas from glucose, and no production of H_2S from GCF medium[59]. There is, however, a taxonomic question of synonyms in the sense that the *A. caviae* from the research by Popoff and colleagues and the *A. punctata* from Schubert's earlier taxonomic work both share the same type strain (ATCC 15468[T]) and both names have standing in the literature. This issue will be addressed towards the end of this chapter in a discussion of current problematic taxonomic issues.

Although these three proposed species of Popoff and colleagues could now be separated on the basis of biochemical characteristics, core oligosaccharide patterns and similarities in DNA, the authors noted[59] that each species contained within itself at least two or three distinct DNA Hybridization Groups (HGs) that could not be separated from one another by phenotypic traits. Phenotypic *A. hydrophila* species (phenospecies) resided among three genospecies (genomic species or genomospecies) referred to as DNA HGs 1–3. DNA HGs 4–6 contained aeromonads that resembled the phenospecies *A. caviae*, and DNA HGs 7–8 resembled *A. sobria* (Table 1.2).

The next numerical taxonomy study was conducted by Lee and Bryant in 1984[60] and involved 163 aeromonad strains from a wide range of environments including natural waters, seafoods, other foods, veterinary specimens, human faeces, other human clinical sources, domestic and hospital environments and a selection of reference strains. Using 119 character traits incubated at 30°C, most strains clustered in phenons corresponding to the previously described *A. hydrophila*, *A. caviae*, *A. sobria* and *A. salmonicida*. Oddly enough, the type strain CIP 7433[T] for *A. sobria* did not cluster within any of these groups. Furthermore, Lee and Bryant found positive lysine decarboxylase activity by Moeller's method to be a useful marker for identifying *A. hydrophila* and *A. sobria*.

With the publication of the First Edition of *Bergey's Manual of Systematic Bacteriology* in 1984[61], the genus *Aeromonas* still resided in the family Vibrionaceae, and the description by the chapter author (Popoff) included the following species: three motile species (*A. hydrophila*, *A. caviae* and *A. sobria*) and one non-motile species (*A. salmonicida*) with three subspecies of *salmonicida*, *achromogenes*, and *masoucida*. Popoff's general description of the genus can be summarized as follows[61]: Aeromonads occur as straight cells that are *rod shaped with rounded ends to coccoid*. Their size range is

8

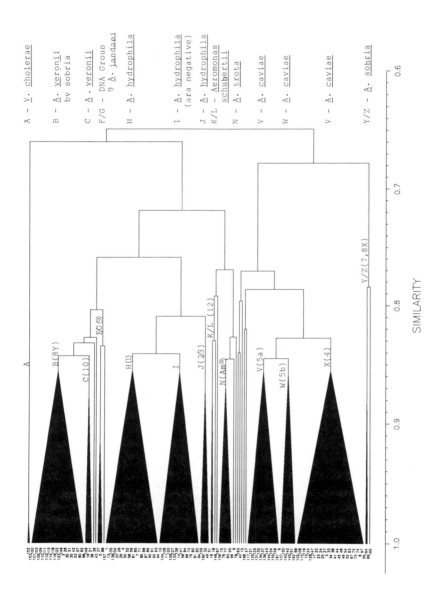

Figure 1.2 Dendrogram based on the S_{SM} coefficient and clustering by UPGMA. Numbers in parenthesis denote DNA hybridization definition strains contained within that cluster. Reprinted with permission of Gustav Fischer Verlag from Reference 90

Table 1.2   Current genospecies and phenospecies within the genus *Aeromonas*

DNA hybridi- zation group	Definition strain	Genospecies	Phenospecies
1	ATCC 7966[T]	*A. hydrophila*	*A. hydrophila*
2	ATCC 14715	Unnamed	*A. hydrophila*
3	ATCC 33658[T]	*A. salmonicida*	*A. salmonicida*[a]
3	CDC 0434-84	Unnamed	*A. hydrophila*[b]
4	ATCC 15468[T]	*A. caviae*	*A. caviae*
5A	CDC 0862-83	*A. media*	*A. caviae*
5B	CDC 0435-84	*A. media*	*A. media*
6	ATCC 23309[T]	*A. eucrenophila*	*A. eucrenophila*
7	CIP 7433[T]	*A. sobria*	*A. sobria*
8X	CDC 0437-84	*A. veronii*	*A. veronii* biovar sobria
8Y	ATCC 9071	*A. veronii*	*A. veronii* biovar sobria[c]
9	ATCC 49568[T]	*A. jandaei*	*A. jandaei*
10	ATCC 35624[T]	*A. veronii*	*A. veronii*
11	ATCC 35941	Unnamed	*Aeromonas* sp. (ornithine-positive)
12	ATCC 43700[T]	*A. schubertii*	*A. schubertii*
13	ATCC 43946	Unnamed	*Aeromonas* Group 501
14	ATCC 49657[T]	*A. trota*	*A. trota*[d]
15??	CECT 4199[T]	*A. allosaccharophila*	*A. allosaccharophila*
16??	CECT 4342[T]	*A. encheleia*	*A. encheleia*

[a]This includes non-motile, psychrophilic strains
[b]This includes motile, mesophilic strains
[c]*A. ichthiosmia* appears to be synonymous with *A. veronii* biovar sobria[106,108,109,138]
[d]*A. enteropelogenes* appears to be synonymous with *A. trota*[105,108,107,139]

0.3–1.0 μm in diameter and 1.0–3.5 μm in length and they exist singly, in pairs or short chains. They are Gram-negative, facultative anaerobes and generally motile by a single polar flagellum, although one species is non-motile. Metabolism of glucose is both fermentative and respiratory. They are oxidase- and catalase-positive, reduce nitrates to nitrites, and utilize carbohydrates with the production of acid, and acid with gas. They are resistant to the vibriostatic agent O/129 (2,4-diamino-6,7-diiso-propylpteridine) with an optimum growth temperature of 22–28°C and a mol% G + C content of 57–63 (Bd, T_m).

Popoff further noted that although DNA homology studies suggested that the subspecies of *A. salmonicida* were not sufficiently different to warrant the status of a subspecies, the three subspecies names were retained in the Approved List of Bacterial Names. Finally, Popoff determined that a halophilic bacterium previously designated as *A. proteolytica* by Merkel and colleagues[62] was sufficiently different to be removed from the genus *Aeromonas* and placed in the genus *Vibrio*, as *V. proteolyticus*.

Janda and colleagues had published in 1983[63] the results of a one-year prospective study in a New York City hospital, which had recovered 32 aeromonad strains (31 clinical and one environmental). Approximately half were faecal isolates (mostly from children under 5 years of age), while most of the wound or blood isolates were from an older population (average age = 56 years) who usually had one of several underlying disorders, including cancer, cirrhosis or renal disease. Most infections appeared to be community acquired as opposed to nosocomial and over 70% were recovered during the summer or autumn months. Seventeen of these strains were further biotyped as either *A. hydrophila, A. caviae* or *A. sobria* using a seven-test battery that was a modification of the original Popoff and Véron scheme with incubation of the tests at 37°C. The predominant species was *A. hydrophila* (47%), with each of the other two species being present in percentages ranging from 24 to 29%. It was noted that two out of three blood isolates were *A. sobria*, the newest species that had just been proposed by Popoff and co-workers.

Janda and colleagues subsequently examined a larger group of 147 *Aeromonas* isolates from diverse clinical and geographical sources using a modified Popoff and Véron biotyping scheme of eight biochemical tests[64]. They found that 93% could be biotyped, with *A. hydrophila* as the predominating species (48%), followed by equal percentages of *A. sobria* and *A. caviae* (25–27%). Additional markers were associated significantly with one or more of these three species, and incorporation of nine additional tests resulted in the accurate identification of 98% of the total isolates. These markers included haemolysis detected on sheep blood agar, lysine decarboxylase, staphylolytic enzyme, lecithinase, arbutin hydrolysis, acid production from cellobiose, mannose, 1-*O*-methyl-α-D-glucopyranosidase in the oxidative–fermentative (O–F) medium, and acid production from lactose in 10% purple agar base medium.

Regardless of the species designation, the most common source among the 147 strains was the gastrointestinal tract (40%). However, *A. sobria* was most often associated with the gastrointestinal tract and bacteraemia (68%), and *A. hydrophila* was most often recovered from wounds and the gastrointestinal tract (50%). *A. caviae* was rarely isolated from blood cultures, suggesting a greater virulence potential for both *A. hydrophila* and *A. sobria* when compared with *A. caviae*. Thus, it was becoming apparent that to understand the pathogenesis and epidemiology of *Aeromonas* infections, researchers and clinical microbiologists had to have a good grasp of the taxonomy of *Aeromonas* so that their studies and clinical identifications would be based on well characterized strains.

Meanwhile, a fifth species, *A. media*, was proposed by Allen and colleagues in 1983[65]. This species had been recovered from river water, and was initially described as non-motile, producing a brown diffusible pigment and growing at an optimum temperature of 35°C, but not implicated in human disease thus far. The type strain (ATCC 33907^T) was found to reside in DNA HG 5 based on Popoff's original DNA hybridization studies. Later reports by the same authors found some *A. media* strains to actually exhibit motility. For differentiation of *A. media* from *A. salmonicida* as well as other aeromonads, readers should consult Table 1.3.

Meanwhile, in an extension of the biochemical and exoenzymatic analyses of clinical *Aeromonas* by Hsu, Waltman and Shotts[66,67], Janda[68] evaluated 127 isolates representing *A. hydrophila*, *A. sobria* and *A. caviae*. This work involved examining each strain for over 70 biochemical tests and exoenzyme activities at 35°C using both conventional biochemical test media and the API-ZYM rapid enzyme strips (BioMérieux Vitek, Inc., Hazelwood, MO). Associations were found between several properties and taxonomically recognizable species, such as positive lysine decarboxylase and Voges–Proskauer reactions and haemolysis on sheep blood agar correlating

Table 1.3 Characteristics to separate *Aeromonas salmonicida* subspecies[a]

Character	*Aeromonas salmonicida* subspecies			
	achromogenes	*masoucida*	*salmonicida*	*smithia*
H$_2$S production	–[b]	+[c]	–	+
Indole production	+	+	–	–
Methyl Red	+	+	+	–
Voges–Proskauer reaction	–	+	–	–
Arginine dihydrolase	+	+	+	–
Aesculin hydrolysis	–	+	+	–
Lipase production	+	+	+	–
Chitin degradation	–	+	–	–
Elastin degradation	–	+	+	–
Acid from:				
D-galactose	+	+	+	–
Lactose	–	+	–	–
Maltose	+	+	+	–
D-mannitol	–	+	+	–
Sucrose	+	+	–	V[d]
Growth at 30°C	–	+	+	–

Character	*A. salmonicida*	*A. media*	Other aeromonads
Motility	–[e]	–[e]	+
Growth at 37°C	–[e]	+	+

[a]Data for table received from Professor Brian Austin
[b]–, negative
[c]+, positive
[d]V, Variable
[e]Positive strains have now been reported

with an identification of either *A. hydrophila* or *A. sobria*, but not *A. caviae*.

Investigators at the CDC in Atlanta, Georgia working on the taxonomy of the genus *Aeromonas* presented an extension of the original DNA homology studies of Popoff and co-workers at both the 1985 American Society of Microbiology (ASM) Annual Meeting in Washington, DC[69] and the First International Workshop on *Aeromonas–Plesiomonas* in Manchester, England in 1986[70]. Using the hydroxyapatite method for DNA–DNA hybridization at 60°C, they studied a large group of reference, environmental and clinical strains and found 9–12 different DNA HGs among these strains. All the non-motile *A. salmonicida* strains were highly related to one another in one single group, but the motile mesophilic aeromonads were extremely heterogenous, with many belonging to species other than *A. hydrophila, A. caviae, A. sobria* or *A. media* as defined by their type strains.

The same authors further proposed a new species for nine ornithine decarboxylase-positive strains (an unusual trait for aeromonads) that had previously been designated as Enteric Group 77. This proposal was to be published in 1987 as *A. veronii* and considered as DNA HG 10, since this group of organisms did not hybridize well enough with any of the known *Aeromonas* type strains tested in this study[71]. Two additional ornithine-positive *Aeromonas* strains, that were not genetically similar to the nine proposed *A. veronii* strains, were reclassified as *Aeromonas* species, ornithine-positive, and most taxonomists now consider them as the unnamed DNA HG 11. Most interestingly, the authors stated in an Addendum in Proof to the paper that the type strain of *A. veronii* (ATCC 35624[T]) was highly related by DNA–DNA hybridization to other *Aeromonas* reference strains from DNA HG 8 (phenotype *A. sobria*) such as ATCC 9071. They had only included the type strain for *A. sobria* (CIP 7433[T]), which actually resides in DNA HG 7, in their original work. This revelation that DNA HG 10 and DNA HG 8 were very similar was somewhat startling since these two genetically similar groups (HGs 8 and 10) were phenotypically very different from one another. *A. veronii* (HG 10) was described as arginine dihydrolase-negative, ornithine decarboxylase-negative and aesculin hydrolysis-positive, while these genetically similar reference strains *A. sobria* from HG 8 were already known to be arginine dihydrolase-positive, ornithine decarboxylase-negative and aesculin hydrolysis-negative. The authors suggested that these two groups of aeromonads might possibly be biogroups of one another within the species *A. veronii* and chose to address this problem in a future publication on the taxonomy of the *Aeromonas* group.

Studies of the genus *Aeromonas* on a molecular level were beginning to appear in the literature and will be mentioned at this point and more extensively explored later in this chapter. Colwell and colleagues[72] in 1986 used 16S rRNA cataloguing, 5S rRNA sequencing, and RNA–DNA hybridization to study members of the family Vibrionaceae. This research resulted in the proposal that the evolutionary history of aeromonads was sufficiently divergent from that of the Vibrionaceae to warrant the proposal of a new family, Aeromonadaceae[72]. Later work by Martinez-Murcia and colleagues[73] using 16S rDNA sequencing and Kita-Tsukamoto and co-workers[74] using 16S rRNA sequencing supported this original proposal to place aeromonads in their own family, the Aeromonadaceae.

Holmberg and colleagues[75], working at the CDC, conducted a retrospective study of 34 faecal isolates of nationwide distribution that included both DNA–DNA hybridization and phenotyping using Popoff and Véron's original scheme. Compared with 68 control subjects, matched by age ranges, the patients from whom the 34 aeromonad isolates were obtained were more likely to have ingested untreated water, specifically well water. Eighteen of the 34 isolates hybridized as DNA HG 4 (phenotype *A. caviae*) with smaller numbers of only one to six isolates residing among a variety of DNA groups including HGs 1 and 3 (phenotype *A. hydrophila*), DNA HG 5 (phenotype *A. media*) and DNA HG 8 (phenotype *A. sobria*). They found no clear correlation between illnesses and any phenotypic or genotypic characteristics. Further, they proposed that some *Aeromonas* strains could be enteropathogenic for the normal host, and could be acquired by drinking untreated water, but did not favour using any species designation at that time.

However, two separate studies published by Brenden and Janda[76,77] in 1987 began to suggest that significant differences may indeed exist between the different phenotypic species, thereby warranting species identification, at least in the clinical laboratory. The first study evaluated 58 clinical *Aeromonas* strains for their ability to produce β-haemolysin in both an agar plate assay and a cell-free assay[76]. They examined only sheep erythrocytes in the agar plate assay, but compared nine different animal erythrocyte cells in the more sensitive cell-free broth assay. Their findings suggested that perhaps haemolysins were related not only to the clinical source of isolation, but also to the aeromonad species from which they were isolated, in particular certain genospecies within a phenotypic species. The second study was a retrospective study of 13 bacteraemia-associated *Aeromonas* isolates collected by Janda over a five-year period[77]. There was a predominance of *A. sobria* among these isolates, with an overall death rate of 31%. Conversely, the few *A. caviae*-associated bacteraemias were always polymicrobic in nature. These findings suggested that patients colonized or locally infected with *A. sobria* and prone to systemic infections, might be at higher risk for the development of more serious disseminated disease.

At around this time, in the late 1980s, most studies evaluating media for the detection of *Aeromonas* were suggesting the use of ampicillin-containing agar to inhibit the overgrowth of normal faecal flora and facilitate the recovery of aeromonads[78,79]. However, a US Navy study of diarrhoeal cases in the Philippines found that 57% of their *Aeromonas* isolates were actually susceptible to ampicillin and would not be recovered when such a medium was employed[80].

A major taxonomic dilemma was brewing in the literature in the late 1980s, since many researchers worldwide were beginning to study aeromonads and publish their results. However, most were simply referring to all their motile isolates as *A. hydrophila* or *A. hydrophila* Group or Complex, and when they did attempt speciation they were not specific in stating their methods, test media and taxonomic scheme in the Materials and Methods sections of these publications. To quote Pazzaglia[81]:

> The scientific literature between 1960 and 1985 is littered with publications which may have used (unreliable or outdated) classifications for identifying *Aeromonas* to the species level. The significant overlap among classification used during this period, and the failure of many authors to detail the methods used for identification, have resulted in a large number of published scientific and clinical reports which identify *Aeromonas* isolates with species designations that are, at best, questionable, and at worst meaningless.

This confusion when deciphering the literature as to just exactly what particular species had been used in published research was exacerbated when growth studies by Statner and colleagues[82] indicated that the optimal growth temperature for 24 motile aeromonads was not always 28°C, as previously stated by Popoff in the *Bergey's Manual of Systematic Bacteriology*. In fact, they found that the optimal growth temperature was often 37°C, even for some environmental strains that presumably had never been exposed to 37°C, either through multiple laboratory passages or by exposure to human body temperatures during infection. Soluble protein profiles by SDS–PAGE at different temperatures also revealed temperature-dependent differences in 68% of the strains tested. Unfortunately, the strains were simply identified as *A. hydrophila* Group using API-20E strips (BioMérieux Vitek, Inc., Hazelwood, MO), so species-related differences could not accurately be determined.

Two more new *Aeromonas* species appeared in the literature in 1988. Hickman-Brenner and co-workers at CDC proposed the name *A. schubertii* for a group of seven mesophilic, motile, mannitol-negative and indole-negative aeromonads found in clinical specimens and considered to reside in DNA HG 12[83]. Two very similar strains (mannitol-negative but indole-positive) were not genetically close enough to be considered *A. schubertii*, but the authors reclassified them from Enteric Group 501 to *Aeromonas* Group 501. Most taxonomists consider this to be DNA HG 13, in the same manner that the two ornithine-positive strains close to *A. veronii* (DNA HG 10) became *Aeromonas* species, ornithine-positive, and reside in the yet unnamed DNA HG 11.

The other new species was proposed by Schubert and Hegazi[84] as *A. eucrenophila*, a mesophilic aeromonad isolated from fresh uncontaminated brooks and rivers. This organism was described as phenotypically resembling *A. punctata*-like or *A. caviae*-like strains that ironically produced gas from glucose, a trait considered negative for all other *A. caviae*. Even more interesting was the fact that this species resided within DNA HG 6, an original Popoff *et al. A. caviae* hybridization group.

Thus, after the proposal of these last two species, researchers and clinical microbiologists were dealing with five species of motile aeromonads isolated from clinical specimens (*A. hydrophila, A. caviae, A. sobria, A. veronii,* and *A. schubertii*) and three non-motile species (*A. salmonicida, A. media, A. eucrenophila*), that had only been isolated from non-human (fish and environmental) sources.

1.3   MAJOR TAXONOMIC STUDIES

Several major taxonomic studies were conducted and published in the late 1980s and early 1990s in an attempt to reconcile the proposed phenotypic species (eight) with the then recognized genotypic species (13 DNA HGs) (Table 1.2). The previously described study by Holmberg and colleagues[75] from CDC had phenotyped 34 diarrhoeal isolates and genotyped 33 diarrhoeal isolates using DNA HG definition strains for HGs 1–8 only. All the strains, except one, were found to reside among only five genospecies: DNA HG 1 (phenotype *A. hydrophila*), DNA HG 3 (motile phenotype *A. hydrophila*), DNA HG 4 (phenotype *A. caviae*), DNA HG 5 (phenotype *A. caviae/A. media*), or DNA HG 8 (phenotype *A. sobria*). The majority of the strains, 17/33, resided in DNA HG 4 (*A. caviae*) and none of the clinical *A. sobria* strains hybridized with the type strain for *A. sobria* (CIP 7433[T]) residing in DNA HG 7. This was the same result that Lee had obtained in 1984, and suggested even more strongly that what were being identified in the clinical laboratory as *A. sobria* (HG 8) were very different from HG 7 *A. sobria* as proposed by Popoff and Véron. Thus, it seemed very likely that the CDC proposal that clinical *A. sobria* (HG 8) might be a biogroup of the genetically similar *A. veronii* (HG 10) was valid for future consideration.

Arduino and colleagues[85] then analysed 132 aeromonad strains from the CDC collection, including definition strains for DNA HGs 1–12 as well as all known ATCC type strains. All 132 strains had previously been genotyped by DNA–DNA hybridization and were now tested for 165 phenotypic traits at 25°C, except for growth tolerances at 42°C and glucose fermentation at 4, 15, 37 and 42°C. The majority of the strains resided in DNA HGs 1, 4 and 8, with the single genospecies majority (37/132) found in DNA HG 4 (phenotype *A. caviae*). Only 41 tests were discriminatory for the 12 groups, and a probability matrix with a computer identification program based on normalized likelihoods accurately placed 85% of the strains in their correct DNA HG.

However, DNA HGs 4, 5 and 6 (phenotypes *A. caviae, A. media* and *A. eucrenophila*) were the most difficult to separate.

Renaud and colleagues[86] in France phenotyped 66 lakewater and 34 various clinical isolates using both the conventional schemes of Popoff and colleagues and Janda and co-workers at 30°C. Each strain was then tested using 147 rapid assimilation tests available on API-50CH, API-AO and API-50AA strips as 49 carbohydrates, 49 organic acids and 49 amino acids at 30°C (BioMérieux Vitek, Inc., Hazelwood, MO). The results were analysed by numerical taxonomy using the S_J coefficient and clusters formed were measured by the mean distance method and presented as both a similarity matrix and a dendrogram. All but 10 strains resided in four clusters designated as A, B, C and D. Groups C and D were divided into two subgroups C_1, C_2, D_1, and D_2. Group A contained *A. sobria* strains and the type strain CIP 7433[T] (DNA HG 7). Group B contained several atypical arabinose-negative *A. hydrophila* strains, and Group C_2 had most of the *A. hydrophila* strains including ATCC 7966[T] (DNA HG 1). Groups C_1, D_1 and D_2 were subgroups of *A. caviae*, but ATCC 15468[T] (DNA HG 4) did not cluster with any group. Of the substrates considered discriminatory, only DL-lactate, gentobiose and diaminobutane were new markers.

The next major study of motile aeromonads to reconcile phenotype to genotype was conducted by Kuijper and colleagues in the Netherlands[87]. This involved a collection of 189 faecal isolates gathered between 1982 and 1986 and analysed for 30 phenotypic traits at 37°C, including most tests from the original Popoff and Véron scheme plus additional tests from the literature. Most of the isolates phenotyped as *A. hydrophila, A. caviae* and *A. sobria*, but when a subset of 142 strains was subjected to DNA–DNA hybridization using HG definition strains for DNA HGs 1–11, the strains resided in DNA HGs 1, 2, 3, 4, 5a and 8. Although the majority resided in DNA HGs 1, 4 and 8, as in previous studies, there were increased frequencies of the other rarer HGs. Most disconcerting was that 8/26 (31%) of the phenotyped *A. hydrophila* strains actually resided in DNA HG 8 (phenotype *A. sobria* suggested as a biogroup of *A. veronii* HG 10). This occurred most probably because of a positive aesculin hydrolysis reaction for DNA HG 8 strains that would phenotypically suggest they belonged to the usual aesculin hydrolysis-positive *A. hydrophila* phenotype.

A large-scale study of 138 motile aeromonads was conducted by Altwegg and colleagues in collaboration with CDC and published in 1990[88]. Their strain collection includes 102 faecal isolates, seven other clinical isolates, three environmental isolates and two unknown isolates from Zurich, Switzerland, nine reference strains from CDC for DNA HGs 1–9, and three faecal, 10 environmental and two miscellaneous strains from Heidelberg, Germany. All 138 strains were tested for 100 phenotypic characters, although only 63 different characters were actually evaluated due to the redundancy of some tests. This final 63 included only 10 conventional tests plus 53 rapid biochemical, carbohydrate and enzyme tests using API-20E, API-50E and API ATB 32GN identification systems (API Systems, La Balme-Les Grottes, France). The similarity values for the strains were analysed using a similarity coefficient similar to S_J, and clustering was by UPGMA. Cluster analysis at an 88% similarity level revealed three major phenons similar to the *A. hydrophila* (20.3%), *A. caviae* (56.6%), and *A. sobria* (18.1%) described by Popoff in *Bergey's Manual of Systematic Bacteriology*. Each phenon contained more than one genospecies and more than one currently named species.

A subset of 126 strains from the original 138 was subjected to DNA–DNA hybridization using the definition strains for DNA HGs 1–9 only, and again a majority of the strains were found to reside in DNA HGs 1, 4 and 8 with slightly larger numbers in groups 2, 3 and 5 than had been reported previously. Like the two recent studies, no strains were found in DNA HG 6 (*A. eucrenophila*), HG 7 (*A. sobria*), HG 11 or HG 12 (*A. schubertii*). Similarly, all but one of the mostly clinical strains that resembled *A. sobria* phenotypically resided in DNA HG 8, not DNA HG 7 which contains Popoff and Véron's type strain CIP 7433[T]. These authors, who included CDC researchers from the original *A. veronii* publication, likewise proposed that the correct solution to the problem that DNA HG 8 (clinical *A. sobria*) and DNA HG 10 (*A. veronii*) are genetically identical but biochemically distinguishable, may be to consider these two groups as either biotypes or subspecies of *A. veronii*.

An interesting distribution was also discovered among the 126 strains that were genotyped in that the DNA HGs were not equally distributed among the clinical and environmental isolates. Most clinical isolates resided in DNA HGs 1, 4, 5a, 5b and 8, whereas more than half of the environmental strains resided in DNA HGs 2, 3 and 5a only. This suggested that strains from certain DNA HGs might have less virulence potential for humans than those from other HGs. A few tests, all from commercial test strips, were suggested as possibly useful for separating the various genotypes within a single phenotypic species. These were: production of either acid from or growth on D-sorbitol (separates HG 3 from 1 and 2 within *A. hydrophila* phenotype), growth on citrate (separates HG 4 from 5a and 5b within *A. caviae* phenotype) and growth on DL-lactate (separates HG 1 from 2 and 3 within *A. hydrophila* phenotype).

Meanwhile, a large numerical taxonomy study of both motile and non-motile fish-associated *Aeromonas* species was published by Austin and colleagues in 1989[89]. They had examined 180 *Aeromonas* strains for 140 unit characters at 25°C and analysed the data by S_{SM} and S_J coefficients with UPGMA clustering. In addition, representative strains from the collection were examined by DNA–DNA hybridization. It was determined that discrete homogeneous clusters of motile *A. hydrophila*, *A. caviae* and *A. sobria* were easily distinguishable from clusters representing the non-motile species *A.*

salmonicida and *A. media*. Further, a new subspecies of *A. salmonicida* known as *A. salmonicida* subspecies *smithia* was elucidated and proposed. Distinguishing characteristics to separate *smithia* from other *A. salmonicida* subspecies can be found in Table 1.3.

It was three years before the next large numerical taxonomy study of the motile aeromonads appeared in the literature. This was an examination by Carnahan and Joseph[90] of 167 motile *Aeromonas* isolates that included clinical (130/167), veterinary (8/167), and environmental (14/167) isolates from several different geographical sources including USA, Bangladesh, India, Sudan, Somalia, Egypt and Indonesia as well as 15 reference aeromonad strains. This was in deference to previous studies by Holmberg, Kuijper and Altwegg where the majority of their isolates were from one major geographical source, i.e. USA, The Netherlands, or Switzerland, respectively. Equally important was the need for this newest study to have a more equitable distribution of clinical isolates from non-faecal sources (35/130) and not just primarily faecal isolates, as was the case in the previously mentioned studies. The bias of clinical and geographical source distribution in previous studies may have resulted in skewed frequencies of species being detected.

The strains were initially examined for 80 phenotypic traits from which 50 discriminatory tests were used for the numerical taxonomic analysis using the SAS/TAXAN[R] program[91]. The definition strains for all 13 known DNA HGs as well as additional reference strains and type strains were included as controls and markers for the various clusters. S_{SM} and S_J coefficients were used to form similarity matrices and dendrograms with clustering by UPGMA. The 50 discriminatory tests covered a range of well documented biochemical traits as well as relatively new biochemical, physiological and antimicrobial resistance tests. All tests, except for growth at 42°C, were run at 36 ± 1°C in an attempt to address both accurate taxonomic classification and possible association with clinically significant virulence-associated markers.

Using a similarity level of 85% for species designation, the analysis of the 167 strains resulted in 12 clusters or phenons plus 13 single-member clusters. Each cluster, with one exception, only contained one DNA HG definition strain or type strain. The large similarity matrix generated using the S_{SM} coefficient (Figure 1.1) demonstrated three distinct clusters which represented *A. hydrophila, A. sobria* and *A. caviae*. However, the cluster of mostly clinical *A. sobria* strains included a DNA HG8Y reference strain (ATCC 9071), but not Popoff and colleagues' *A. sobria*-type strain (CIP 7433[T]) or the DNA HG8X reference strain, which is the actual definition strain for HG8. Also of interest is the fact that this large cluster of HG8 strains was very close to (merging at a similarity value of 84%) a smaller cluster containing the *A. veronii* (HG10) type strain, thereby strongly suggesting that these two biochemically distinct clinical groups were indeed distinct biovars of the same species as other taxonomists had proposed. Joseph and colleagues officially proposed[92], in a later case-study involving an *A. veronii* infection, that ornithine-positive *A. veronii* (HG 10) should be considered as *A. veronii* biovar veronii, and ornithine-negative clinical *A. sobria* strains (HG 8) be regarded as *A. veronii* biovar sobria. This meant that all previous references to clinical *A. sobria* isolates, starting with the isolation by Joseph and colleagues in 1979[57], were actually members of the *A. veronii* biovar sobria species (DNA HGs 8/10), and were not genotypically or phenotypically related to the original environmental *A. sobria* species initially proposed by Popoff and Véron in 1979. Joseph and colleagues have confirmed that the original 1979 NMRI-7 isolate from the diver as well as other earlier strains published as *A. sobria*, are indeed members of *A. veronii* biovar sobria as represented by the strain ATCC 9071[93]. These suggested nomenclatural changes have now received wide acceptance in the literature and have even been adopted by the databases of several major rapid identification systems.

A dendrogram derived from the numerical taxonomy study by Carnahan and Joseph[90] gives a good representation of all the clusters uncovered with this research (Figure 1.2). Each cluster consisted of mostly clinical strains, but also included a few environmental and/or veterinary strains. One small cluster of five strains contained the DNA definition strains for both DNA HGs 2 and 3 (phenotype *A. hydrophila*), suggesting that they were extremely difficult to separate, as had been previously noted in other studies. Other small clusters or single strains contained the definition strain for HG 6 (*A. eucrenophila*), HG 7 (phenotype *A. sobria*), DNA HG 9 (unnamed group resembling *A. sobria*), DNA HG 10 (phenotype *A. veronii*), DNA HG 11 (*Aeromonas* sp. ornithine-positive), DNA HG 12 (phenotype *A. schubertii*), and DNA HG 13 (*Aeromonas* Group 501). Clusters containing DNA HGs 2, 3 and 7 were almost exclusively environmental or veterinary strains, whereas the strains in clusters containing DNA HGs 9, 10 and 12 were exclusively clinical isolates from extraintestinal sources.

Most interesting was the discovery that two small clusters of atypical strains contained no DNA definition strain or type strain. The first was a cluster of three strains that merged at the 84% similarity (S) level with the DNA definition strain for HG 9 (CDC 0787-80). This DNA HG 9 had been established in the large DNA–DNA hybridization study at CDC in the early 1980s, but there had been insufficient numbers of these strains to designate a new species at that time. Carnahan, Joseph and Fanning took their four strains from the numerical taxonomy study and combined them with five similar strains from various collections, and further genotypic and phenotypic study resulted in the proposal of a new species, *A. jandaei*, that was negative for sucrose fermentation and aesculin hydrolysis. This was first presented by Carnahan as an Oral Paper Presentation at the 3rd International Symposium on *Aeromonas* and *Plesiomonas* in Helsingør,

Denmark in September 1990, and subsequently published and validated in 1991[94,95].

The other atypical cluster from their initial taxonomic study consisted of seven ampicillin-susceptible *A. sobria*-like strains from one specific locality, Southeast Asia. These strains were evaluated more extensively and combined with six other similar strains from Indonesia and the USA. After phenotyping and DNA–DNA hybridization, it was determined that this cluster was also a new species of *Aeromonas*, namely *A. trota*, which was negative for aesculin hydrolysis and the Voges–Proskauer reaction, and susceptible to ampicillin (an unusual trait for most aeromonads). This new species was first proposed and presented at the 3rd International Symposium on *Aeromonas* and *Plesiomonas* in Helsingør, Denmark in September 1990, and subsequently published and validated in 1991[96]. The authors considered *A. trota* to exist as the new DNA HG 14, since they already considered the *Aeromonas* Group 501 strains to reside in the unnamed DNA HG 13 based on the original *A. schubertii* paper. However, some authors do not consider *Aeromonas* Group 501 as HG 13 and therefore often refer to *A. trota* as HG 13, not HG 14 as was initially proposed (see Table 1.4). It should be noted that *Aeromonas* researchers at CDC, currently consider *A. trota* to reside in DNA HG 14.

The ultimate goal of the taxonomic study of motile aeromonads by Carnahan and Joseph was to use the frequency matrix of the SAS/TAXAN[R] analysis to derive a subset of discriminatory conventional tests to identify clinical aeromonads. These tests should be readily available in most clinical diagnostic laboratories, and their use should retain the identification power

Table 1.4 Selected biochemical properties of *Aeromonas* HGs, expressed as percentage positive responses

Test[a]	HG[b]												
	1	2	3	4	5	6	7	8	9	10	11	12	13[c]
LDC	100	50	67	0	0	0	100	100	100	100	50	82	100
ODC	0	0	0	0	0	0	0	0	0	100	100	0	0
ADH	100	88	67	100	91	86	0	100	100	0	0	91	100
Aesculin	95	100	100	93	91	100	0	0	0	100	50	0	8
Gas (glucose)	91	75	50	0	0	100	50	89	100	80	50	0	69
VP	91	75	67	0	0	0	0	94	91	100	0	18	0
Arabinose	86	100	100	100	100	86	0	17	0	0	0	0	0
Mannitol	95	100	100	100	100	100	100	100	100	100	100	0	69
Sucrose	100	100	100	100	100	71	100	100	0	100	100	0	23

[a]Abbreviations: LDC, lysine decarboxylase; ODC, ornithine decarboxylase; ADH, arginine dihydrolase; VP, Voges–Proskauer reaction
[b]Each number indicates percentage of isolates positive for each indicated trait
[c]HG 13 represents *A. trota* (originally proposed as HG 14)
(Table reprinted with permission from publisher and authors from reference 102)

of the entire frequency matrix. Eighteen such discriminatory tests were chosen from the original 50 and used to construct two multi-level dichotomous keys using a BASIC computer program (FLOABN) by Behram and colleagues[97]. FLOABN was written as a modification of a PASCAL 'greedy algorithm' program previously developed and published for the identification of bacterial species[98]. The first key, 'Aerokey I', was based on clusters and single strains representing all currently known DNA HGs 1–14, and used 12 of the 18 discriminatory tests to differentiate these HGs from one another. However, because several HGs were only represented by single strains or only a couple of strains, this key was never published and its significance awaits future studies examining larger numbers of these rare genospecies. The second dichotomous biotyping key, 'Aerokey II', was based on the clusters that contained the major species found in clinical specimens: *A. hydrophila* (arabinose-positive and arabinose-negative clusters), *A. veronii* biovar sobria, *A. caviae*, *A. veronii* biovar veronii, *A. schubertii*, *A. jandaei*, and *A. trota*. Seven of the original 18 discriminatory tests were used to construct Aerokey II, and its validity was confirmed by accurately identifying 97% of 60 blinded clinical aeromonad isolates from an independent reference laboratory, and 100% of 18 ATCC reference strains for the newest species[99]. Aerokey II has already been somewhat modified by other microbiologists in the field, resulting in the publication of Aeroscheme[100]. (For further information on identification of aeromonads see Chapter 3.)

Two more large studies involving the relationship between phenotype and genotype among aeromonads emerged in 1992. The first, conducted by Kämpfer and Altwegg[101], examined 176 aeromonads strains representing all currently recognized genospecies, except that *Aeromonas* Group 501 was not included and the authors chose to consider *A. trota* as HG 13 and not HG 14 as originally proposed. Each of the 176 strains was examined for 329 different characters in standard microtitration plates at 30°C. Using S_{SM} and S_J as measures of similarity with a UPGMA computer algorithm, they found 16 clusters and 7 unclustered strains at the 93.5% similarity level. HGs 1, 4, 5, 6, 7, 9, 12 and 13 (*A. trota* in this research) were mostly represented by a single cluster or phenon with DNA HGs 2 and 3 in very closely related phenons. The phenotypic data correlated well with the taxa as defined by DNA–DNA hybridization and multilocus enzyme analysis (see discussion later in this chapter). A matrix was constructed for probabilistic identification of all genospecies using 30 diagnostic characters. Using a Willcox probability of >0.99, this matrix achieved a correct identification rate of 71.51%; and when a Willcox probability of >0.90 was used, a correct identification rate of 83.7% was achieved.

The second study, by Abbott and colleagues[102], involved biochemical analysis of 133 strains of *Aeromonas* from clinical (102/133), veterinary (16/133), and environmental (15/133) sources that had already been identified to DNA hybridization group. Out of the 58 biochemical, morphological or growth properties examined at 35°C, between nine and 16 tests were shown to be capable of identifying 132/133 (99%) of the strains to their correct DNA HG. The authors selected eight standard biochemical tests as a 'starting point' for separating the DNA HGs 1–13 (Table 1.4) followed by smaller test batteries for closely related HGs that often involved biochemical tests not readily available in most laboratories. Note that these authors included two *Aeromonas* Group 501 strains (considered HG 13) as reference strains for HG 12 *A. schubertii* and, therefore, considered *A. trota* to be HG 13 in their Table, not HG 14 as originally proposed.

1.4   UNNAMED DNA HYBRIDIZATION GROUPS

It should be obvious from all the above studies that the genus *Aeromonas*, at least as far as motile species are concerned, is very diverse. We have yet to name all of the known DNA HGs such as HG 3 (motile subspecies), HG 5a, HG 11 (*Aeromonas* species, ornithine-positive), and HG 13 (*Aeromonas* Group 501) and there are certainly more HGs to be discovered. Future work needs to consider conducting biochemical tests on such strains at both 35–37°C and 25–30°C in order to determine any temperature-dependent reactions that may exist among the different species, especially those from the environment.

To this end, there has been some progress made in terms of published tests that might be useful in distinguishing closely related groups of aeromonads. One such related group includes DNA HGs 1, 2 and 3 (phenotype *A. hydrophila*), of which HG 3 contains motile, mesophilic strains from human sources as well as the established non-motile, psychrophilic *A. salmonicida* subspecies. As previously mentioned, work by Altwegg and colleagues suggested acid from or growth on D-sorbitol, and DL-lactate utilization as helpful with these three groups[88]. The study by Abbott and colleagues[102] suggested that acid production from D-rhamnose, salicin and lactose, oxidation of gluconate, elastase production and production of phenylpyruvic acid were discriminatory for these groups. Phenotypic and genotypic studies of this *A. hydrophila* Complex (HGs 1, 2, 3) by Hänninen[103] involving analysis of 22 strains suggest that separation of these groups can best be accomplished using DL-lactate and urocanic acid utilization at 30°C, acid production from D-sorbitol and D-rhamnose at both 30°C and 35°C and a new marker known as 'growth temperature gradient'. The mean maximum growth temperature (T_{max}) was significantly higher for HG 1 strains (41.1 ± 0.56°C, $p < 0.001$) than that of DNA HG 2 (38.2 ± 0.79°C) or HG 3 (38.6 ± 0.75°C) and may explain why different biochemical test results are obtained with these HGs depending on the temperature of incubation used (30°C or 37°C). Ali and colleagues [104] have recently proposed *Aeromonas bestiarum* for DNA HG2 based on DNA/DNA hybridization, ribotyping and utilization of DL-lactate, acid from D-sorbitol and utilization of urocanic acid.

Another closely related group includes DNA HGs 4, 5 and 6, where HG 5b includes the type strain for *A. media* and HG 6 includes *A. eucrenophila*. Altwegg and colleagues[88] suggested citrate utilization as helpful, whereas Abbott and co-workers[102] additionally suggested acid from glycerol and D-mannose as well as H_2S from cysteine, haemolysis, pyrazinamidase and production oi phenylpyruvic acid. However, both these studies are based on small numbers of available strains for these particular HGs, and larger numbers of strains will need to be examined before definitive differences can be established and new species names designated. Also remaining to be named are the strains referred to as *Aeromonas* species, ornithine-positive (HG 11) and *Aeromonas* Group 501 (HG 13). Once again, when sufficient numbers of these HGs are examined by rigorous phenotypic and genotypic standards, such studies could result in their proposal as new 'bona fide' species within the genus *Aeromonas*.

1.5   TAXONOMIC ISSUES TO BE CONSIDERED

Returning to the aforementioned problems with *A. caviae* and *A. punctata* as synonyms for one another and sharing the same type strain, there are two currently opposing opinions as to how this issue should be resolved. Hickman-Brenner and colleagues from CDC stated in their paper proposing *A. veronii*[71] that they considered *A. punctata* the older and more established of the two synonyms on The Approved List of Bacterial Names. They further suggested resolving this dilemma by submitting a 'Request for an Opinion' to the Judicial Commission of the International Committee on Systematic Bacteriology. However, Schubert, who proposed *A. punctata* as a valid species in the Eighth Edition of *Bergey's Manual of Determinative Bacteriology* in 1974 disagrees with these assertions, and considers that a 'Request for an Opinion' is not necessary at present to resolve this issue. Schubert and Hegazi stated in their paper proposing *A. eucrenophila*[84] that, although they too consider *A. caviae* as a later and illegitimate synonym of *A. punctata*, the trivial term *A. caviae* should be temporarily retained in order to 'secure continuity in the literature', especially for strains isolated from cases of paediatric diarrhoea.

In terms of the proposal that DNA HGs 8 and 10 are genetically identical, but contain phenotypically different biovars, the names *A. veronii* biovar veronii and *A. veronii* biovar sobria as proposed by Joseph and colleagues[92] are currently being used throughout most of the recent literature. This was also proposed as a means of keeping a 'secure continuity in the literature' between what used to be considered as clinical *A. sobria* isolates and what are now known to be genetically similar, but biochemically distinct biovars within the species *A. veronii*. Their official acceptance will most probably be a discussion item for a future Taxonomic Working Group on *Aeromonas* currently being established within the ICSB Vibrionaceae subcommittee[105].

There are a few additional controversial issues within *Aeromonas* taxonomy to be considered. Within a few months of the proposal of *A. jandaei* and *A. trota* at the 3rd International Symposium on *Aeromonas* and *Plesiomonas* in Helsingør, Denmark (September 1990) and just prior to their publication and validation, two new species of *Aeromonas* were published based on limited DNA–DNA hybridization studies and limited biochemical analyses. The first new species proposed was *A. enteropelogenes*, isolated from human faecal specimens from India, and negative for aesculin hydrolysis and the Voges–Proskauer[106] reaction. The second new species proposed was *A. ichthiosmia*, a group of arabinose-negative and sucrose-positive isolates from mainly environmental sources[107]. Both these species were likewise validated in early 1991, despite the fact that their brief, initial description appeared very similar to *A. trota* and *A. veronii* biovar sobria, respectively. In fact, early analysis of the type strains of *A. enteropelogenes* and *A. ichthiosmia* found them to be identical to *A. trota* and *A. veronii* biovar sobria by phenotyping, fatty acid profile by gas liquid chromatography (GLC) and to have nearly identical ribotyping patterns[108]. Further proof that they are most likely not two new species is the fact that one of the strains published by Schubert and co-workers as *A. ichthiosmia* (ATCC 11163) was in the original 1985 DNA–DNA hybridization study from CDC[69] and had already been stringently examined by DNA–DNA hybridization, and belonged to DNA HG 8, which is now generally considered as *A. veronii* biovar sobria. Lastly, a study of the genera *Aeromonas* and *Plesiomonas* by small subunit 16S rDNA sequencing by Collins and co-workers[109] revealed that *A. enteropelogenes* and *A. ichthiosmia* were indeed identical to the earlier proposed *A. trota* and *A. veronii* biovar sobria, respectively. Although *A. trota* and *A. veronii* biovar sobria appear to be widely accepted in the literature, most researchers do not realize that *A. enteropelogenes* and *A. ichthiosmia* may be synonyms for *A. trota* and *A. veronii* biovar sobria. These taxonomic issues will no doubt also be topics for discussion at future meetings of the Taxonomic Working Group on *Aeromonas*.

Finally, two additional new species have been proposed recently. The first, *Aeromonas allosaccharophila*, was proposed by Martinez-Murcia and colleagues[110] as a new mesophilic aeromonad species based on phenotypic studies and 16S rRNA gene sequence analysis of three atypical aeromonad strains. Oddly enough, one of the three strains was an original DNA HG 11 strain (ATCC 35942) published as an *Aeromonas* species, ornithine-positive by Hickman-Brenner and co-workers[71] in their *A. veronii* paper in 1987. Although this small group of three strains proposed as *A. allosaccharophila* appeared to be distinct by 16S rRNA gene sequencing, the recommendations from the Ad Hoc Committee on Reconciliation of Approaches to Bacterial

Systematics by Wayne and colleagues[111] suggested that DNA–DNA reassociation or hybridization represents the best procedure for establishing new species at this time. The committee further recommended that a distinct genospecies that cannot be differentiated from another genospecies on the basis of any known phenotypic property or properties shared by this new species, should not be named until specific phenotypic differentiation can be made. Herein lies the problem with *A. allosaccharophila*. The species is so named because each one of the three strains ferments a different unusual sugar, namely D-raffinose, D-melibiose and D-rhamnose, but none of the three ferment any useful combination of these sugars. Only one of the three is ornithine decarboxylase-positive (ATCC 35942 from HG 11). Therefore, in the average laboratory, it would be very difficult to distinguish *A. allosaccharophila* from other aeromonads, since occasional strains in many other DNA HGs will ferment these sugars. The further lack of DNA–DNA hybridization data to support this new species proposal was very recently rectified in a Note published in the April 1995 issue of the *International Journal of Systematic Bacteriology*[112]. However, the authors did not include the definition strain for DNA HG 2 in their DNA–DNA hybridization studies, and proposed *A. allosaccharophila* as DNA HG 14, when in fact *A. trota* had already been proposed as HG 14, making HG 15 the next available HG group. It is also a little troubling to taxonomists in general when a new species is proposed on very limited numbers of strains, such as the three used to describe *A. allosaccharophila*. Hence, when Esteve and colleagues proposed *A. encheleia* in July 1995 as a new aeromonad species isolated from European eels, and based on only four strains, some taxonomic issues of concern were raised. This new mesophilic species has a phenotypic profile that includes an inability to use D-gluconate, L-arginine and L-glutamine as sole carbon and energy sources. However, there are several instances where the results for these tests for other known established *Aeromonas* species are described as ND (no data available) in their own Table 3, and their DNA–DNA hybridization data once again did not include the reference strain for DNA HG 2. In light of the already tortuous and confusing path that *Aeromonas* taxonomy has taken, it is critical that any new species should be composed of a number of similar strains that have been extensively phenotyped and genotyped against all known species type strains and definition strains for all known HGs.

1.6   OTHER TAXONOMIC METHODS

Almost any method available for subtyping *Aeromonas* has been used to approach taxonomic problems within this genus[114]. While some of these exhibited excellent correlation with the various genospecies as defined by DNA–DNA hybridization data, others like bacteriophage typing[115], protein fingerprinting[116,117] and serotyping[118-120] have very little or no taxonomic value at all, but may be very valuable tools for epidemiological investigations. Because quantitative DNA–DNA hybridization, as the only single method accepted for defining new species, is not readily available in most laboratories, combinations of other methods have been used in a polyphasic approach to attempt to group strains and identify them to the species level[121]. These methods will be briefly reviewed in the following paragraphs, and more extensively discussed in Chapter 4.

1.6.1 ENZYME ELECTROPHORESIS

Picard and Goullet[122] established eight species-specific zymotypes among the three species *A. hydrophila, A. sobria* and *A. caviae*. This was accomplished using the isoelectric points of malate dehydrogenase for these three phenospecies as well as the mobility of lactate dehydrogenase from *A. sobria*, and there was good correlation with the previously established DNA HGs of Popoff and Véron[54]. Later, Altwegg and colleagues[123] quantitatively analysed data obtained by the multilocus enzyme electrophoresis (MEE) method as published by Selander *et al.*[124], and again found an excellent, although not perfect, correlation with genomic species as defined by DNA–DNA hybridization. Based on the relative frequencies of alleles at only four loci selected from the 11 enzymes analysed (Table 1.5), 48 clinical *Aeromonas* strains were examined to detect their DNA HG. Only one single strain was not grouped correctly[125]. While four enzymes are sufficient for separating DNA hybridization groups 1 to 12 (Table 1.2), fewer enzymes are needed to distinguish between genomic groups within a given phenotypic species. For the *A. hydrophila* group, the enzymes nucleoside phosphorylase (NSP, E.C 2.4.2.1) and malate dehydrogenase (MDH, E.C.1.1.1.37.) allow complete separation of DNA HGs 1 to 3. A combination of the three enzymes NSP, MDH and NAD-dependent glyceraldehyde phosphorylase (G3PD, E.C. 1.2.1.12.) provides excellent separation of DNA HGs 4, 5a, 5b and 6 of the *A. caviae* group, which are also more difficult to separate biochemically than groups within the *A. hydrophila* group. The remaining DNA HGs 7, 8/10, 11 and 12 can also be separated on the basis of only two enzymes, MDH and G3PD.

1.6.2 SEQUENCES OF 16S RIBOSOMAL RNA GENES

The genes coding for the RNA of the small subunits of ribosomes (16S rDNA) have a relatively conserved nucleotide sequence and most precisely reflect bacterial phylogeny[126]. The respective sequences have been determined for most genomic species of the genus *Aeromonas* by Martinez-Murcia and colleagues[73,110], and phylogenetic trees have been constructed based on these data. The various species exhibited very high levels of overall

Table 1.5 Frequency of alleles at four selected enzyme loci which allow separation of DNA hybridization groups 1 to 12

Locus and allele	Frequency of allele in DNA hybridization group (number of strains analysed) (from Altwegg et al.[123])											
	1 (15)	2 (7)	3 (9)	4 (43)	5A (16)	5B (8)	6 (2)	7 (2)	8 (32)	9 (2)	11 (2)	12 (2)
MDH												
1	0.07	1.00	–	–	–	–	–	–	–	–	–	–
2	0.86	–	1.00	–	0.06	–	1.00	–	0.03	–	1.00	–
3	0.07	–	–	1.00	0.94	0.12	–	1.00	0.97	1.00	–	–
4	–	–	–	–	–	0.88	–	–	–	–	–	–
5	–	–	–	–	–	–	–	–	–	–	–	1.00
NSP												
1	–	–	–	–	–	0.12	–	–	–	–	–	–
2	–	1.00	1.00	–	0.75	0.88	–	1.00	0.94	1.00	–	1.00
3	1.00	–	–	1.00	0.25	–	1.00	–	0.06	–	1.00	–
AKD												
1	–	–	–	–	–	–	–	–	0.19	–	–	–
2	–	–	–	–	–	–	–	–	0.75	1.00	–	–
3	–	–	–	–	0.06	–	–	1.00	0.06	–	–	–
4	0.93	1.00	1.00	–	–	–	1.00	–	–	–	–	–
5	–	–	–	–	0.75	–	–	–	–	–	1.00	–
6	0.07	–	–	1.00	0.19	1.00	–	–	–	–	–	–
7	–	–	–	–	–	–	–	–	–	–	–	1.00
G3PD												
1	–	1.00	–	0.02	–	0.88	1.00	–	–	–	–	–
2	–	–	0.89	–	1.00	0.12	–	1.00	–	–	1.00	–
3	0.20	–	0.11	0.98	–	–	–	–	–	1.00	–	–
4	0.07	–	–	–	–	–	–	–	–	–	–	1.00
5	0.73	–	–	–	–	–	–	–	0.09	–	–	–
6	–	–	–	–	–	–	–	–	0.91	–	–	–

Abbreviations: MDH, malate dehydrogenase; NSP, nucleoside phosphorylase; ADK, adenylate kinase; G3PD, NAD-dependent glyceraldehyde-phosphate dehydrogenase

sequence similarity (98–100%) to each other and formed a distinct evolutionary line with the gamma subclass of the *Proteobacteria*[73]. However, several of the relationships derived from these sequences were in marked disagreement with previously published DNA–DNA hybridization results.

For example, *A. salmonicida* subspp. *achromogenes* and *masoucida* (DNA HG 3) have the same sequence as the reference strain for *Aeromonas* HG 2, but differ in two nucleotides from the sequence for *A. salmonicida* subsp. *salmonicida*. This suggests that strains belonging to DNA HGs 2 and 3 are very closely related and might even be considered as one single species.

However, combining these two hybridization groups into one contradicts the fact that mesophilic strains of HGs 2 and 3 can clearly be separated by ribotyping and multilocus enzyme electrophoresis[103,104,125]. DNA–DNA hybridization of such strains has been contradictory[121]. In addition, the rDNA sequence of mesophilic HG 3 strains has not been determined and may also shed more light on the true relationship among these groups. Ribosomal DNA sequences of HG 4 (*A. caviae*) and HG 14 (*A. trota*) differ by only one single nucleotide, although these species are quite different biochemically and share a DNA–DNA hybridization homology of only 30%[96]. Likewise, HG 1 (*A. hydrophila*) and HG 5 (*A. media*) are biochemically different, but their rDNA sequences differ by only three nucleotides.

Identical 16S rDNA sequences were found for *A. veronii* biovar veronii and *A. veronii* biovar sobria, providing further evidence of the genetic identity of the previously established HGs 8 and 10 despite marked biochemical differences. In contrast *A. sobria* (HG 7) and *A. veronii* biovar sobria (HG 8) were only distantly related to each other while having a DNA–DNA homology of 60–65% as originally determined by Popoff and colleagues[59] under optimal reassociation conditions. As previously mentioned, identical rDNA sequences also confirmed that *A. enteropelogenes* is most probably the same species as *A. trota*[109], since they are identical or nearly identical phenotypically and by ribotyping, respectively[108]. For similar reasons, *A. ichthiosmia* seems to be a synonym for *A. veronii* biovar sobria[105,106] (Table 1.2).

Differences in the nucleotide sequences of 16S rDNA were not found to be evenly distributed over the entire molecule, but rather to cluster in two 'hypervariable' regions corresponding to the regions around positions 160 and 460, respectively, of the *Escherichia coli* sequence described by Brosius *et al.*[127] (Table 1.6). These differences provided a basis for the development of amplification/hybridization systems that allow specific identification of *A. sobria*[128], *A. hydrophila* and *A. veronii*[129], and DNA HG 11[130].

1.6.3   RIBOTYPING

In addition to being a versatile tool for epidemiological investigations[114], ribosomal RNA gene restriction patterns have been shown to correlate very well with DNA–DNA hybridization data[131]. When using restriction enzyme *Sma*I for digesting the chromosomal DNA and plasmid pGML1 (which contains a 832 bp insert representing part of the 16S rRNA gene of *E. coli*) for hybridization, low-molecular-weight bands (<4 kb) gave distinct banding patterns with relatively little variation for the various DNA HGs. The main problems associated with this type of analysis are that for some of the species only two to four strains have been available for analysis, and that some HGs (i.e. *A. veronii* and *A. jandaei*) are very closely related and cannot be separated reliably. However, it has been shown recently that ribotyping and

Table 1.6 16S rDNA sequences of *Aeromonas* strains at two variable regions corresponding to positions 154–167 and 457–476 of the *E. coli* 16S rDNA sequence according to Brosius *et al.*[127]

Species, strain	DNA group	Region 1 (position 154–167)	Region 2 (position 457–476)
A. hydrophila ATCC 7966[T]	1	AGUUGGAAACGACU	UGAUGCCUAAUACGUAUCAA
A. salmonicida subsp. salmonicida NCIMB 1102[Ta]	3	AGUUGGAAACGACU	UGGCGCCUAAUACGUGUCAA
A. caviae ATCC 15467[b]	4	AGUUGGAAACGACU	CAGUAGCUAAUAUCUGCUGG
A. media ATCC 33907[Tc]	5B	AGUUGGAAACGACU	UGAUGCCUAAUACGCAUCAG
A. eucrenophila NCIMB 74[T]	6	AGUUGGAAACGGCU	UGAUGCCUAAUACGCAUCAG
A. sobria NCIMB 12065[T]	7	AGUUGGAAACGACU	UGGCAGCUAAUAUCUGUCAG
A. jandaei ATCC 49568[T]	9	UACUGGAAACGGUA	CAGUAGCUAAUAUCUGCUGG
A. veronii NCIMB 13015[Td]	8/10	UACUGGAAACGGUA	UGGUAGCUAAUAAACUGCCAG
Aeromonas sp. ATCC 35941	11	AGUUGGAAACGACU	UGGUCGCUAAUAAACGGCCAA
A. schubertii DSM 4882[T]	12	UACUGGAAACGGUA	UGGUGGUUAAUACCUGCCAG
A. trota ATCC 49657[T]	14	AGUUGGAAACGACU	CAGUAGCUAAUAUCUGCUGG
A. allosaccharophila	15	AGUUGGAAACGACU	UGGUAGCGAAUAAACUGCCAG

Data from Martinez-Murcia *et al.*[73,110]

[a]Sequence at these positions identical to *A. salmonicida* subsp. *achromogenes* NCIMB 1110[T] and *A. salmonicida* subsp. *masoucida* NCIMB 2020[T]

[b]Sequence at these positions identical to *A. caviae* NCIMB 13016[T] and NCIMB 9235

[c]Complete sequence identical to NCIMB 2237[T] and ATCC 33950

[d]Complete sequence identical to *A. veronii* ATCC 35622, and to *A. veronii* biogroup sobria ATCC 9071 and NCIMB 37

analysis by amplified fragment length polymorphisms (AFLP) give almost identical results when applied to the clustering of strains (see section 1.6.5).

1.6.4   CELLULAR FATTY ACID COMPOSITION

Gas liquid chromatographic analysis of cellular fatty acid methyl esters (FAMEs) again showed a very high similarity among all *Aeromonas* strains and groups analysed[132]. Nevertheless, minor quantitative variations could be used to differentiate phenospecies and/or DNA HGs. For example, the phenotypic complexes *A. hydrophila, A. sobria* and *A. caviae* were basically grouped into distinct FAME clusters. These data are in good agreement with earlier studies[133,134] that did not use strains of DNA HGs. Minor differences in FAME data between different studies can be explained by different culture conditions and different sets of strains used. In 1994, another study[135] again found very similar correlations between FAMEs and DNA–DNA hybridization data. The utility of FAME analysis for assigning *Aeromonas* strains to particular phenospecies or HGs was documented with strains isolated from five Flemish drinking water production plants[136].

1.6.5   AMPLIFIED FRAGMENT LENGTH POLYMORPHISMS (AFLP)

This recently developed genomic fingerprinting technique, which detects DNA polymorphisms by selective amplification of restriction fragments, has successfully been applied to *Aeromonas* strains[137]. Basically, most AFLP groups corresponded to the various DNA hybridization groups of the strains used in that study. Comparison of the complex banding patterns is facilitated by computer analysis, and this also allows quantitation and, consequently, the generation of dendrograms which essentially result in homogeneous groups of strains. The utility of this method has been shown when larger numbers of strains were analysed[138]. Unexpectedly, two subgroups were found among *A. eucrenophila* strains, one of which was much more closely related to DNA HG 11 strains than to the other subgroup of *A. eucrenophila*. These results were fully confirmed by quantifying differences in ribotyping patterns[139]. The significance of these findings is not clear at the moment but questions the original DNA hybridization results obtained with these strains[84] which therefore should be repeated.

1.7   CONCLUSIONS

This is where the taxonomy of the aeromonads stands at the present time, with an illustrious, albeit somewhat confusing, past and a hopeful future, with many questions still to be resolved and many new ones to be asked. It cannot be stressed strongly enough how important it is for persons working with aeromonad strains to know, as can best be determined, the true species of the strains with which they are working. There have been several papers published, even in recent years, on clinical cases as well as detailed molecular studies where the researcher was simply told at the outset that the strain they were to use was simply *A. hydrophila*. Subsequently, the research was published with significant results and with the strain listed as *A. hydrophila* in the Materials and Methods section of the paper. Often when such strains were re-examined by taxonomists, they were found to actually be *A. veronii* biovar sobria, *A. trota*, or even a member of an as yet unnamed DNA hybridization group[140]. Especially in the areas of clinical epidemiology, pathogenesis research, antimicrobial susceptibility studies and human volunteer trials, it is paramount to make a valiant effort to use the most up-to-date methods with which to classify aeromonads prior to attempting to publish valid results with such strains. The aeromonads are a truly fascinating but enigmatic group of aquatic microorganisms that hold many more conundrums besides their taxonomic status within their realm.

1.8   REFERENCES

1   Hazen TC, Flierman CB, Hirsch RP, Esch GW. Prevalence and distribution of *Aeromonas hydrophila* in the United States. *Appl Environ Microbiol* 1978; **36**: 731–8.

2   Seidler RJ, Allen DA, Lockman H, Colwell RR, Joseph SW, Daily OP. Isolation, enumeration and characterization of *Aeromonas* from polluted waters encountered in diving operations. *Appl Environ Microbiol* 1980; **7**: 1010–18.

3   Kaper JB, Lockman H, Colwell RR, Joseph SW. *Aeromonas hydrophila*: ecology and toxigenicity of isolates from an estuary. *J Appl Microbiol* 1981; **50**: 359–77.

4   Burke V, Robinson R, Gracey M, Peterson D, Partridge K. Isolation of *Aeromonas hydrophila* from a metropolitan water supply: seasonal correlation with clinical isolation. *Appl Environ Microbiol* 1984; **48**: 361–6.

5   Van der Kooj, D. Properties of aeromonads and their occurrence and hygienic significance in drinking water. *Zentralbl Bakt Hyg B* 1988; **187**: 1–17.

6   Alonso JL, Botella MS, Amoros I, Alonso MA. The occurrence of mesophilic aeromonads species in marine recreational waters of Valencia (Spain). *J Environ Sci Health* 1994; **3**: 615–28.

7   Sanarelli G. Über einen neuen Mikroorganismus des Wassers, welcher für Thiere mit veränderlicher und konstanter Temperatur pathogen ist. *Zentralbl Bakt Parasitenk* 1891; **9**: 222–8.

8   Hill KR, Caselitz FH, Moody LM. A case of acute metastatic myositis caused by a new organism of the family *Pseudomonodaceae*: a preliminary report. *W Ind Med* 1954; **3**: 9–11.

9   Mathewson JJ, Dupont HL. *Aeromonas* species: role as human pathogens. In Remington JS, Swartz MN (eds) *Current Clinical Topics in Infectious Diseases* Vol 12e, Cambridge: Blackwell Scientific, 1992; 26–36.

10 Lederberg J, Shope RE, Oaks SC Jr. *Emerging Infections: Microbial Threats to Health in the United States.* Committee on Emerging Microbial Threats to Health, Institute of Medicine. Washington, DC: National Academy Press, 1992; 199–201.

11 Altwegg MA, Geiss HK. *Aeromonas* as a human pathogen. *Crit Rev Microbiol* 1989; **16**: 253–86.

12 Janda JM. Recent advances in the study of the taxonomy, pathogenicity, and infectious syndromes associated with the genus *Aeromonas. Clin Microbiol Rev* 1991; **4**: 397–410.

13 Farmer JJ, Arduino MJ, Hickman-Brenner FW. The genera *Aeromonas* and *Plesiomonas.* In Balows A, Trüper HG, Dworkin M, Harder W. Schleifer K-H (eds) *The Prokaryotes,* Second Edition, Vol 3. New York: Springer Verlag, 1992: 3012–45.

14 Joseph SW, Carnahan AM. The isolation, identification, and systematics of the motile *Aeromonas* species. In *Annual Review of Fish Diseases.* New York: Elsevier, 1994; 315–43.

15. Chester FD. A preliminary arrangement of the species of the genus *Bacterium.* Contribution to determinative bacteriology. Part I. *9th Annu Rep Delaware College Agric Exp Sta* 1897: 92.

16 Chester FD. *A Manual of Determinative Bacteriology.* New York: Macmillan, 1901: 235.

17 Zimmerman OER. Die Bakterien unserer Trink- und Nutzewässer. *Berl Naturwiss Ges Chemnitz* 1890; **11**: 86–7.

18 Ernst P. Die Frühjahrsseuche der Frösche und ihre Abhängigkeit von Temperatureinflüssen. *Beitr Pathol Anat* 1890; **8**: 203–20.

19 Marcus LC. Infectious diseases of reptiles. *J Am Vet Med Assoc* 1971; 629–31.

20 Schotts EB, Gaines JL, Prestwood AK. *Aeromonas*-induced deaths among fish and reptiles in an eutrophic inland lake. *J Am Vet Med Assoc* 1972; **161**: 603–7.

21 Gray SJ. *Aeromonas hydrophila* in livestock: incidence, biochemical characteristics and antibiotic susceptibility. *J Hyg Camb* 1984; **92**: 365–75.

22 Shane SM, Harrington KS, Montrose MS, Roebuck RG. The occurrence of *Aeromonas hydrophila* in avian diagnostic submissions. *Avian Dis* 1984; **28**: 804–7.

23 Kluyver AJ, van Niel CB. Prospects for a natural system of classification of bacteria. *Zentralbl Bakteriol Parasitenk Infektionskr Hyg Abt II* 1936; **94**: 369–403.

24 Stanier RY. A note on the taxonomy of *Proteus hydrophilus. J Bacteriol* 1943; **46**: 213.

25 Snieszko SF. Genus IV. *Aeromonas* Kluyver and van Niel, 1936. In Breed RS, Murray EGD, Smith NR (eds) *Bergey's Manual of Determinative Bacteriology,* 7th Edn. Baltimore: Williams and Wilkins, 1957: 189.

26 Emmerich R, Wiebel C. Über eine durch Bakterien erzeugte Seuche unter den Forellen. *Arch Hyg* 1894; **21**: 1–21.

27 Lehmann KB, Neumann R. *Atlas und Grundriss der Bakteriologie.* Munich: JF Lehmann, 1896.

28 Zimmerman OER. Die Bakterien unserer Trink- und Nutzwässer, Part 1. *Berl Naturwiss Ges Chemnitz* 1890; **1**: 38–9.

29 Marsh MC. *Bacterium truttae,* a new bacterium pathogenic to trout. *Science* 1902; **16**: 706.

30 Duff DCB. Dissociation in *Bacillus salmonicida* with special reference to the appearance of a G form of culture. *J Bacteriol* 1937: 49–67.

31 Griffin PJ, Snieszko SF, Friddle SB. A more comprehensive description of *Bacterium salmonicida. Trans Am Fish Soc* 1952; **82**: 129–38.

32 Griffin PJ, Snieszko SF, Friddle SB. A new adjuvant in the diagnosis of fish furunculosis caused by *Bacterium salmonicida*. *Vet Med* 1953; **48**: 280–2.

33 O'Leary WM, Panos C, Helz GE. Studies on the nutrition of *Bacterium salmonicida*. *J Bacteriol* 1956; **72**: 673–6.

34 Smith IW. The classification of *Bacterium salmonicida*. *J Gen Microbiol* 1963; **33**: 263–74.

35 MacInnes JI, Trust TJ, Crosa JH. Deoxyribonucleic acid relationships among members of the genus *Aeromonas*. *Can J Microbiol* 1979; **25**: 579–86.

36 Aiken RS, Barlin B, Miles AA. A case of botulism. *Lancet* 1936; **ii**: 780.

37 Miles AA, Halnan ET. A new species of microorganism (*Proteus melanovogenes*) causing black rot in eggs. *J. Hyg* 1937; **37**: 79–97.

38 Caselitz FH. Eine neue Bacterium der Gattung: *Vibrio* Muller, *Vibrio jamaicensis*. *Z Tropenmed Parasitol* 1955; **6**: 52–63.

39 Ewing WH, Hugh R, Johnson JH. *Studies on the* Aeromonas *Group*. Atlanta, GA: Centers for Disease Control 1961; 1–37.

40 Eddy BP. Cephalotrichous, fermentative Gram-negative bacteria: the genus *Aeromonas*. *J Appl Bacteriol* 1960; 23: 216–49.

41 Eddy BP, Carpenter KP. Further studies on *Aeromonas*. II. Taxonomy of *Aeromonas* and C27 strains. *J Appl Bacteriol* 1964; **27**: 96–109.

42 Ewing WH, Johnson JG. The differentiation of *Aeromonas* and C27 strains from *Enterobacteriaceae*. *Int Bull Bacteriol Numen Taxon* 1960; **10**: 223–30.

43 McCarthy DH. The bacteriology and taxonomy of *Aeromonas liquefaciens*. Technical Report Series 2. Weymouth, Dorset: 1975; 2.

44 Schubert RHW. Über die biochemischen Eigenschaften von *Aeromonas hydrophila*. *Zentralbl Bakteriol Hyg* 1963; **188**: 62–9.

45 Schubert RHW. Die Systematik der Aeromonaden. *Arch Hyg* 1967; 150: 681–8.

46 Schubert RHW. Über den Aminosäureverwendungsstoff wechsel bei Angehörigen des Genus *Aeromonas*. *Zentralbl Bakteriol I Orig* 1969; **211**: 402–5.

47 Schubert RHW. Infrasubspezifische Taxonomie von *Aeromonas hydrophila*. *Zentralbl Bakteriol Abt I Orig* 1969; **211**: 407–9.

48 Habs H, Schubert RHW. Uber die biochemischen Merkmallüe und die taxonomische Stellung von *Plesiomonas shigelloides* (Bader). *Zentralbl Bakteriol Abt I Orig* 1962; **186**: 316–27.

49 Schubert RHW. Zur Taxonomie von *Aeromonas salmonicida* subsp. *achromogenes* (Smith 1963) Schubert 1967 und *Aeromonas salmonicida* subsp. *masoucida* Kimura 1969. *Zentralbl Bakteriol I Orig* 1969; **211**: 413–7.

50 Schubert RHW. The taxonomy and nomenclature of the genus *Aeromonas* Kluyver and Van Niel 1936. I. Suggestions on the taxonomy and nomenclature of the aerogenic *Aeromonas* species. *Int J Syst Bacteriol* 1967; **17**: 23–37.

51 Schubert RHW. The taxonomy and nomenclature of the genus *Aeromonas* Kluyver and Van Niel 1936. II. Suggestions on the taxonomy and nomenclature of the anaerogenic *Aeromonas* species. *Int J Syst Bacteriol* 1967; **17**: 273–9.

52 Schubert RHW. Genus II. *Aeromonas* Kluyver and Van Niel, 1936. In Buchanan RE, Gibbons NE (eds) *Bergey's Manual of Determinative Bacteriology*, 8th Edn. Baltimore: Williams and Wilkins, 1974; 345–8.

53 Von Graevenitz A, Mensch AH. The genus *Aeromonas* in human bacteriology. *N Engl J Med* 1968; **278**: 245–9.

54 Popoff M, Véron M. A taxonomic study of the *Aeromonas hydrophila–Aeromonas punctata* group. *J Gen Microbiol* 1976; **94**: 11–22.

55 Sneath PHA, Sokal RR. *Numerical Taxonomy*. San Francisco: WH Freeman, 1973.

56 Austin B, Priest FG. *Modern Bacterial Taxonomy*. United Kingdom: Van Nostrand Reinhold, 1988.

57	Joseph SW, Daily OP, Hunt WS, Seidler RJ, Allen DA, Colwell RR. *Aeromonas* primary wound infection of a diver in polluted water. *J Clin Microbiol* 1979; **10**: 46–9.

58	Shaw D, Hodder HJ. Lipopolysaccharides of the motile aeromonads; core oligosaccharide analysis as an aid to taxonomic classification. *Can J Microbiol* 1978; **24**: 864–8.

59	Popoff MY, Coynault C, Kiredjian M, Lemelin M. Polynucleotide sequence relatedness among motile *Aeromonas* species. *Curr Microbiol* 1981; **5**: 109–14.

60.	Lee JV, Identification of *Aeromonas* in the routine laboratory. *Experientia* 1987; **43**: 355–6.

61	Popoff M, Genus III. *Aeromonas*. Kluyver and Van Niel 1936,398[AL]. In Krieg NR, Holt JG (eds) *Bergey's Manual of Systematic Bacteriology*, Vol 1. Baltimore, MD: Williams and Wilkins, 1984: 545–8.

62	Merkel JR, Traganza ED, Mukherjee BB, Griffin TB, Prescott JM. Proteolytic activity and general characteristics of a marine bacterium, *Aeromonas proteolytica*, sp.n. *J Bacteriol* 1964; **73**: 247–52.

63	Janda JM, Bottone EJ, Reitano M. *Aeromonas* species in clinical microbiology: significance, epidemiology, and speciation. *Diag Microbiol Infect Dis* 1983; **1**: 221–8.

64	Janda JM, Reitano M, Bottone EJ. Biotyping of *Aeromonas* isolates as a correlate to delineating a species-associated disease spectrum. *J Clin Microbiol* 1984 **19**: 44–7.

65	Allen DA, Austin B, Colwell RR. *Aeromonas media*, a new species isolated from river water. *Int J Syst Bacteriol* 1983; **33**: 599–604.

66	Hsu TC, Waltman WD II, Shotts EB. Correlation of extracellular enzymatic activity and biochemical characteristics with regard to virulence of *Aeromonas hydrophila*. *Int Symp Fish Biol* 1981; **49**: 101–11.

67	Waltman WD II, Shotts EB, Hsu TC. Enzymatic characterization of *Aeromonas hydrophila* complex by the API ZYM system. *J Clin Microbiol* 1982; **16**: 692.

68	Janda JM. Biochemical and exoenzymatic properties of *Aeromonas* species. *Diagn Microbiol Infect Dis* 1985; **3**: 223–32.

69	Fanning GR, Hickman-Brenner FW, Farmer JJ III, Brenner DJ. DNA relatedness and phenotypic analysis of the genus *Aeromonas*. *Abstr Annu Meet ASM* 1985; **C116**: 319.

70	Farmer JJ III, Hickman-Brenner FW, Fanning GR, Arduino MJ, Brenner DJ. Analysis of *Aeromonas* and *Plesiomonas* by DNA–DNA hybridization and phenotype. *1st Int Workshop Aeromonas/Plesiomonas*, Manchester UK, 1986; Abst. P1.

71	Hickman-Brenner FW, MacDonald KL, Steigerwalt AG, Fanning GR, Brenner DJ, Farmer JJ III. *Aeromonas veronii*, a new ornithine decarboxylase-positive species that may cause diarrhoea. *J Clin Microbiol* 1987; **25**: 900–6.

72	Colwell RR, MacDonell MR, De Ley J. Proposal to recognize the family *Aeromonadaceae* fam. nov. *Int J Syst Bacteriol* 1986; **36**: 473–7.

73	Martinez-Murcia AJ, Benlloch S, Collins MD. Phylogenetic interrelationships of members of the genera *Aeromonas* and *Plesiomonas* as determined by 16S ribosomal DNA sequencing: lack of congruence with results of DNA–DNA hybridizations. *Int J Syst Bacteriol* 1992; **42**: 412–21.

74	Kita-Tsukamoto K, Oyaizu H, Nanba K, Simidu U. Phylogenetic relationships of marine bacteria, mainly members of the family *Vibrionaceae*, determined on the bases of 16S rRNA sequences. *Int J Syst Bacteriol* 1993; **43**: 8–19.

75	Holmberg SD, Schell WL, Fanning GR, Wachsmuth IK, Hickman-Brenner FW, Blake PA, Brenner DJ, Farmer JJ III. *Aeromonas* intestinal infections in the United States. *Ann Intern Med* 1986; **105**: 683–9.

76 Brenden R, Janda JM. Detection, quantitation and stability of the beta-hemolysin of *Aeromonas* spp. *J Med Microbiol* 1987; **24**: 247–51.

77 Janda JM, Brenden R. Importance of *Aeromonas sobria* in *Aeromonas* bacteremia. *J Infect Dis* 1987; **155**: 589–91.

78 Mishra S, Nair GB, Bhadra RK, Sikder SN, Pal SC. Comparison of selective media for primary isolation of *Aeromonas* species from human and animal feces. *J Clin Microbiol* 1987; **25**: 2040–3.

79 Kelly MT, Stroh EMD, Jessop J. Comparison of blood agar, ampicillin blood agar, MacConkey-Ampicillin-Tween agar, and modified cefsulodin-irgasan-novobiocin agar for isolation of *Aeromonas* spp. from stool specimens. *J Clin Microbiol* 1988; **26**: 1738–40.

80 Kilpatrick ME, Escamilla J, Bourgeois AL, Adkins HJ, Rockhill RC. Overview of four US Navy overseas research studies on *Aeromonas*. *Experientia* 1987; **43**: 365–7.

81 Pazzaglia G. Studies of virulence and mechanisms of pathogenesis in *Aeromonas* species. ScD Dissertation, Johns Hopkins University, MD, 1993: 401.

82 Statner B, Jones MJ, George WL. Effect of incubation temperature on growth and soluble protein profiles of motile *Aeromonas* strains. *J Clin Microbiol* 1988; **26**: 392–3.

83 Hickman-Brenner FW, Fanning GR, Arduino MJ, Brenner DJ, Farmer JJ III. *Aeromonas schubertii*, a new mannitol-negative species found in human clinical specimens. *J Clin Microbiol* 1988; **26**: 1561–4.

84 Schubert RHW, Hegazi M. *Aeromonas eucrenophila* species nova *Aeromonas caviae* a later and illegitimate synonym of *Aeromonas punctata*. *Zentralbl Bakteriol Hyg A* 1988; **268**: 34–9.

85 Arduino MJ, Hickman-Brenner FW, Farmer JJ III. Phenotypic analysis of 132 *Aeromonas* strains representing 12 DNA hybridization groups. *J Diarrh Dis Res* 1988; **6**: 137.

86 Renaud F, Freney J, Boeufgras JM, Monget D, Sedaillan A, Fleurette J. Carbon substrate assimilation patterns of clinical and environmental strains of *Aeromonas hydrophila*, *Aeromonas sobria*, and *Aeromonas caviae* observed with a micromethod. *Zentralbl Bakteriol Hyg A* 1988; **269**: 323–30.

87 Kuijper EJ, Steigerwalt AG, Schoenmakers BSCIM, Peeters MF, Zanen HC, Brenner DJ. Phenotypic characterization and DNA relatedness in human fecal isolates of *Aeromonas* spp. *J Clin Microbiol* 1989; **27**: 132–8.

88 Altwegg M, Steigerwalt AG, Altwegg-Bissig R, Lüthy-Hottenstein J, Brenner DJ. Biochemical identification of *Aeromonas* genospecies isolated from humans. *J Clin Microbiol* 1990; **28**: 258–64.

89 Austin DA, McIntosh D, Austin B. Taxonomy of fish associated *Aeromonas* spp., with the description of *Aeromonas salmonicida* subsp. *smithia* subsp. nov. *Syst Appl Microbiol* 1989; **11**: 277–90.

90 Carnahan AM, Joseph SW, Systematic assessment of geographically and clinically diverse aeromonads. *Syst Appl Microbiol* 1993; **16**: 72–84.

91 Jacobs D. Displaying taxonomic data analyses using SAS/GRAPH[R]. *Proc 2nd Annu Northeast SAS User Group Regional Conference*, 1988: 132–8.

92 Joseph SW, Carnahan AM, Brayton PR, Fanning GR, Almazan R, Zbick C, Trudo EW Jr, Colwell RR. *Aeromonas jandaei* and *Aeromonas veronii* dual infection of a human wound following aquatic exposure. *J Clin Microbiol* 1991; **29**: 565–9.

93 Joseph SW, Carnahan AM, Rollins D, Walker DI. *Aeromonas* and *Plesiomonas* in the environment: value of differential biotyping of aeromonads. *J Diarrh Dis Res* 1988; **6**: 80–7.

94   Carnahan AM, Joseph SW. *Aeromonas* update: new species and global distribution. 3rd International Workshop on *Aeromonas* and *Plesiomonas*. *Experienta* 1991; **47**: 402–3.

95   Carnahan AM, Fanning GR, Joseph SW. *Aeromonas jandaei* (formerly genospecies DNA group 9 *A. sobria*), a new sucrose-negative species isolated from clinical specimens. *J Clin Microbiol* 1991; **29**: 560–4.

96   Carnahan AM, Chakraborty T, Fanning GR, Verma D, Ali A, Janda JM, Joseph SW. *Aeromonas trota* sp. nov.: an ampicillin-susceptible species isolated from clinical specimens. *J Clin Microbiol* 1991; **29**: 1206–10.

97   Behram S, Grauzlis N, Carnahan AM, Joseph SW. A PC-based procedure for automated classification and identification of populations with binary characteristics. *Binary* 1993; **5**: 199–206.

98   Matousek J, Schindler J. Selecting a small well-discriminating subset of tests from a frequency matrix. *Binary* 1989; **1**: 19–28.

99   Carnahan AM, Behram S, Joseph SW. Aerokey II: A flexible key for identifying clinical *Aeromonas* species. *J Clin Microbiol* 1991; **29**: 2843–9.

100 Furuwatari C, Kawakami Y, Akahane T, Hidaka E, Okimura Y, Nakayama J, Furihata K, Katsuyama T. Proposal for an Aeroscheme (modified Aerokey II) for the identification of clinical *Aeromonas* species. *Med Sci Res* 1994; **22**: 617–9.

101 Kämpfer P, Altwegg M. Numerical classification and identification of *Aeromonas* genospecies. *J Appl Bacteriol* 1992; **72**: 341–52.

102 Abbott SL, Cheung WKW, Kroske-Bystrom S, Malekzadeh T, Janda JM. Identification of *Aeromonas* strains to the genospecies level in the clinical laboratory. *J Clin Microbiol* 1992; **30**: 1262–6.

103 Hänninen ML. Phenotypic characteristics of the three hybridization groups of *Aeromonas hydrophila* complex isolated from different sources. *J Appl Microbiol* 1994; **76**: 455–2.

104 Ali A, Carnahan A, Altwegg M, Lüthy-Hottenstein J, Joseph S. *Aeromonas bestiarum* sp. nov. (formerly genomo-species DNA group 2 *A. hydrophila*), a new species isolated from non-human sources. *Med. Microbiol Lett* 1996; **5**: 156–65.

105 Holmes B, personal communication.

106 Schubert RHW, Hegazi M, Wahlig W. *Aeromonas enteropelogenes* species nova. *Hyg Med* 1990; **15**: 471–2.

107 Schubert RHW, Hegazi M, Wahlig W. *Aeromonas ichthiosmia* species nova. *Hyg Med* 1990; **15**: 477–9.

108 Carnahan AM. *Aeromonas* taxonomy: A sea of change. 4th International Workshop on *Aeromonas/Plesiomonas* Atlanta, GA. *Med Microbiol Newslett* 1993; **2**: 206–11.

109 Collins MD, Martinez-Murcia AJ, Cai J. *Aeromonas enteropelogenes* and *Aeromonas ichthiosmia* are identical to *Aeromonas trota* and *Aeromonas veronii* respectively, as revealed by small-subunit rRNA sequence analysis. *Int J Syst Bacteriol* 1994; **43**: 855–6.

110 Martinez-Murcia AJ, Esteve C, Garay E, Collins MD. *Aeromonas allosaccharophila* sp. nov., a new mesophilic member of the genus *Aeromonas*. *FEMs Microbiol Lett* 1992; **91**: 199–206.

111 Wayne LG, Brenner DJ, Colwell RR, Grimont PAD, Kandler O, Krichevsky MI, Moore LH, Moore WEC, Murray RGE, Stackebrandt E, Starr MP, Trüper HG. Report of the Ad Hoc Committee on Reconciliation of Approaches to Bacterial Systematics. *Int J Syst Bacteriol* 1987; **37**: 463–4.

112 Esteve C, Gutierrez MC, Ventosa A. DNA relatedness among *Aeromonas allosaccharophila* strains and DNA hybridization groups of the genus *Aeromonas*. *Int J Syst Bacteriol* 1995; **45**: 390–1.

113 Esteve C, Gutierrez MC, Ventosa A. *Aeromonas encheleia* sp. nov., isolated from European eels. *Int J Syst Bacteriol* 1995; **45**: 462–6.
114 Altwegg M. Subtyping methods. In Austin B, Altwegg M, Gosling PJ, Joseph S (eds) *The Genus* Aeromonas. Chichester: John Wiley, 1996: 109–125.
115 Altwegg M, Altwegg-Bissig R, Demarta A, Peduzzi R, Reeves MW, Swaminathan B. Comparison of four typing methods for *Aeromonas* species. *J Diarrh Dis Res* 1988; **6**: 88–94.
116 Millership SE, Want SV. Characterization of strains of *Aeromonas* spp. by phenotype and whole-cell protein fingerprint. *J Med Microbiol* 1993; **39**: 107–13.
117 Mullar R, Millership S. Typing of *Aeromonas* spp. by numerical analysis of immunoblotted SDS–PAGE gels. *J Med Microbiol* 1993; **39**: 325–33.
118 Sakazaki R, Shimada T. O-serogrouping for mesophilic *Aeromonas* strains. *Jpn J Med Sci* 1984; **37**: 247–55.
119 Cheasty R, Gross RJ, Thomas LV, Rowe B. Serogrouping of the *Aeromonas hydrophila* Group. *J Diarrh Dis Res* 1989; **6**: 95–8.
120 Thomas LV, Gross RJ, Cheasty T, Rowe B. Extended serogroups scheme for motile, mesophilic *Aeromonas* species. *J Clin Microbiol* 1990; **28**: 980–4.
121 Altwegg M. A polyphasic approach to the classification and identification of *Aeromonas* strains. *Med Microbiol Lett* 1993; **2**: 200–5.
122 Picard B, Goullet Ph. Comparative electrophoretic profiles of esterases, and of glutamate, lactate and malate dehydrogenases, from *Aeromonas hydrophila, A. caviae*, and *A. sobria*. *J Gen Microbiol* 1985; **131**: 3385–91.
123 Altwegg M, Reeves MW, Altwegg-Bissig R, Brenner DJ. Multilocus enzyme analysis of the genus *Aeromonas* and its use for species identification. *Zentralbl Bakteriol Int J Med Microbiol* 1991; **275**: 28–45.
124 Selander RK, Caugant DA, Ochman H, Musser JM, Gilmour MN, Whittam TS. Methods of multilocus enzyme electrophoresis for bacterial population genetics and systematics. *Appl Environ Microbiol* 1986; **51**: 873–84.
125 Altwegg M, Steigerwalt AG, Janda JM, Brenner DJ. Identification of *Aeromonas* species by isoenzyme analysis. *Abstracts of the 89th Annual Meeting of the American Society for Microbiology*, New Orleans 1989, Abstract C-253.
126 Woese CR. Bacterial evolution. *Microbiol Rev* 1987; **51**: 221–71.
127 Brosius J, Palmer JL, Kennedy JP, Noller HF. Complete nucleotide sequence of 16S ribosomal RNA gene from *Escherichia coli*. *Proc Natl Acad Sci USA* 1978; **75**: 4801–5.
128 Ash C, Martinez-Murcia AJ, Collins MD. Molecular identification of *Aeromonas sobria* by using a polymerase chain reaction-probe test. *Med Microbial Lett* 1993; **2**: 80–6.
129 Dorsch M, Ashbolt NJ, Cox PT, Goodman AE. Rapid identification of *Aeromonas* species using 16S rDNA targeted oligonucleotide primers: a molecular approach based on screening of environmental isolates. *J Appl Bacteriol* 1994; **77**: 722–6.
130 Martinez-Murcia AJ, personal communication.
131 Martinetti Lucchini G, Altwegg M. rRNA gene restriction patterns as taxonomic tools for the genus *Aeromonas*. *Int J Syst Bacteriol* 1992; **42**: 384–9.
132 Huys G, Vancanneyt M, Coopman R, Janssen P, Falsen E, Altwegg M, Kersters K. Cellular fatty acid composition as a chemotaxonomic marker for the differentiation of phenospecies and hybridization groups in the genus *Aeromonas*. *Int J Syst Bacteriol* 1994; **44**: 651–8.
133 Canonica FP, Pisano MA. Gas-liquid chromatographic analysis of fatty acid methyl esters of *Aeromonas hydrophila, Aeromonas sobria*, and *Aeromonas caviae*. *J Clin Microbiol* 1988; **26**: 681–5.
134 Hansen W, Freney J, Labbe M, Renaud F, Yourassowsky E, Fleurette J. Gas-

liquid chromatographic analysis of cellular fatty acid methyl esters in *Aeromonas* species. *Zentralbl Bakteriol* 1991; **275**: 1–10.

135 Kämpfer P, Balsczyk K, Auling G. Characterization of *Aeromonas* genomic species by using quinone, polyamine, and fatty acid patterns. *Can J Microbiol* 1994; **40**: 844–50.

136 Huys G, Gersters I, Vancanneyt M, Coopman R, Janssen P, Kersters K. Diversity of *Aeromonas* sp. in Flemish drinking water production plants as determined by gas-liquid chromatographic analysis of cellular fatty acid methyl esters (FAMEs). *J Appl Bacteriol* 1995; **78**: 445–55.

137 Huys G, Gersters I. Vancanneyt M, Coopman R, Janssen P, Vestraete W, Kersters K. Chemotaxonomic analysis and genomic fingerprinting of *Aeromonas* sp. isolated from Flemish drinking water production plants. *Abstracts of the 5th International* Aeromonas–Plesiomonas *Symposium*, Edinburgh, 1995.

138 Huys G, Coopman R, Janssen P, Kersters K. High resolution genotypic analysis of the genus *Aeromonas* by AFLP fingerprinting. *Int J Syst Bact* 1996; **46**: 572–80.

139 Altwegg M, unpublished results.

140 Carnahan A, unpublished results.

2 Isolation and Enumeration of Aeronomads

NELSON P. MOYER

Hygienic Laboratory, Iowa City, USA

2.1 INTRODUCTION

Aeromonas hydrophila and the motile, mesophilic aeromonads which cause human disease are distinctly differentiated by growth temperature from the psychrophilic, non-motile aeromonads causing diseases in fish. The media and methods required for their isolation differ markedly. To further complicate studies of aeromonads, they are present in aquatic, terrestrial and intestinal habitats, together with a myriad other microorganisms. Because of the diversity of habitats, variety of samples and the numbers of competing microflora, representative isolation and enumeration of aeromonads are among the most challenging exercises microbiologists face. Investigators have a bewildering choice of media and methods, and they are left to choose those which are most appropriate to their needs.

Aeromonads grow readily on most eugonic media; however, to recover them from polymicrobic sources such as stool, sewage, sludge, surface water and food requires the use of selective and differential media, sometimes with prior enrichment. Media for the isolation of *Aeromonas* spp. from contaminated sources were first borrowed from clinical microbiology and later modified for environmental use. Foods present yet another set of difficulties for microbiologists who must add enumeration, resuscitation of injured cells and enrichment techniques to their armamentarium of culture methods. Media used in the clinical laboratory are highly selective and wholly unsuitable for detecting low numbers of injured organisms in foods and drinking water.

Several publications detailing isolation of aeromonads from foods have appeared, and methods for aeromonads are included in some editions of the US Food and Drug Administration *Bacteriological Analytical Manual* and the *Compendium of Bacteriological Methods for Analysis of Foods* published by the American Public Health Association[1-4]. However, few comprehensive reviews of media and methods for isolation of aeromonads from environmental sources have been published[5-7]. The purpose of this chapter is to present the options available to microbiologists for isolation and enumeration of aeromonads in the many and varied environments in which they occur.

The Genus Aeromonas. Edited by B. Austin, M. Altwegg, P.J. Gosling and S. Joseph
© 1996 John Wiley & Sons Ltd

2.2 BACTERIOLOGY

While clinical laboratories prefer to use the temperature range 35–37°C to effect a rapid recovery of the aetiologic agent to expedite patient diagnosis, there is an advantage in using lower temperatures for environmental samples since aeromonads grow rapidly at 28–30°C, while competing members of the Enterobacteriaceae prefer a higher growth temperature. Selective media for isolation of aeromonads must be read and counted after 18–24 h incubation at 28–30°C for accurate results. Holding plates longer than 24 h before the first reading risks inaccuracies from colour changes around colonies, which render them unrecognizable as presumptive aeromonads from overgrowth of competing microbial flora.

The lower limit of growth of aeromonads is usually considered to be pH 5.2 while they tolerate alkaline pH up to 9.8, a feature exploited in the use of alkaline peptone water (APW), pH 8.6–9.1, for their enrichment and enumeration using the most probable number technique[8].

Aeromonads produce a variety of hydrolytic extracellular enzymes, haemolysins and cytotoxins[9]. Differential media have exploited the variety of hydrolytic enzymes produced by aeromonads to facilitate their isolation. Substrates commonly used include gelatin (gelatinase), starch (amylase), DNA (DNAse), elastin (elastase), dextrin (dextrinase), and egg yolk (lecithinase).

2.2.1 DIFFERENTIAL REACTIONS FOR THE GENUS *AEROMONAS*

Some strains of *Aeromonas*, including all strains of *A. caviae*, rapidly die in broth culture containing glucose after 12 h incubation at 37°C. This 'suicide phenomenon' results from the accumulation of acetic acid from glucose metabolism which lowers the pH of the broth to 5.2[10,11]. This phenomenon has been suggested as a differential characteristic for identification of aeromonads when used in conjunction with aesculin hydrolysis and production of gas from glucose[12,13]. To test for the suicide phenomenon, inoculate a tube of nutrient broth statically and incubate it at 30°C for 24–36 h. A pellet in the bottom of the tube over a clear broth is a positive test, while a turbid culture is a negative test. Subculture a loopful of the sediment to a blood agar plate to test viability.

Aeromonads differ from the Enterobacteriaceae in their inability to ferment inositol and xylose; hence, these carbon sources are frequently incorporated into culture media together with a pH indicator as differential ingredients. Colony types of xylose-fermenting and xylose-non-fermenting bacteria are difficult to distinguish in mixed cultures[14]. Media which rely upon carbohydrates for differentiation of aeromonads must be made highly selective to facilitate recognition of *Aeromonas* colony types. Xylose and lactose-containing media are inhibitory to some strains of *Aeromonas*. The mechanism is not known, but it is postulated that autoclaving of media containing these carbon sources produces toxic products which interfere with growth of aeromonads. The toxicity of xylose is not apparent in media such as xylose desoxycholate citrate (XDC) agar, which are boiled instead of autoclaved[15]. Xylose 'intoxication' of culture media may also be prevented by aseptic addition of filter sterilized xylose after the basal medium has been autoclaved and cooled below 50°C.

Several carbohydrate reactions have value for rapid differentiation of aeromonads. Most *Aeromonas* spp. are inositol-negative and mannitol-positive, while *Plesiomonas* is inositol-positive and mannitol-negative. *A. schubertii* is mannitol-negative. *A. jandaei* is sucrose-negative. *A. eucrenophila* and *A. caviae* are non-aerogenic. *A. encheleia* is gluconate-negative and may be missed when starch glutamate ampicillin penicillin (SGAP-10C) agar is used to isolate aeromonads from environmental samples. Fermentation of glucose with or without gas production separates aeromonads from pseudomonads.

Altwegg and colleagues[16] noted temperature- and media-dependent decarboxylase reactions among aeromonads and suggested that reference laboratories should use Fay and Barry's formulation to determine ornithine, lysine and arginine reactions after incubation at 29°C for 24 h. Clinical laboratories frequently use motility indole ornithine medium (MIO) to screen colonies picked from enteric plates. It was suggested that two tubes of MIO, incubated at 22°C and 37°C, should be used to screen picks from Yersinia Selective Agar (Difco) plates. Additionally, colonies are streaked on blood agar for oxidase testing; *Yersinia* being differentiated from aeromonads by exhibiting motility at 22°C, but not at 37°C, and by a negative oxidase test. *A. veronii* is arginine-negative and ornithine-positive, whereas other aeromonads are ornithine-negative.

Aeromonas species are differentiated from *Vibrio furnissii* and *V. fluvialis* by failure to grow at 4°C and in 6–8% NaCl broth. These reactions are extremely important for correct identification of environmental isolates of these organisms, which closely resemble each other[17,18].

Other tests recommended for recognition of aeromonads and differentiation among aeromonads include cephalothin susceptibility[19], pyrazinamidase[20] and the CAMP test[21].

2.2.2 SCREENING MEDIA

Kaper and co-workers[22] modified MIO semi-solid agar to develop a medium for screening presumptive *Aeromonas* colonies picked from selective media. This medium permits determination of mannitol and inositol fermentation, ornithine decarboxylation, indole production, motility, H_2S and gas production in a single tube. Typically, three to ten colonies per sample are stabbed into tubes of *A. hydrophila* (AH) Medium, which is then incubated at 30°C for 24 h. Organisms, which ferment mannitol and do not decarboxylate ornithine, produce a yellow butt with a purple band at the top of the tube. Organisms, which are ornithine-positive and mannitol-positive or -negative, give an alkaline reaction throughout the tube. Tryptose is added to the medium to facilitate a test for indole by addition of Kovac's reagent. Motility is discernible as turbidity throughout the butt. Cultures which produce an alkaline reaction in the top centimetre of the tube and an acid butt are tested for oxidase production. Motility and indole reactions are not particularly helpful in the presumptive identification of *Aeromonas* spp.

An evaluation of AH medium, also known as Kaper's multi-test medium, was published by Toranzo *et al.*[23], who preferred incubation at 25°C for 48 h. They found false negative indole reactions, and reported that differentiation of aerogenic from anaerogenic strains was problematic. Overall, Kaper's multi-test medium is the best single tube method available for screening colonies picked from selective media. Differential characteristics of aeromonads and coliforms on Kaper's medium are shown in Table 2.1.

Hugh-Leifson or a similar medium may be used to determine whether glucose metabolism is oxidative or fermentative. While the OF reaction does not provide as much information as Kaper's multi-test medium, OF media are readily available commercially.

Triple sugar iron agar (TSI) and lysine iron agar (LIA) are frequently used to screen colonies picked from stool cultures in clinical laboratories. On TSI agar, aeromonads produce K/A or A/A reactions, with or without gas. Reactions on LIA are K/K or K/A. H_2S is not produced on either medium. The spot oxidase test may be performed on growth from the LIA slant. Oxidase-positive cultures are tested further to confirm that they are aeromonads.

De Ryck and colleagues[24] suggested a method for screening aeromonads from Yersinia Selective Agar (Difco) plates. Mannitol-negative colonies were

Table 2.1 Reactions on Kaper's multi-test medium

Organism	Top	Butt	Motility	H_2S	Indole
Aeromonas hydrophila	K	A	+	−	+
Klebsiella pneumoniae	A	A	−	−	−
Klebsiella oxytoca	A	A	−	−	+
Escherichia coli	K	K or A	+ or −	−	+
Salmonella spp.	K or A	K or A	+	+	−
Enterobacter spp.	K or N	K or N	+	−	−
Proteus spp.	R	K or A	+	+ or −	+
Yersinia enterocolitica	K or N	K or N	−	−	+ or −
Citrobacter spp.	K	A or K	+	+	−
Serratia spp.	N or K	N or K	+	−	−

K, alkaline; A, acid; N, neutral; R, red (deamination reaction)

inoculated onto plates of Phenol Red Broth Base (Difco) containing xylose, aseptically added after autoclaving and cooling, at a final concentration of 0.4%, pH 7.8, by streaking a thin line approximately 1 cm long on the plate (spot inoculation). Up to 16 colonies could be inoculated per plate. Plates were incubated and xylose-negative, oxidase-positive colonies were considered presumptive aeromonads.

Nutrient agar plates containing gelatin and O/129 for determination of oxidase and gelatinase production, and sensitivity to O/129 have been used to screen colonies picked from selective media[25]. This system permits presumptive identification of aeromonads to genus with only two media.

Slade and colleagues[26] proposed a new medium, gelatin arginine dihydrolase (GAD) agar, for screening of aeromonads picked from selective media. After incubation, GAD medium was placed at 4°C for 1–2 h and a positive gelatinase reaction was indicated if the medium remained liquid. Arginine dihydrolase reactions were recorded if the medium turned orange to red from the original cerise colour after passing through an intermediate yellow stage. Most aeromonads isolated from bile salts brilliant green (BBG) or xylose desoxycholate citrate (XDC) media did not demonstrate typical positive reactions on GAD until 48 h of incubation. Colonies picked from BBG and XDC yielded false negative GAD results 14.8% and 18.2% of the time, respectively. Further refinement of GAD as an alternative to Kaper's multi-test medium was suggested.

Pessoa and Da Silva[27] described a screening medium for presumptive identification of *Aeromonas* and *Vibrio* which permitted determination of motility, sucrose fermentation, gas production from glucose, H_2S indole and urease production, and tryptophane deaminase and lysine decarboxylase reactions in a single tube. Growth from the slant was suitable for oxidase reaction and determination of β-galactosidase and gelatinase reactions and for antigenic analyses. This test has not been evaluated recently, and no data are available on its performance with newly identified aeromonads.

Kalina and colleagues[28] described a medium (A-2) using gelatin as the sole nitrogen source and starch as the sole carbon source for differentiation of *Vibrio* and *Aeromonas*. Other organisms common to faeces, sewage and surface water failed to grow in the medium. This medium has not been independently evaluated.

2.3 APPROACHES TO THE ISOLATION OF AEROMONADS

2.3.1 SAMPLE COLLECTION METHODS

2.3.1.1 Environmental samples

Water samples may be collected in sterile, screw-capped glass or plastic bottles or Whirl-Pak bags (Nasco, Fort Atkinson, WI). A sample volume of 200–1000 ml is sufficient for most investigations. Sample containers used to collect chlorinated water should contain 24 mg/l sodium thiosulphate to inactivate free residual chlorine. Versteegh *et al.*[29] recommended addition of EDTA (50 mg/l) to sample bottles to neutralize toxic effects of heavy metals leached from plumbing. Concentrations of EDTA up to 250 mg/l have been reported[30].

Bravo and colleagues[31] used Moore swabs enriched in alkaline peptone water (APW) pH 8.6, to isolate aeromonads from the sewage system of a hospital. Spira bottles, a modification of the Moore swab, have been used by the author to isolate aeromonads from groundwater pumped from wells[32,33]. Empty one-pound (454 g) plastic media bottles are used by cutting a 2 cm hole in the bottom and packing the bottle with 30 cm of sterile cotton gauze in a manner which prevents channelling of water. Spira bottles may be placed in zip-lock bags until used for sample collection. At the sample site, approximately 10 litres of water are passed through the bottle and allowed to flow back to the source. The bottles are then returned to the bag for transport to the laboratory.

Sediment and sludge samples may be collected using commercially available sample or core devices. Usually, a 200 g sample is sufficient. Samples are placed in screw-capped containers or Whirl-Pak bags. When sampling equipment must be used repeatedly to obtain environmental samples, it is permissible to disinfect the apparatus with sodium hypochlorite (18 mg/l), followed by a thorough rinse in the next water to be sampled.

Environmental samples should be placed in insulated containers with ice packs for transport to the laboratory and stored at 2–8°C. Where enumeration is important, samples must be processed within 6–8 h of collection. All samples should be processed within 24 h of collection.

2.3.1.2 Fish

Fish samples are taken by swabbing the gills, lesions and external surfaces and inoculating plates directly and placing swabs into transport or enrichment medium. Tissue, faeces and organ samples of 1–2 g each are aseptically placed into sterile containers for later processing. Smears for Gram stain should be made from external lesions, gills, fins and internal organs (kidney and liver) if septicaemia is suspected[34,35].

Blood for serological studies is collected by caudal venipuncture without use of anticoagulant. One drop of blood may be added to 3 ml of TSB if culture is desirable.

2.3.1.3 Food

Whenever possible, a food sample should weigh at least 25 g. Ideally, 100–200 g of sample should be collected. A 25 g sample is weighed and added to 225 ml of 0.1% peptone water, followed by homogenization by blending in a mechanical blender or in a stomacher for 1–2 min. Further dilution in peptone water or phosphate-buffered saline is necessary if enumeration is desirable. Plates are inoculated with 0.1 ml of dilutions and the sample is spread across the surface of the plates with a bent glass rod which has been surface sterilized by dipping it into 95% ethanol and flaming. Automated systems are available for laboratories that process large numbers of samples.

Since aeromonads are capable of growing at 5°C in foods, samples must be refrigerated in transit and processed as soon as possible upon arrival in the laboratory. Some strains may be sensitive to acid pH below 5.5, so neutral buffers must be used to prepare sample dilutions. Aeromonads have been isolated from frozen oysters held for one year and stabilized cell suspensions are recoverable after many years of storage at −70°C, suggesting that freezing and thawing has little effect on recovery from frozen foods, although enrichment techniques may be necessary when aeromonads are present in low numbers.

Sampling of environmental surfaces is an important adjunct to investigations of foodborne illness. Surfaces may be wiped with swabs or with a 10 × 10 cm gauze pad moistened with saline or 0.1% peptone in saline, unless the site is already moist. The swabs or pads are placed in a sterile plastic bag, transported refrigerated, and processed the same day. Swabs are placed into enrichment media and incubated overnight at 25°C prior to being subcultured to plating media. Gauze pads are placed into a bottle containing 10 ml of APW or TSBA enrichment broth and processed in the same way as the swabs.

2.3.1.4 Humans and animals

Specimens of wound exudates and body fluids from normally sterile sites may be aspirated or collected on swabs and submitted in sterile containers. Blood cultures are inoculated into commercial blood culture bottles and incubated following the usual protocol for handling blood cultures in the clinical laboratory.

Faecal samples for *Aeromonas* culture may be submitted in Cary–Blair transport medium at ambient temperature, and held at refrigerator temperatures until processed. Approximately 0.1 g of faeces is inoculated onto each plate with a swab and the plate is streaked for isolated colonies with a wire loop, If enrichment broth is used, approximately 1 g of stool is inoculated into 10 ml of medium. Rectal swabs may be used to collect faecal samples from infants and animals, but faecal specimens are preferred.

2.3.2 NON-SELECTIVE MEDIA

Aeromonads grow readily on eugonic media, and blood agar is preferred for culturing clinical specimens from normally sterile sites. The choice of a non-selective medium depends upon the nature of the sample, the organisms which may interfere with recognition of aeromonads, and the purpose of the investigation.

2.3.3   DIFFERENTIAL AND SELECTIVE MEDIA

Use of a selective medium is a compromise and incorporation of dyes, bile salts and antibiotics into isolation media always reduces the yield of the desired organisms. While selective media reduce the counts of aeromonads in environmental samples, non-selective media are subject to overgrowth, reducing the ability to recognize colours and zones on differential media. Virtually all inhibitory agents incorporated into selective media for isolation of aeromonads will prevent recovery of a small percentage of strains.

Dyes such as basic fuchsin, crystal violet, brilliant green and eosin are incorporated into culture media to inhibit Gram-positive bacteria and facilitate recovery of members of the Enterobacteriaceae. While most aeromonads tolerate concentrations of dyes used in culture media, some strains are inhibited.

Because aeromonads are frequently found together with coliform organisms, media containing bile salts intended for isolation of enteric pathogens have been used for the isolation of aeromonads from contaminated environmental and intestinal sources. Bile salts were not found to be inhibitory at concentrations of 2.5 g/l regardless of the form used[36]. Difco Bile Salts proved inhibitory at this concentration for 4% of strains. Different brands of bile sales[14] and formulations of MacConkey agar[37] may affect colony morphology and recovery rates. In one study, 25% of aeromonads failed to grow on TCBS agar, and 2% failed to grow on MacConkey agar[36].

Addition of ampicillin to culture media for recovery of aeromonads from clinical specimens is based upon their relative resistance to that antibiotic. The concentration of ampicillin for optimal recovery with minimal inhibition of sensitive strains is a matter of controversy. Investigators have recommended concentrations of 10 mg/l, 20 mg/l, 30 mg/l and 40 mg/l, but no comparative study has been published which directly compares all four concentrations of ampicillin on a variety of samples. Environmental and food microbiologists use ampicillin-containing media for contaminated environmental samples since environmental strains share the same resistance characteristics to ampicillin as clinical strains. Ampicillin sensitivity is variable with *A. caviae, A. schubertii* and *A. veronii* subsp. sobria, *A. media, A. sobria* (HG 7), *A. eucrenophila*, HG 11 (unnamed) and *A. trota* cannot be recovered on media containing ampicillin[38]. The MIC for ampicillin for most aeromonads (88% of strains tested) range between 10 and ≥80 mg/l, but some strains of *A. caviae* are sensitive to ampicillin concentrations of 2.5 mg/l[39,40].

Want and Millership[14] examined the effects of incorporating ampicillin, bile salts and carbohydrates into media for recovery of aeromonads from stools. Most data support the use of ampicillin (10 mg/l) for isolation of clinically significant aeromonads. Less than 5% of *Aeromonas* strains tested are sensitive to ampicillin[41,42].

Ampicillin is relatively unstable in culture media and the initial concentration is reduced by 25% within one week after preparation[15]. Media containing ampicillin must be stored at 2–8°C for no longer than one week before use. The success of investigators using media containing 30 mg/l suggests that plates may have been stored longer than one week with a progressive loss of potency of the ampicillin which allowed *A. caviae* and other moderately susceptible aeromonads to grow. Stock solutions of ampicillin must be prepared, dispensed in unit volumes containing the desired concentration for a litre of medium, and frozen at −20°C to maintain the activity of the antibiotic.

Approximately 14% of aeromonads are resistant to cefsulodin at 4 mg/l and 4% of aeromonads are resistant to novobiocin at 4 mg/l[43]. Some strains of *A. media* are resistant to cephalothin. Cefsulodin sensitivity has been suggested as the reason some strains of *A. sobria* and *A. hydrophila* are inhibited on Yersinia Selective Agar (Difco). *A. jandaei* is resistant to colistin.

A list of differential and selective media, their abbreviations, the application for which they are used and the literature citation containing their formulation is presented in Table 2.2.

2.3.4 ENRICHMENT TECHNIQUES

2.3.4.1 Environmental samples

Enrichments should not be used for ecological studies since the predominant strain(s) will quickly overgrow other strains which may be present. Enrichments are only useful as a presence–absence test for aeromonads in food and drinking water, or for monitoring marine populations in shellfish harvesting areas.

2.3.4.2 Faecal samples

The use of enrichment for recovery of aeromonads from stools is one of the more contentious issues facing the clinical microbiologist. It is acknowledged that sensitive enrichment methods obviate the ability to produce quantitative results and may select asymptomatic carriers. The use of highly selective media in a search for specific enteric pathogens is widely practised in clinical microbiology. Because of the enormous variety of faecal flora, it has been necessary to couple the use of selective media with enrichments and

Table 2.2   Media for isolation of aeromonads

Medium	Abbreviation	Sample	Method	Reference
Aeromonas Agar (Difco)	XIBBG	Stool	Streak plate	52a
Aeromonas Medium (Ryan's medium)	AM	Stool, food, water	Streak plate	†
Alkaline Peptone Water	APW	Stool, water, food	Enrichment, MPN	3
Ampicillin Dextrin Agar	ADA	Water	Spread plate, MF	15
Bile Salts Brilliant Green Agar	BBG	Stool	Streak plate	52
Bile Salts Brilliant Green Starch Agar	BBGS	Food, water	Spread plate	160
Blood Ampicillin Agar	BAP	Clinical specimens	Streak plate	36
Butzler Campylobacter Selective Agar	BCSA	Stool	Streak plate	175
Congo Red Agar	CRA	*A. salmonicida*	Spread plate	137
Coomassie Blue Agar	CBA	*A. salmonicida*	Spread plate	133
Desoxycholate Citrate Agar	DCA	Stool	Streak plate	107, 153
Dextrin Fuchsin Sulphite Agar	DFS	Water	Spread plate, MF	67
DNAse Methyl Green Bile Agar	DMGB	Stool	Streak plate	176
DNAse Toluidine Blue Ampicillin Agar	DNTA	Stool	Streak plate	144
Glutamate Starch Penicillin Agar	GSP	Water	Spread plate	66
Inositol Brilliant Bile Greeen Salts Agar	IBG	Stool	Streak plate	145
MacConkey Trehalose Agar	MT	Water	Spread plate	86
MacConkey Tween Ampicillin Agar	MTA	Stool	Streak plate	147
Meso-Inositol Xylose Agar	MIX	Water	MF	100
Modified Ampicillin Dextrin Agar (O/129)	MADA	Seawater	MF	87
Modified Peptone Beef Extract Glycogen Agar	PBG	Water	Pour plate	58
Modified PXA	MPXA	Stool	Streak plate	47
Modified Rimler–Shotts Medium	MRS	Water	MF, MPN	59, 73
Modified Starch Ampicillin Agar (Lachica)	MSA	Food	Spread plate	3
Paranitrophenyl Glycerine Agar	PNPG	Stool	Streak plate	36
Pril Xylose Ampicillin Agar	PXA	Stool	Streak plate	68
Rimler–Shotts Agar	RS	Water	Spread plate, MF	62
Rippey–Cabelli Agar	mA	Water	MF	65
Starch Ampicillin Agar	SA	Food	Spread plate	158
Starch DNA Ampicillin Agar	SDA	Food	Spread plate	3
Starch Glutamate Ampicillin Penicillin Agar	SGAP-10C	Water	Spread plate	70

Table 2.2   (*continued*)

Medium	Abbreviation	Sample	Method	Reference
Tryptic Soya Ampicillin Broth	TSBA	Food	Enrichment, MPN	3
Tryptone Xylose Ampicillin Agar (Modified PXA)	TXA	Water	MF	15
Xylose Ampicillin Agar	XAA	Water	MF	103
Xylose Desoxycholate Citrate Agar	XDC	Stool	Streak plate	49
Yersinia Selective Agar (Modified CIN)	YSA	Stool	Streak plate	43

† unpublished formulation, Oxoid CM833

occasionally even pre-enrichments for recovery of specific pathogens. Whenever possible, concurrent use of non-selective media without enrichment is encouraged.

Robinson and colleagues[44,45] reported comparative studies of several enrichment broths to enhance recovery of aeromonads from stools. They found that nutrient broth gave the best recovery; however, they contended that enrichment allows isolation of aeromonads in the low faecal concentrations likely to be found in convalescent patients, carriers and those with subclinical infection. They argued that routine use of enrichment for isolation of faecal aeromonads, by detecting *Aeromonas* spp. in low numbers in patients without diarrhoea, is likely to confuse interpretation of epidemiological studies seeking to clarify the relationship between *Aeromonas* spp. and acute diarrhoea.

Several investigators recommend alkaline peptone water (APW) adjusted to pH 8.6–9.8 for enrichment of faecal specimens to be cultured for aeromonads[46–50]. The basic formulation contains 1% peptone in water at pH 8.6, but some investigators have increased the pH to as high as 9.8, added ampicillin (10–40 mg/l) and/or 0.5% desoxycholate to increase selectivity[51]. The most favourable incubation temperature for APW enrichments is also contentious. Millership and Chattopadhyay[52] found better growth at 25°C while Price and Hunt[52a] found no difference between 25°C and 37°C. Aeromonads have been isolated from GN broth enrichments, but at about half the efficiency of APW[46]. Moyer *et al.*[48] found that enrichment procedures increased the number of *A. caviae* isolates, suggesting that APW may be detecting transient colonizers, rather than frank pathogens.

In the author's experience, only 19% of *Aeromonas* isolates are recovered only from APW enrichments and 8% of isolates are recovered from cold enrichment broth intended for recovery of *Yersinia enterocolitica*. In this laboratory, 72% of aeromonads were isolated on blood agar containing ampicillin (10 mg/l) and only 6% were recovered using the usual media for

isolation of *Salmonella, Shigella* and *Campylobacter*. Clearly, reliance on
non-selective blood agar and routine enteric media will miss clinically
significant aeromonads. Until typing systems are available to distinguish
between virulent and avirulent strains of aeromonads, clinical laboratories
must continue to use sensitive methods for detection of aeromonads in stools
of patients with diarrhoea, and leave the question of clinical relevance to the
physician.

2.3.4.3   Fish

Typical and atypical strains of *A. salmonicida* have been recovered from fish
samples by enrichment. The decision whether or not to use enrichment
procedures depends upon the purpose of the investigation, the nature of the
sample, the acceptable sensitivity of recovery and whether enumeration of
aeromonads is desirable.

2.3.4.4   Food

When small numbers of aeromonads are anticipated and where there is the
probability of cell injury from environmental pressures, samples may be
placed in enrichment broths. Two broths, tryptic soy ampicillin (30 mg/l)
broth (TSAB) and APW are most frequently used, although several
formulations have been proposed. APW has been used without bile salts and
with or without ampicillin for enrichment and enumeration of aeromonads
from foods by the most probable number (MPN) technique.

2.3.5   ENUMERATION METHODS

Enumeration of aeromonads poses herculean challenges to microbiologists.
While they are ubiquitous in aqueous environments and grow well on most
laboratory media, they are also accompanied by myriad other bacteria with
similar growth requirements. To further complicate their recovery, they have
wide varying susceptibility to the selective agents incorporated into culture
media so that no medium has been developed that will recover all
Aeromonas spp. and strains reliably and quantitatively from the
environment. Until recently, the taxonomic confusion within the genus
Aeromonas was a major limitation to a systematic study of the distribution of
environmental aeromonads.

Plating methods using differential and selective media necessarily
underestimate cell populations. The limit of detection of aeromonads from
stools has been estimated at 2–3 $\log_{10}$. Below this level, they are overgrown by
competing flora or diluted below detection by laboratory manipulations[53].
Staining methods are tedious and may overestimate cell populations, and
instrumental methods such as flow cytometry and fluorometric detection of

dye-stained cells are just beginning to be applied for enumeration of environmental microorganisms.

Direct inoculation of specimens is common in clinical laboratories and enumeration is rarely necessary. Microbiologists may report a semi-quantitative estimate of growth based upon the number of quadrants showing growth. Spread plate and pour plate techniques, using decimal dilutions of samples, are used primarily for food, sediment and sludge samples. Selection of dilutions for quantitative species determination is critical. With low dilutions, aeromonads may be overgrown or suppressed, while at higher dilutions, they may be diluted to extinction. Spread plates are subject to overgrowth by *Bacillus* and swarming *Proteus* unless selective ingredients are incorporated into the medium. Water samples are usually processed using the membrane filtration method and the membranes are placed onto the surface of differential and selective media for incubation. Where the number of aeromonads is estimated to be low, membranes may be placed into enrichment broth for a presence–absence test. The efficiency of membrane filtration methods depends upon the properties of the medium selected and the organisms in the sample. Aeromonads in food and water samples may be enumerated using the MPN technique described in *Standard Methods for the Examination of Water and Wastewater* and the *Compendium of Methods for the Microbiological Examination of Foods*. Alkaline peptone water and tryptic soy broth containing ampicillin (30 mg/l) are the most frequently used media for MPN methods.

2.3.5.1 Biotyping

Microbiologists responsible for the isolation of aeromonads from environmental and intestinal sources encounter a staggering number of biotypes[54]. The author has isolated up to nine biotypes from water drawn from a 150 foot (~45 m) drilled, steel-cased well, while Gray reports at least seven biotypes from a single sample of animal faeces. This situation presents a challenge to the epidemiologist attempting to understand the relationship of clinical isolates with their environmental source.

2.3.5.2 Stock cultures

Working cultures of *Aeromonas* spp. are grown on nutrient media which do not contain fermentable carbohydrate, placed at 5°C, and subcultured every 2–3 weeks. Effendi and Austin[55] suggest a maintenance medium containing 0.5% peptone, 0.1% yeast extract, 2.4% NaCl, 0.7% Mg_2SO_4, 0.075% KCl and 1.5% agar for marine strains. Cultures may be stored in 15% glycerol at -20°C for short periods (months), but lyophilization or freezing in serum inositol broth (consisting of 25 g of inositol dissolved in 50 ml of distilled

water, filter sterilized, and aseptically added to 450 ml of sterile calf serum) at
−70°C is preferred for long-term storage of stock cultures.

2.3.5.3 Supplemental tests

Whenever studies of *Aeromonas* populations in the environment are
contemplated, it is prudent to perform additional microbiological and
physicochemical analyses which may provide insight into the interpretation
of *Aeromonas* culture results. The number of aeromonads in natural waters
and sediments shows a correlation to the presence of faecal coliforms,
therefore tests for enumeration of total and faecal coliform bacteria should
be included in any survey of unknown waters. Heterotrophic plate count data
and knowledge of the predominant bacterial flora may guide investigators to
selection of an appropriate selective medium for optimal recovery of
aeromonads. Besides microbiological tests, turbidity, salinity, temperature,
pH, dissolved oxygen, biological oxygen demand (BOD), assimilatable
organic carbon, and free chlorine residual may be useful data in some
investigations.

2.4   ISOLATION MEDIA AND METHODS

2.4.1   ISOLATION OF AEROMONADS FROM ENVIRONMENTAL SAMPLES

Aeromonads may be present in sufficient numbers to permit direct detection
in samples collected during summer months. One loopful (10 μl) may be
streaked directly onto culture media.

Moore swabs and gauze removed from Spira bottles is placed into a
container of APW which is incubated at 25°C overnight and subcultured to
plating media for isolation of aeromonads.

Selective media are required for isolation of aeromonads from stool,
sewage, sludge and heavily polluted waters. Counts in sewage average 10^5
CFU/ml, counts in polluted river water are in the range 10^3–10^5 CFU/ml and
counts in clean river water in the range 10^0–10^2 CFU/ml. High counts are
associated with high BOD of water[56].

Peptone Beef Extract Glycogen Broth (PBG)[57] was modified by addition
of bromothymol blue, sodium lauryl sulphate and agar for isolation and
enumeration of aeromonads from aquatic animals and their environment[58].
Pour plates were prepared, incubated for 4–5 h at room temperature, overlaid
with 2% agar in water, and reincubated at room temperature for three to
four days. Aeromonads appear as large yellow colonies which can be picked
and screened using a spot oxidase test. *Citrobacter, Proteus* and *Serratia*
colonies may resemble *Aeromonas* but are differentiated by being oxidase-

negative. PBG has been used to enumerate aeromonads in faecal suspensions from aquatic animals and aquarium water. Seidler and co-workers[59] used PBG inoculated by the pour plate method and incubated at 37°C for 24 h to enumerate aeromonads in polluted river water.

Schubert[60,61] recommends dextrin fuchsin sulphite (DFS) agar in conjunction with membrane filtration for enumeration of aeromonads from clear waters. The membrane is placed upside down on DFS agar and the plate is overlaid with melted and tempered DFS and incubated at 30°C for 24 h. This procedure is reported to prevent overgrowth of pseudomonads in water samples. When red colonies appear, they are counted and the membrane is cut from the plate and individual colonies are subcultured for identification.

Shotts and Rimler[62] developed a differential and selective medium for recognition of aeromonads in water samples. Maltose served as the carbon source, lysine and ornithine were added for detection of decarboxylase reactions, cysteine was added to detect hydrogen sulphide production, novobiocin and desoxycholate were added as selective agents and bromo thymol blue served as the pH indicator. Inoculum was spread across the surface of plates, and they were incubated at 37°C for 18–24 h. *Aeromonas* and H_2S-negative *Citrobacter* appeared as yellow colonies, while H_2S-positive colonies of *Citrobacter* and *Proteus* appeared yellow with a black centre. Maltose-negative *Citrobacter, Salmonella* and *Edwardsiella* appear green with a black centre, *Pseudomonas, Salmonella, Enterobacter, Escherichia, Klebsiella, Shigella, Plesiomonas* and some *Proteus* spp. appear greenish-yellow to green. Some *Proteus* and *Plesiomonas* were inhibited. Plates must be read before 26 h or the yellow acid reaction of maltose fermenters will begin to fade and gradually change to green. The medium has been used with membrane filtration but it has been replaced in most laboratories by more recent formulations. The chief criticisms of Rimler Shotts (RS) medium are its lack of selectivity, especially for estuarine samples[63]. *Klebsiella* sometimes produces yellow colonies which are indistinguishable from aeromonads on primary plates[64].

Seidler and colleagues[59] modified RS agar by omission of L-lysine, L-ornithine, sodium thiosulphate and agar to produce Modified Rimler Shotts Medium (MRSM). When samples yielded fewer than 10 colonies per millilitre, aeromonads were enumerated by an MPN method using membrane filters which were placed directly into tubes of MRSM and incubated at 37°C for 24 h. Tubes showing acid reactions at 24 h were streaked onto MacConkey agar plates and plates were incubated at 35°C for 24 h. Lactose-negative colonies were picked to *Aeromonas* multi-test medium and gelatin agar and those giving typical reactions were identified to species.

Rippey and Cabelli[65] formulated a medium which employed trehalose as the fermentable carbohydrate and ampicillin and ethanol as selective agents, for detection and enumeration of aeromonads in natural waters. The medium, designated mA, was used in conjunction with membrane filtration.

Filter membranes were placed upright onto the surface of the medium and incubated aerobically for 20 h at 35–37°C. Aeromonads appeared as circular, convex, yellow colonies, 1–3 mm in diameter. Other organisms appeared blue to green. *Klebsiella* were inhibited by the ethanol. After presumptive aeromonads were counted, the filter membrane was transferred onto a plate of mannitol medium and incubated at 37°C for 2–3 h. Colonies which remain yellow, indicating mannitol fermentation, are scored and the membrane is transferred to a filter pad saturated with phosphate-buffered saline for 60 s to neutralize organic acid end-products which could result in a false negative oxidase test. The filters were next transferred to a filter pad saturated with oxidase reagent and colonies which turned purple within 10–15 s were counted as presumptive aeromonads. The method resulted in a 2% false positive rate and a 11% false negative rate. The upper counting limit of 70 colonies per membrane requires preparation of 0.5 log dilutions or use of replicate filters to avoid overcrowding of membranes. The mA medium was found to be superior to RS and PBG for recovery of aeromonads from water. This membrane filtration procedure is tedious and technically challenging but it is still used by some investigators for enumeration of aeromonads in environmental water samples. As with RS medium, mA results in the misidentification of *Vibrio* as *Aeromonas* in estuarine and marine samples[63].

Havelaar and colleagues[15] evaluated seven selective media for enumeration of aeromonads by membrane filter using pure cultures and natural water samples. The media comparisons included GSP[66], DFS[67], RS[62], mA[65] and XDC[49]. Additionally, Pril xylose ampicillin (PXA) agar[68] was modified by deleting Pril to produce tryptone xylose ampicillin (10 mg/l and 30 mg/l) agar (TXA 10 and TXA 30). Of these media, only DFS and mA allowed quantitative recovery of aeromonads compared to non-selective TSA control media. The best qualities of both of these media were combined to produce a new medium, ampicillin dextrin agar (ADA). This new medium exploits the high specificity of dextrin fermentation[67] to limit growth of organisms in environmental samples. Ampicillin (10 mg/l) is added as a selective agent. After filtration of samples, membranes are placed directly onto the agar surface and plates are incubated aerobically at 30°C for 24–48 h. Colonies, 1–2 mm in diameter and appearing bright yellow, were counted as presumptive aeromonads. Optimal colony definition occurred with a bromo thymol blue concentration of 80 mg/l and a final pH of 7.8 ± 0.2 at 25°C[69]. This medium is recommended for enumeration of aeromonads in water samples using membrane filtration or spread plate techniques.

Ampicillin Dextrin Broth[15] consists of ADA without agar and this medium is suitable for enrichment or enumeration of aeromonads using the MPN method.

Huguet and Ribas[70] modified glutamate starch phenol red (GSP) agar[66] by adding a trace amount of glucose to enhance recovery of stressed cells and ampicillin (20 mg/l) to reduce the growth of pseudomonads and other background flora. This new medium, SGAP-10C, enhanced recovery of aeromonads from river water samples when incubated at 28°C for 48 h. Colonies of aeromonads appear yellow while other organisms appear pinkish in colour.

Monfort and Baleux[71] compared RS and PXA for recovery of aeromonads from various kinds of waters. Difficulty in discrimination colour of colonies and inability to distinguish some vibrios from aeromonads were the primary drawbacks of RS. RS was found to inhibit some strains of *Aeromonas*. PXA was found to be an acceptable medium for isolation of aeromonads from polluted and non-polluted marine and fresh water[72].

Arcos and colleagues[73] conducted the most comprehensive evaluation to date of media for isolation of aeromonads from natural water samples. Inositol brilliant green bile (IBB), bile brilliant green (BBG), RS, XDC and DFS agars allowed growth of several microorganisms common to sewage and these media were considered unsatisfactory for isolation of aeromonads from water. mA, modified PXA, and DNAse toluidine blue ampicillin (DNTA) agars effectively inhibited growth of *Edwardsiella, Salmonella, Staphylococcus* and *Streptococcus*; however, modified PXA proved inhibitory to some *A. hydrophila* strains, and it was considered unsatisfactory for environmental use. Based upon data for selectivity, specificity, recovery efficiency and quality of growth, mA proved to be the preferred medium for quantitative recovery of aeromonads from water. When the top four media included in this evaluation were ranked, the highest ranking was accorded to mA, followed by DNTA, MacConkey Tween 80 (MACT), and starch bile (SB) agars.

Knøchel[74] compared starch agar (SA), DFS and blood ampicillin (10 mg/l) agar (BA10) and blood ampicillin (30 mg/l) agar (BA30) for enumeration of aeromonads from environmental samples and determined that SA provided the highest number of confirmed picks at 85% of presumptive colonies. Colonies were screened on SA plates with half-strength Lugol's iodine solution.

Poffé and Op de Beeck[75] reported that unpublished evaluations of SGP, RS, PBG, SA, DFS, bile salts brilliant green starch (BBGS) agar, mA and ADA suggested that SA and mA may be useful for enumeration of aeromonads from surface water and wastewater. Subsequent studies revealed that overgrowth of *E. coli* and *P. fluorescens* presented problems of colony recognition on SA, and mA was chosen for the environmental survey of aeromonads using membrane filtration. By filtering tenfold dilutions, it was possible to enumerate *Aeromonas* populations as high as 10^6 CFU/ml of sample.

Ribas and colleagues[76] compared SGAP-10C to ADA and mA for recovery of aeromonads from water from a variety of sources using membrane filtration in conjunction with mA and spread plates for the other media. They expressed a preference for SGAP-10C despite a requirement for 48 h incubation. Recoveries were essentially the same for all three media.

These investigators cited the technically demanding methodology for recognition of presumptive aeromonads on mA and the breakthrough of background organisms on ADA to support their preference.

Bernagozzi and co-workers[77] compared mA, RS and Ryan's *Aeromonas* Medium with ampicillin (5 mg/l) for isolation of aeromonads from surface waters using membrane filtration. Plates were incubated aerobically at 30°C for 24–36 h. *Alcaligenes, Pseudomonas, Salmonella, Shigella, Vibrio* and *Yersinia* spp. grew on all media. *Escherichia, Streptococcus* and *Proteus* were inhibited on mA but not RS or Ryan's medium. None of these media gave a 2 log reduction of background flora. Despite the high concentration of ampicillin (30 mg/l) in mA, this medium was preferred for both freshwater and seawater samples.

Hazen and colleagues[78] used filter pads soaked with RS medium for membrane filter enumeration of aeromonads in a variety of surface waters. Sample volumes of 100, 10, 1, 0.1, 0.01 and 0.001 ml or equivalents were filtered. This method gave a 2.5% false positivity rate. When colony counts exceed 100 colonies on a membrane, use of half-log dilutions produces countable plates. Sample volumes of 5, 1, 0.5 or 0.1 ml shift the number of colonies per membrane into the countable range (30–100 colonies).

Abeyta and co-workers[79] compared tryptic soy ampicillin broth (TSBA) and modified Rimler-Shotts broth (MRSB) in conjunction with Rimler-Shotts agar (RSA) and PBG for enrichment of aeromonads from oysters and water from oyster beds. All media and enrichments were incubated at 35°C for 24 h. TSBA gave the highest enrichment while the combination of TSBA and PBG gave the best recovery from both water and oyster samples.

Venkateswaran and co-workers[80] placed membrane filters onto pads pre-soaked with APW and incubated them at 30°C for 6 h before transferring the filters to selective and differential media. This pre-enrichment method improved recovery of aeromonads by 15.6% in direct comparisons with non-enriched filters. The method provides a convenient means of resuscitating starved or dormant cells, since studies demonstrated the ability to recover cells damaged by exposures to heat and cold. The APW formulation used contained 1% peptone, 0.5% yeast extract, and 0.05% sodium deoxycholate at pH 8.6. This pre-enrichment method had been previously used to enumerate aeromonads in environmental samples in conjunction with modified Pril xylose ampicillin agar[81].

Abeyta and colleagues[82] surveyed the shellfish harvesting estuaries of the California coast and the Pacific northwest using TSBA and APW enrichments. For all samples, 25 g was added to 225 ml of broth and incubated at 35°C for 24 h. Broth cultures were streaked onto MacConkey and PBG agar plates incubated at 35°C for 24 h for recovery of aeromonads. Typical colonies were picked to multi-test screening medium and presumptive aeromonads were identified to species. Overgrowth of marine bacteria limited recognition of aeromonads in warm marine waters

(16–26°C), but did not interfere with isolation of aeromonads from cold marine waters (10–16°C). APW was not as effective as TSBA in enrichment of aeromonads from marine waters and the combination of TSBA and PBG was the best combination in this trial. Aeromonads were isolated from waters with salinity from 0–48‰, and there was an inverse relationship between salinity and the number of aeromonads isolated. The highest recovery rates were obtained at sites where the faecal coliform levels were also high, but aeromonads were also isolated from waters without detectable faecal coliforms.

Venkateswaran and colleagues[82a] used APW and modified PXA to study environmental aeromonads. After filtering samples, the membranes were placed onto filter pads soaked in alkaline peptone water containing 0.05% sodium deoxycholate for 30 min and incubated for 6 h at 37°C before being placed onto the surface of modified PXA and incubated at 37°C for 18–24 h. These investigators also placed membrane filters into three tube sets of APW for enumeration by the MPN method. Tubes were incubated at 37°C for 18 h and streaked onto modified PXA plates which were incubated at 37°C for 18–24 h. Tubes producing aeromonads on modified PXA plates were scored positive and the MPN tables were used to determine counts of aeromonads in samples. Abbey and Etang[83] used APW at a final pH of 9.1 to isolate aeromonads from a variety of environmental sources. Other media used for enumeration of aeromonads by the MPN technique include APW[84,85], TSBA[1], modified RS medium[86] and XDC[25].

Alonso and Garay[87] modified ADA and SA by addition of O/129 (50 mg/l) and ampicillin (16 mg/l) and used these media (mADA and mSA) for membrane filter enumeration of aeromonads in seawater. Presumptive colonies were screened in Kaper's multi-test medium and 95% of presumptive colonies picked from both media proved to be aeromonads.

Recently, studies of river water marine samples using early media such as mA[88] and modified PXA[81,89] have been published.

Paniagua and colleagues[90] used SA, originally developed to isolate aeromonads from food, to study fish and river samples and found 76% *A. hydrophila*, 12% *A. caviae* and 11% *A. sobria* in water and sediments of rivers in Spain. Mateos *et al.*[91] also used SA to recover aeromonads from 0.1 ml of water and sediment samples plated directly on the medium.

Enteric culture media have been used to isolate aeromonads from highly polluted waters. Nazer and colleagues[92] used APW and DCA (SS agar) to culture aeromonads from canal water in London, and Chowdhury *et al.*[93] used MacConkey agar and BBGS for freshwater estuarine and marine samples. BBGS was somewhat more inhibitory to background microorganisms.

mA, ADA, SA, PXA and SGAP-10C are the most widely used plating media for enumeration of aeromonads. Personal experience suggests that SGAP-10C may be too selective, and underestimate aeromonad populations, and there are several disadvantages to its use. Colonies develop slowly, yellow colour changes of the media from starch utilization are difficult to detect in some cultures and plates must be incubated for 48 h before results are observable. No studies have compared these media directly so it is not possible to recommend an optimal recovery medium. Media selection must be made from knowledge of the other bacteria present in samples, their numbers relative to aeromonads, and the physical nature of the sample.

2.5   METHODS FOR ISOLATION OF AEROMONADS FROM DRINKING WATER

Aeromonads are sensitive to chlorine, yet they have been isolated from chlorinated water systems, where they persist as a component of biofilm[94,95]. The number of aeromonads is generally <10 CFU/ml in distribution system water.

For enrichment of aeromonads from clear water samples and drinking water samples, filter 50–100 ml through membrane filters and place the filters into 10–25 ml of APW. Incubate cultures overnight at 25°C and subculture to plates for detection of aeromonads. Another method for isolation of aeromonads from clear waters with low numbers of organisms is to pass 4–40 litres of water through a 0.45 μm in-line filter (Gelman Mini-Capsule filter or equivalent) from a pressurized source. The ends of the filter are plugged with sterile rubber stoppers for transport to the laboratory. The residual water in the filter is aseptically decanted into a sterile container and saved if desired, and APW is then placed into the in-line filter which is then incubated at 25–35°C overnight. Samples for subculture may be removed by attaching a syringe (without needle) to the port on the filter and 0.1 ml is added to plating media for recovery of aeromonads. The residual water in the filter may be cultured directly, or passed through a 0.45 μm filter membrane using a Swinnex holder. The membrane may be placed into APW or placed directly onto culture medium. This method has been useful for isolation of aeromonads from wells of patients with aeromonas gastroenteritis[96].

Burke *et al*.[97,98] isolated aeromonads from chlorinated and unchlorinated water supplies, including ground water, using membrane filtration and teepol broth (Oxoid) modified to contain 0.2% teepol. Water temperature and chlorine residual were the variables found to affect aeromonad recovery.

Millership and Chattopadhyay[99] used APW subcultured to XDC and BBG and membrane filtration using BBG incubated anaerobically to isolate aeromonads from chlorinated drinking water sources. Aeromonads were recovered from 25% of samples in summer and 7% of samples in winter. *Pseudomonas* overgrowth proved to be a problem with these media, hence the recourse to anaerobic incubation.

Cunliffe and Adcock[100] used meso inositol xylose agar (MIX) incubated anaerobically to isolate aeromonads from drinking water using the membrane filter technique. Plates were incubated at 30°C for 18 h in an atmosphere of 10% hydrogen and 90% nitrogen and then incubated aerobically at 35°C for 24 h. Anaerobic incubation suppressed non-aeromonad growth and gave a 92% confirmation rate. Carbon dioxide was avoided in the gas mix to prevent surface acidity of plates which could have reduced recovery of aeromonads.

Havelaar and colleagues[101] sampled raw, treated and distribution system water in The Netherlands and enumerated aeromonads using ADA incubated at 30°C for 24 h. Treated water usually contained less than 10 CFU/100 ml. When regrowth of aeromonads occurred in distribution systems, counts ranged from 10–3300 CFU/100 ml.

Moyer and colleagues[102] isolated aeromonads from the water treatment plant and distribution system of a small community using MF with ADA and APW. Duplicate water samples were filtered through 0.45 μm membranes. One membrane was aseptically placed onto the surface of ADA which was incubated at 35°C for 24 h. The second membrane was aseptically placed into a bottle containing 10 ml of APW and incubated at 25°C for 24 h. APW enrichments were subcultured to BA10 plates which were incubated at 35°C for 24 h and oxidase-positive colonies were identified. APW recovered aeromonads from a well sample which was negative on ADA, but there was agreement between APW and ADA results with all other samples. Positive samples ranged between <1–11 CFU/ml on ADA.

Holmes and Sartory[103] compared ADA, xylose ampicillin agar (XAA), Ryan's Aeromonas Medium (Oxoid), and Aeromonas Agar (Difco) for enumeration of aeromonads from drinking water. Samples were filtered, the membrane was placed onto the agar surface, and plates were incubated at 30°C for 24 h. ADA and Ryan's medium gave the desired selectivity and sensitivity; however, colony characteristics were more consistent on Ryan's medium.

Mascher et al.[104] used BA10 in conjunction with MF to isolate aeromonads from drinking water. Mariottini and Mingoia[105] used mA, BA20 and APW to isolate aeromonads from drinking water and characterized 128 strains. Ghanem et al.[106] isolated enterotoxigenic aeromonads from tapwater using membrane filtration and SGP medium.

2.5.1 PRIVATE WELLS

Bhat and colleagues[107] isolated aeromonads from well water implicated in an outbreak of *Aeromonas* gastroenteritis using filtration and enrichment of the filter in DCA medium without the sugars and agar. Seitz filter pads were cultured in 25 ml of the broth which was subcultured to plating media after overnight incubation.

Krovacek and colleagues[108] isolated aeromonads from well water in

Sweden in association with the investigation of long-term gastroenteritis in a child of 1.5 years. Using membrane filtration with ADA, cell counts of well water were 70–46 000 CFU/100 ml. *A. hydrophila* strains isolated produced enterotoxin and haemolysin characteristic of the faecal isolate.

2.5.2   BOTTLED WATER

Aeromonads have been isolated from bottled waters by investigators in Canada, Saudi Arabia, Spain and elsewhere[109–111]. Warburton and co-workers[112] developed a method for detection of aeromonads in bottled waters as a means of enforcing the proposed Canadian regulations of 0 CFU/100 ml for aeromonads. The method incorporated mA–trehalose agar[113] with the hydrophobic grid filters and Ryan's medium. The resulting procedure includes a 4 h resuscitation step and use of both mA–trehalose and Ryan's medium[114].

2.5.3   INTERFERENCE WITH COLIFORM METHODS

Clark and Vlassoff[115], Grabow and DuPrez[116], and Lupo and colleagues[117] reported that lactose-fermenting aeromonads interfered with coliform enumeration by MF and MPN. Clark[118] reported that aeromonads comprise 12% of bacteria isolated by MF and PA from drinking water. This phenomenon is recognizable as light turbidity in laural sulphate broth tubes with or without a tiny bubble of gas in the inverted tube. When PA methods are used for coliform analysis, tubes incubated at 35°C showing turbidity only are streaked to MacConkey agar and colourless colonies are screened for gelatinase production and oxidase reaction. Turbid tubes at 35°C which are clear at 44.5°C are suggestive of aeromonads.

2.6   METHODS FOR ISOLATION OF AEROMONADS FROM FISH

Diagnosis of diseases in fish cannot be adequately determined from study of a single fish. Furthermore, examination of dead fish is to be avoided, since proliferation of bacteria from endogenous sources may give misleading culture information. Methods must include a search for both *A. salmonicida*, which causes furunculosis and septicaemia, and *A. hydrophila*, which causes ulceration, haemorrhagic lesions, dropsy and septicaemia.

The clinical pathology of fish disease is variable, depending upon the virulence of the bacterial strain, the age of the fish and environmental factors. Subclinical infections and carrier states occur with the intestine as the primary site of colonization[119]. It is not possible to differentiate between furunculosis and diseases caused by other Gram-negative organisms on clinical signs alone. Culture examination and antigen detection, by serological methods or molecular detection of pathogens, are necessary for a definitive diagnosis. Antigen detection and molecular methods are also useful for detecting asymptomatic infections when cell populations are present below the detection limit of cultural methods.

Both non-selective and selective media are generally used to isolate aeromonads[120]. Böhm et al.[121] added 10% serum and ampicillin (50 mg/l) to tryptone soya agar in addition to furunculosis agar (Difco) for isolation of A. salmonicida from diseased fish. Furunculosis agar (Difco) contains tryptone (10 g/l), yeast extract (5 g/l), L-tyrosine (1 g/l), NaCl (2.5 g/l) and agar (15 g/l). Plates were incubated at 25°C for up to six days. Both typical and atypical strains of A. salmonicida may be isolated by inoculating up to 1 g of fish tissue into 10 ml of tryptic soy broth and incubating at 20°C for 24 h, followed by addition of horse blood to a final concentration of 5%, and extending the incubation time another 48–72 h. Rimler-Shotts agar has also been widely used for isolation of aeromonads from fish[122,123].

Non-inhibitory plating media for other fish pathogens besides A. salmonicida should be incubated at 20–25°C to retard the growth of bacteria with temperature optima in the mesophilic range above 25°C. A disadvantage of lower incubation temperature is the need for longer incubation times and the combination of incubation time and temperature selected is dependent upon the organisms desired and the media used.

A. salmonicida produces a brown soluble pigment on media containing 0.1% tyrosine or phenylalanine. Many authors have reported atypical pigment production of strains of A. salmonicida causing disease[124–127].

A. salmonicida exhibits different colony types depending upon the state of the A-layer of the cell envelope, and these colony types have been designated rough, smooth and G-phase[128]. The presence of A-layer outer membrane proteins correlates with roughness, virulence, autoagglutination, adhesion to fish cells, the ability to bind Congo red and Coomassie brilliant blue dyes, and hydrophobicity. The A-layer of A. salmonicida is a $50\,000\,M_r$ protein (A-protein) which confers virulence by providing organisms with a protective barrier against the host defences. Loss of virulence follows repeated subculture in the laboratory and the maximal temperature which supports growth is increased from 18–26°C for A$^+$ strains to 30°C for A$^-$ strains[129]. The A-layer has been associated with autoagglutination and adherence in addition to virulence. Congo red binds to cells exhibiting A-layer and these colonies appear dark red when grown on tryptone soya agar containing 30 mg/l of Congo red. Congo Red Agar (CRA) is used to differentiate A$^+$ and A$^-$ strains of A. salmonicida in cultures and also serves as an indicator of cell surface integrity.

Udey[130] reported that incorporation of Coomassie blue into nutrient growth medium was a convenient marker for presence of the A-layer.

The medium was modified by Wilson and Horne[131] for differentiation of A[+] and A[-] strains of *A. salmonicida*, and Cipriano and Bertolini[132] used it for that purpose. Markwardt *et al.*[133] used the Udey form of the medium to recover *A. salmonicida* from fish. The medium consisted of tryptone soya agar containing 100 mg of Coomassie Brilliant Blue Dye per litre. Plates were inoculated with samples in dilution and incubated for 48 h at 21°C. Colonies producing A-layer protein stained medium to dark blue while those without A-layer protein stained blue green or light blue. Some strains of *A. hydrophila* which also produce A-layer protein will stain medium blue. *Yersinia ruckeri, Vibrio* spp. and coliform organisms stain light blue. This medium provides a rapid and simple means of determining A-layer-producing clones upon primary isolation of *A. salmonicida*.

Cipriano and Bertolini[132], Cipriano and Ford[134], and Teska and Cipriano[135] used dilution counts on Coomassie blue agar to improve the sensitivity of culture for *A. salmonicida* by $2-3 \log_{10}$ over streak plate methods. Kidney tissue from infected fish were macerated in phosphate buffer at a 1:10 (W/V) dilution and 0.01 ml of decimal dilutions were plated onto Coomassie brilliant blue (CBB) agar. Plates were incubated at ambient temperature for up to 48 h and examined for dark blue colonies. Dark blue colonies which were non-motile, oxidase-positive, gelatinase-positive, indole-negative, Gram-negative rods giving a K/A reaction in TSI agar were presumptively counted as *A. salmonicida*. This method is used extensively in the United States as a non-invasive way to culture *A. salmonicida* from mucous samples of hatchery fish[136]. The choice of incubation temperature for Congo red and Coomassie blue agars enables investigators to select for A[+] or A[-] cell populations.

Ishiguro and colleagues[137] observed a loss of virulence in *A. salmonicida* grown at 30°C; indeed, at that growth temperature, virulent strains represented <10% of the total population. The attenuated strains did not autoaggregate and did not possess the A-layer. Johnson *et al.*[138] reported isolation of A-layer-negative strains of *A. salmonicida* which retained the ability to cause disease in fish. These strains would be missed on CBB unless they were checked for oxidase reaction and pigment production.

Chapman and colleagues[139] isolated an oxidase-negative strain of *A. salmonicida* from coho salmon with furunculosis using CBB agar. Oxidase-positive strains were simultaneously isolated. The importance of subculturing several colonies from different fish for identification and antibiotic sensitivity tests cannot be overemphasized. The oxidase-negative strains would probably have been overlooked on other differential media but the combination of blue colour on CBB and production of a brown soluble pigment prompted further characterization.

2.7 METHODS FOR ISOLATION OF AEROMONADS FROM HUMANS

2.7.1 EXTRAINTESTINAL INFECTIONS

Altwegg[140] reviewed the isolation procedures for extraintestinal infections. Aeromonads grow well on blood agar and MacConkey agar, which are routinely used in most clinical laboratories, and no special media are generally required for specimens from normally sterile or polymicrobic sites. Ampicillin-containing media are not recommended since ampicillin-sensitive aeromonads have been reported from clinical materials.

2.7.2 INTESTINAL INFECTIONS

Aeromonads are usually recognized in clinical laboratories because of their strong β-haemolysis on blood agar and their positive oxidase reaction, although not all pathogenic strains are haemolytic and false negative oxidase tests occur with growth taken from media containing fermentable carbohydrate. Among the traditional enteric plating media, MacConkey agar (MAC), desoxycholate citrate agar (DCA) and bile brilliant green agar (BBG) permit growth of aeromonads; however, they are frequently overlooked unless they are present in high numbers and the microbiologist performs an oxidase test on subcultures from medium without fermentable carbohydrate.

Cary–Blair transport medium is widely used for submitting stools for aeromonas culture when immediate culture of the specimen is not possible[96,141]. The use of enrichments for specimens held in transport media is not recommended by some investigators since aeromonads may grow in storage, even at 4°C[142,143]. This phenomenon may result in detection of transitory carrier states in the absence of intestinal disease.

DNA toluidine blue ampicillin agar was proposed to facilitate isolation of aeromonads from stools in clinical laboratories[144]. Production of DNAse is shared by members of the genera *Serratia, Enterobacter, Proteus* and *Pseudomonas* in addition to *Aeromonas* so ampicillin (30 mg/l) was incorporated to improve selectivity of the medium. Aeromonads resistant to ampicillin and producing DNAse exhibit a clear zone around the colony, while organisms which do not produce DNAse have no zone.

Inositol Brilliant Green Bile Salts (IBG) agar was developed for isolation of *Plesiomonas shigelloides* from stool[145]. Inositol non-fermenting *Plesiomonas* and *Aeromonas* colonies appear whitish to pinkish and colonies can be picked for performance of spot oxidase tests. Millership and Chattopadhyay[52] deleted inositol from IBG to produce brilliant green bile salts agar (BBG) and this medium has been used for isolation of *Aeromonas* and *Plesiomonas* from stools. Plates are incubated at 37°C and examined for growth at 18 and 48 h. Plates are screened for oxidase reaction by flooding the plates and immediately subculturing oxidase-positive colonies for screening tests to confirm their identity as aeromonads.

In an effort to improve the selectivity of culture media for isolation of aeromonads from stools, Rogol *et al.*[68] formulated a new medium, Pril Xylose Ampicillin agar (PXA), using nutrient agar as the base with 1% xylose, phenol red (25 mg/l), ampicillin (30 mg/l), and the quaternary ammonium detergent Pril (0.02%). Colonies of *Aeromonas* and *Plesiomonas* appear colourless while coliforms produce yellow colonies. Pril had previously been incorporated into culture media to inhibit swarming of *Proteus*. PXA was recommended for enteric cultures[50], but the difficulty of obtaining the detergent has limited its use outside Europe. Studies have shown that 44.6% of *A. caviae* and 17.8% of *A. hydrophila* strains were inhibited on PXA.

Blood agar containing 430 mg/l of paranitrophenyl glycerine (PNPG) and/or blood ampicillin (10 mg/l) agar has been used to isolate aeromonads from stools of children with diarrhoea[36]. Colonies are 2–3 mm in diameter and may or may not be surrounded by a zone of β-haemolysis. Oxidase tests may be performed directly on the plate or growth picked from a colony may be screened using the spot oxidase test. Simultaneous recovery of *A. hydrophila* and *A. caviae* from a patient with watery diarrhoea was attributed to use of BA10 because both haemolytic and non-haemolytic oxidase-positive colony types were recognized[39]. Blood ampicillin agar has become a standard medium for isolation of aeromonads from clinical specimens[43,44,146].

Unfortunately, there has been little agreement concerning the concentration of ampicillin to be incorporated into the medium for maximal suppression of contamination while providing optimal recovery of aeromonads. Ampicillin concentrations of 10, 15, 20 and 30 mg/l each have their proponents Because it is now known that some aeromonads are sensitive to ampicillin, most investigators opt for a concentration of 10 mg/l for media used for isolation of aeromonads from clinical specimens. Use of a non-selective differential medium together with a moderately selective medium is necessary for recovery of all *Aeromonas* spp. which may be pathogenic for humans.

Clinical laboratories have fortuitously recovered aeromonads on MacConkey agar when they have been present in high numbers. MacConkey agar with trehalose substituted for lactose and MacConkey agar with Tween 80 (MTA), with or without ampicillin (10–30 mg/l) were advocated to improve isolation of aeromonads from stools. Comparative studies of MTA with other selective media have not supported its use[147]. MTA recovers less than half of the total aeromonad isolates in most studies and does not recover strains which do not grow on other selective media.

Altwegg and Jöhl[148] found that aeromonads could be recovered from Yersinia Selective Agar (Difco) plates used to isolate *Yersinia* spp. from stools. Both *Yersinia* and *Aeromonas* appear as clear colonies with a dark rose centre termed 'bulls eyes'. Colonies must be transferred to media without fermentable carbohydrate for oxidase testing.

Misra et al.[175] fortuitously recognized growth of large, grey, haemolytic colonies on Butzler's Campylobacter Selective Agar (BCSA) as Aeromonas spp. despite incubation of plates in a microaerophilic environment at 42°C. When they compared the growth of aeromonads recovered from stool specimens on BCSA and Sheep Blood Agar containing 30 mg/l ampicillin (BAA30) incubated at 37°C, 26.7% of their isolates were detected on both media, 43.7% of the isolates were detected on BAA30 only and 29.6% of the isolates were recovered on BCSA only. When isolates recovered on only one of the two media were subcultured to the other medium to determine if its selectivity prevented primary isolation, all isolates grew upon both media. Incubation at 42°C is above the usual temperature range for isolation of aeromonads. A. hydrophila was recovered more frequently than A. sobria or A. caviae, suggesting that 42°C is moderately inhibitory for aeromonads. Since some strains of A. caviae are sensitive to ampicillin concentrations of 30 mg/l in culture media, it is not possible to determine whether reduced recovery of A. caviae resulted from 42°C incubation, high ampicillin concentration, or both selective pressures.

While the ability to recover aeromonads from stool specimens using Butzler's medium for isolation of Campylobacter spp. is an interesting observation, this medium should not be relied upon as a sole method for isolation of aeromonads from stools. The antibiotic combinations in the Campylobacter selective media of Skirrow or Blaser are not suitable for recovery of all aeromonads since they contain trimethoprim and polymyxin B, antibiotics to which most aeromonads are sensitive. The advantage of using Butzler's medium for Campylobacter isolation, instead of other formulations, is the ability to detect potentially pathogenic aeromonads in stool specimens without the expense of adding several plates to the enteric pathogen culture schema which are specifically designed for Aeromonas detection.

Others have reported recovery of A. hydrophila from stools that were inoculated onto blood agar, MacConkey agar and Butzler's medium for isolation of Campylobacter[150]. The hanging drop method for motility determination, together with TSI slants and O/129 discs were used to differentiate Campylobacter and Aeromonas taken from Butzler's medium.

Starch agar (SA), originally developed for recovery of aeromonads from food, has been applied to isolation of aeromonads from stools of humans and animals[48,53] with varying success. Stern et al.[53] preferred it to MacConkey ampicillin, blood ampicillin agar and Yersinia Selective Agar (Difco), while the author has found SA difficult to read and messy to work with because of the need to flood the plates with Lugol's iodine solution to visualize the zones resulting from amylase activity. Colonies must be subcultured to screening media before the oxidase test is performed.

Price and Hunt[52a] and Hunt *et al.*[151] prepared Xylose Irgasan Bile Salts Brilliant Green Agar (XIBBG) for isolation of aeromonads from stool. This formula is now commercially available as *Aeromonas* Agar (Difco). Aeromonads produce colourless colonies while xylose fermenters appear pink.

Ryan modified the formula of XLD in an attempt to develop a universal medium for recovery of *Aeromonas* and *Plesiomonas* in addition to other members of the Enterobacteriaceae. When ampicillin (5 mg/l) is added to the medium, the performance is improved for isolation of aeromonads. The medium has been recommended for clinical specimens and environmental samples. Plates are inoculated and incubated aerobically at 30–35°C for 24 h. Aeromonads appear as dark green colonies, 0.5–1.0 mm in diameter with darker centres. Pseudomonads produce pinpoint to 0.25 mm blue/grey translucent colonies. *Aeromonas* Medium Base (Ryan) and Ampicillin Selective Supplement are available from Oxoid (CM833 and SR136 respectively).

Ghanem and colleagues[106] used SGAP containing 0.4 ml of 2.5% pimaricin per litre to culture aeromonads from stool suspensions in saline by the pour plate method. Aeromonads appear yellow while other organisms produce pink colonies.

The large number of publications recommending different media formulations for isolation of aeromonads resulted in publication of numerous media comparative studies.

Von Graevenitz and Bucher[50] evaluated DFS, DNTA, IBB, PBG, PXA, RS, RC, SSXLD and XDC for recovery of aeromonads from stool and recommended four solid media for isolation of aeromonads – IBB, XDC, DFS and PXA. They also evaluated TSBA and APW enrichment broths and recommended APW for enrichment of *Aeromonas* spp.

Moulsdale[47] modified PXA to contain 25 mg/l PNPG and 20 mg/l ampicillin as the selective agents and used it for isolation of aeromonads from stool. This medium, together with MAC, DCA, XDCA and BA, was evaluated by Robinson *et al.*[45], who concluded that BA with 10 mg/l ampicillin was the most satisfactory of the media evaluated for isolation of aeromonads from stool.

The usual enteric media designed to detect *Salmonella, Shigella* and *Campylobacter* are not very effective for isolation of aeromonads from stools. Kay *et al.*[152] reported that aeromonads were cultured from 19 of 1248 (1.5%) of specimens using MacConkey, SS, XLD, HE, CIN, TCBS and Butzler's selective medium for *Campylobacter*. When these investigators adopted BA (10 mg/l) and APW (both incubated at 35°C) for specific isolation of aeromonads, their recovery rate increased to 15.3% the first year and they recommend this combination of media over all others they have evaluated. Interestingly, their recovery of aeromonads on the routine enteric media increased from 1.5% to 5%, suggesting that familiarity with an organism is indispensable to its recognition on plating media, regardless of the media used.

Figura[153] reported the results of stool culture using DNAT, APW with XDC as recommended by Shread *et al.*[49] and *Salmonella Shigella* Agar modified by the addition of 2% sodium desoxycholate (SSSD). Upon finding more *Aeromonas* isolates on SSSD agar, he developed a selective enrichment semisolid medium (SESAM) composed of APW with 2% sodium desoxycholate and 0.17% agar at pH 8.6 and a selective differential medium (SDAM) composed of XDCA with a final concentration of 2% sodium desoxycholate, 0.065 g/l bromo thymol blue and 0.1 g/l acid fuchsin instead of neutral red with a final pH of 7.7. These latter media were compared with DNTA, SSSD and APW/XDCA for isolation of aeromonads from stools. The combination of SESAM/SDAM yielded the highest recovery overall, but some strains were inhibited by the high concentration of sodium desoxycholate.

Desmond and Janda[154] evaluated eight routine enteric media for their efficiency in recovering aeromonads from stools. BG, BS, DC, HE, SS, MAC, XLD and EMB were included in the survey and blood agar was included as the sole non-selective medium. Plating efficiencies were determined using cell suspensions of known concentrations. Some strains of aeromonads were significantly inhibited (1 log reduction) compared to counts on blood agar and BG failed to support growth of *A. caviae* and *A. sobria* strains. DC, MAC and XLD were the least inhibitory media and they may be satisfactory for recovery of aeromonads from stools in conjunction with a blood agar plate in laboratories which cannot include media specifically formulated for recovery of aeromonads.

Misra and co-workers[149] compared five selective media for isolation of aeromonads from human and animal faeces. Faecal samples or swabs of animal faeces were submitted in Cary–Blair transport medium and inoculated onto plates of BG, XDC, DNTA (containing 10 mg/l ampicillin), BA (10 mg/l) and BA (30 mg/l). Plates were incubated at 37°C and examined for growth at 24, 48 and 72 h. Colonies suggestive of *Aeromonas* spp. were picked for spot oxidase tests except from XDC, where colonies were too small to pick and the plates were flooded with oxidase reagent (1% tetramethyl-*p*-phenylenediamine dihydrochloride). Oxidase-positive colonies were screened using the multi-test medium of Kaper modified as slants instead of deeps. Colonies exhibiting alkaline slant/acid butt reactions were presumptively considered aeromonads. Colonies were also screened for O/129 resistance. Final confirmation was accomplished using the API system. Recovery of aeromonads was best using BA30, followed by DNTA, BA10, XDC and BBG. XDC and BBD were the least inhibitory for usual faecal flora and poor recovery of aeromonads probably resulted from overgrowth by competing organisms. *A. sobria* and *A. caviae* were not recovered on BBG and *A. sobria* was not recovered on XDC. BA30 and DNTA were recommended as the best media for stool cultures.

Kelly and co-workers[147] compared blood agar, BA20, MacConkey Ampicillin Tween 80 Agar (MAT), YSA (modified CIN) for isolation of aeromonads from stools transported in Cary–Blair medium and found that BA20 provided the highest recovery of any single plate but optimal recovery of aeromonads required use of more than one medium. The combination of BA20 and YSA resulted in 100% recovery of aeromonads isolated in this study.

Moyer and colleagues[48] reported the results of an international multilaboratory study to establish the optimal culture media, incubation time and incubation temperature for recovery of aeromonads from stool specimens. This study compared BA10 incubated at 25°C and 37°C, APW incubated at 25°C and 35°C, YSA and SAA, both incubated at 25°C. BAA incubated at 37°C and YSA together with enrichment in APW, incubated at 25°C and subcultured to BA10 and *Yersinia* Selective Agar (Difco), detected 89.9% of the total isolates. Optimal recovery of aeromonads from stools is more dependent upon the number and kinds of primary plates used for screening stools than upon the enrichment method employed. APW streaked to BA10 gave the highest recovery of any single plate. The recommended methods for isolation of aeromonads from stools is shown in Figure 2.1.

Wilcox and colleagues[155] compared *Aeromonas* Agar (Difco) to XDC and BA30 for recovery of aeromonads from stools. They recommend *Aeromonas*

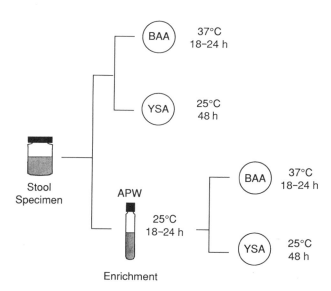

Figure 2.1 Recommended culture media and methods for isolation of *Aeromonas* spp. from stools. BAA, blood ampicillin agar (10 mg/l); YSA, *Yersinia* selective agar; APW, alkaline peptone water, pH 8.6

Agar (Difco), containing bile salt, irgasan, and brilliant green as selective agents, incubated at 37°C, for optimal recovery of aeromonads. APW enrichment was not recommended as it increased the yield of aeromonads but the clinical significance of these isolates in the absence of isolation on primary plates was doubtful. A single plate of *Aeromonas* Agar (Difco) should be used in addition to the usual enteric media for diagnosis of gastroenteritis.

Pascual and colleagues[156] published a comparison of GSP and BA10 on stools cultured in their laboratory from January 1992 to April 1993. BA10 recovered all 37 of the aeromonads isolated while GSP recovered only 15 of 37 isolates and only 3 of 13 *A. hydrophila* strains. GSP was not recommended for stool culture for aeromonas.

Clinical laboratories are under extreme pressure to minimize costs while attempting to isolate a wide variety of aetiologic agents from human specimens, thus multi-purpose media are particularly appealing to clinical microbiologists. One such medium is Cefsulodin-Irgasan-Novobiocin Agar (CIN) for isolation of *Yersinia enterocolitica* from stools. A commercially available modification of the original formula is marketed as Yersinia Selective Agar (YSA). This formula differs from original CIN by reducing the concentration of cefsulodin to 4 mg/l instead of 15 mg/l. Altorfer *et al.*[43] fortuitously isolated aeromonads from YSA and this formulation has proven to be extremely useful for simultaneous detection of *Aeromonas* and *Yersinia* spp. Plates are streaked with faecal material and incubated at 25°C for 48 h.

These results are typical of other media comparative studies where recovery efficiencies are always enhanced by using several media, incubated at different temperatures. No single schema has emerged with universal acceptance for isolation of aeromonads from stools.

Studies of incubation temperatures suggest that 35–37°C increases the yield of positive results over incubation at 30°C[48,155].

Aeromonads are occasionally isolated concurrently with other enteric pathogens such as *Campylobacter, Salmonella* or *Yersinia*, and more than one *Aeromonas* spp. may be present in a stool specimen. Microbiologists must be alert for multiple pathogens in polymicrobic specimens if they are to provide relevant identifications of bacteria in clinical materials.

As the number of media formulations for recovery of aeromonads from stools increases, the question of clinical relevance must be addressed. Transitory colonization probably occurs dependent upon seasonal fluctuations of aeromonad populations in surface water, in shellfish, on vegetables eaten uncooked and upon sanitation practices during production and preparation of meat, dairy products and other processed foods. Culture of aeromonads from stools in the absence of gastrointestinal illness carries little significance. An exhaustive search for aeromonads using a variety of differential and selective media with enrichments is costly and time-consuming. Clinical laboratories wishing to screen stools for aeromonads should add one plate of BA10 agar in addition to the YSA plate used for isolation of *Yersinia*. Enrichments are not necessary for routine specimens.

2.8 METHODS FOR ISOLATION OF AEROMONADS IN FOODS

Aeromonads have an association with meat and vegetable products and exhibit a psychrotrophic nature which could contribute to its role in food spoilage and as a foodborne pathogen. As the aetiologic role of aeromonads as causes of gastrointestinal disease became apparent, several investigators began to search for them in foods. Early attempts to isolate aeromonads from foods relied upon culture media designed for clinical specimens. These media contained bile salts, desoxycholate, brilliant green, Pril, ampicillin, novobiocin, cefsulodin or irgasan as selective agents and xylose, inositol, mannitol or sheep blood. Because media developed for stool culture were, of necessity, highly selective, they were not suitable for recovery of aeromonads from food products. The natural flora of food products posed quite different challenges from those encountered with clinical materials or water. Food microbiology requires quantitative culture methods for isolation and enumeration of organisms.

Methods for isolation of aeromonads from foods are the only ones which approach any degree of standardization[3]. Detection of aeromonads in foods has depended upon direct cultural inoculation of diluted samples using the pour or smear plate techniques or enumeration based upon determination of most probable numbers using multiple tube sets of TSBA30 or APW. For direct plating, 10–25 g samples were homogenized for 1–2 min in diluent using a Colworth stomacher as described by Peterkin and Warburton[113]. Decimal dilutions are prepared and inoculated onto SA, BA30 or *Aeromonas* Agar (Difco) plates, which are then incubated aerobically at 35°C for 24 h. Common diluents include phosphate-buffered saline, 0.1% peptone water, peptone saline, peptone/Tween 80 diluent, tryptone soya broth containing ampicillin (10–30 mg/l) or APW. A detection limit of 100 CFU/g has been reported for food.

When enrichment is desirable, a 25 g portion is placed into 225 ml of enrichment broth. If sample is limited, 25 ml of the original 1 : 10 dilution may be added to 225 ml of enrichment broth and incubated at 28°C overnight. APW is generally used for enrichment, incubated at 28°C for 18–24 h, and 0.1 ml transferred to BA30 and BIBG plates which are incubated at 35°C for 24 h. Counts typically range from below the detection limit to $>10^6$ CFU/g.

Ground meat and chicken samples may be enriched by washing 10 g samples in 90 ml of 0.1% peptone water and adding 10 ml of the rinse water to 90 ml of TSBA30 incubated at 30°C for 24 h. A loopful of culture is streaked onto SA, ADA or other differential and selective media, which is also incubated at 30°C for 24 h. Gobat and Jemmi[157] cultured aeromonads from 10 g samples of raw and ready-to-eat fish and meat products, suggesting that sample size may be selected based upon the expected level of product contamination.

If enumeration is desirable, portions of the dilutions or enrichment broths are inoculated into three or five tube sets of tryptone soya broth with ampicillin (30 mg/l) or APW for quantitation by the MPN method. A loopful of growth from each positive tube is inoculated onto plating media and read as positive or negative for typical aeromonads. The MPN tables are used to determine the number of aeromonads per gram of sample.

Palumbo and colleagues[158] added starch (0.1%) and ampicillin (10 mg/l) to phenol red agar base (Difco) to produce starch ampicillin agar (SA) and used this medium for quantitative recovery of aeromonads from foods[159]. Plates are inoculated with 0.1 ml of tenfold dilutions of samples by the spread plate technique and incubated at 28°C for 24 h. *Aeromonas* colonies are 3–5 mm in diameter and appear yellow to honey-coloured. The plates are flooded with 5 ml of Lugol's iodine solution and colonies surrounded by a clear zone are subcultured for additional testing. Gram-negative organisms which are amylase-, oxidase-, catalase-positive and DNAse-positive and resistant to O/129 are presumptively considered to be *Aeromonas* spp. Disadvantages of the medium are the necessity of flooding the plate with Lugol's iodine solution, which obviates the ability to perform a direct oxidase test, and its inability to differentiate between *Vibrio* and *Aeromonas*; the latter limitation is a serious deficiency for quantitative estimation of aeromonads in fish, shellfish or estuarine samples.

SA, modified SA (Lachica) and BBGS have been most frequently used for isolation of aeromonads from foods, although many other formulations have been advocated. SA and BBGA are incubated at 28°C overnight (24 h maximum) and plates are flooded with 5 ml of Lugol's iodine solution. On SA, amylase-producing colonies are yellow, grow 3–5 mm in diameter, and are surrounded by a clear zone in the medium. Where overcrowding and growth of background flora make it difficult to count presumptive *Aeromonas* colonies with clear zones, growth may be washed or scraped from the plates and the zones may be counted instead of colonies. On modified SA, *A. hydrophila* colonies are surrounded by a light halo against a blue background and Lugol's iodine is not needed.

BBG medium[52] was modified for recovery of aeromonads from foods and water samples by adding starch[160]. Spread plates are inoculated with 0.1 ml from tenfold dilutions of sample and plates are incubated at 30°C for 24–48 h. Plates were flooded with 0.5 ml of Lugol's iodine solution and examined for colonies surrounded by a clear zone indicating amylase production. Presumptive colonies were picked to a multiple test medium[22] and motile strains which produced an alkaline surface and acid butt are further tested for oxidase and indol production, gelatin hydrolysis and sensitivity to O/129 to presumptively identify them as members of the genus *Aeromonas*. This modified medium, brilliant green bile starch agar (BBGS), facilitates quantitative recovery of aeromonads when inoculated with food, water and stool samples[85].

Farber and co-workers[161] used the hydrophobic grid membrane filter method of Rippey and Cabelli to isolate aeromonads from foods packaged under a modified atmosphere. After filtering 1 ml of each dilution through the membranes, they were placed onto the surface of mA-trehalose agar which was incubated at 35°C for 20 h. Trehalose-positive colonies (yellow) were marked and the membrane was placed onto the surface of mA-mannitol agar and incubated for 2–3 h at 37°C. Colonies which remained yellow were marked and the membrane was transferred to a filter pad soaked with phosphate-buffered saline, pH 7.0, for 1 min and then to a filter pad saturated with tetra-*p*-phenylenediamine dihydrochloride and left for 15 s before the oxidase reaction was scored. Trehalose-, mannitol- and oxidase-positive colonies were counted as members of the *A. hydrophila* complex.

Knøchel and Jeppesen[162] surveyed drinking water and foods in Denmark for aeromonads using membrane filtration of 100 ml samples and placing the membranes onto SA plates. After incubation, plates were flooded with half-strength Lugol's iodine solution and colonies with a clear zone were counted as presumptive aeromonads. Yellow colonies on membranes were counted since the zones produced by amylase activity were difficult to read under the membranes. One to four presumptive aeromonad colonies were confirmed using biochemical tests. Aeromonads were recovered in salads with a pH of 5.8 ± 0.52 suggesting that aeromonads survive at lower pH than was previously thought[162]. In all, 36% of raw foods and 6.5% of processed foods grew aeromonads. Altogether, 116 *Aeromonas* isolates were recovered and 86% of these were *A. hydrophila*. Less than a third of the drinking water samples were positive for aeromonads and the numbers per sample ranged from 1–40 CFU/100 ml.

Ibrahim and MacRae[164] isolated aeromonads from red meat and milk samples using APW and SA and Ryan's agar with ampicillin (5 mg/l). Both plating media gave equivalent recovery of aeromonads, but Ryan's medium was occasionally overgrown by swarming *Proteus* in pork samples. *Aeromonas* contamination was widespread in beef, lamb, pork and milk samples.

Pin and colleagues[165] used APW enrichment with subculture of serial dilutions of Oxoid *Aeromonas* medium to sample fish, shellfish, raw meat, cheese and pre-prepared salads purchased from retail consumer outlets. Poultry had the highest incidence of positive samples (100%), followed by lamb (60%), shellfish, fish, pork and beef (40%) and cheese (20%). *A. hydrophila* accounted for most isolates and *A. caviae* was not recovered. Further studies demonstrated that 89% of *Aeromonas* strains isolated from food possess virulence factors[166]. Human strains isolated usually possess haemagglutinating ability and produce haemolysin, and it was suggested that these characteristics may help to determine whether food strains are pathogenic.

Ciufecu and co-workers[167] compared isolation of aeromonads on 15 media (TSA, DFS, ADA, BBGS, Ovotrip Agar in six variations and Egg Yolk Sulphite Agar in five variations). DFS and EYS (AGOS) in the unmodified version demonstrated the desired selectivity for reliable isolation of aeromonads from foods. Counts in fish and meat were 10^3–10^6 CFU/g. *Vibrio* spp. may be confused with aeromonads on EYS plates since both organisms are lecithinase-positive; however, incorporation of O/129 into the medium completely inhibited vibrios. Oxidase-, lecithinase-positive non-fermenters resulted in a false positive rate of 5.8% on EYS agar.

Fricker and Tompsett[168] compared BA10, MacConkey agar and bile salts irgasan brilliant green agar (BIBG) for isolation of aeromonads from foods. APW was used for enrichments. Samples (10 g) were placed into 100 ml of APW and incubated at 37°C for 24 h. A loopful of culture was streaked onto each plate and plates were incubated at 37°C for 24 h. Colonies were screened using oxidase reaction, resistance to O/129, inability to grow in broth containing 6% NaCl and ability to ferment glucose. At least one isolate from each food source was identified to species using routine biochemical tests incubated at 30°C. MacConkey plates were frequently overgrown with *E. coli*, while BA10 plates were occasionally overgrown with swarming *Proteus*. BIBG agar frequently yielded a pure culture of *Aeromonas* from samples of poultry, beef, pork, fish, cooked meats and pre-prepared salads. This medium is commercially available as *Aeromonas* Agar (Difco) and it is recommended for recovery of aeromonads from foods.

Gobat and Jemmi[169] evaluated seven selective agar media and two enrichment broths for isolation of aeromonads from meat, fish and shellfish samples. Plating media used in this evaluation included YSA (modified CIN), Bile salts irgasan brilliant green agar (BIBG), starch ampicillin DNA Agar (SADA), *Aeromonas* Medium (Ryan) with ampicillin (5 mg/l), Blood Ampicillin Agar (BA10, BA20 and BA30). Samples were prepared in decimal dilution and 0.1 ml was spread onto each plate. Plates were incubated at 35°C and examined at 18–24 h. Enrichment broths included tryptic soy broth with ampicillin (10 mg/l) and APW pH 8.7 incubated for 18 h at 28°C or 35°C. These media were challenged with *Escherichia coli, Klebsiella pneumoniae, Proteus vulgaris, Pseudomonas aeruginosa, Salmonella typhimurium, Yersinia enterocolitica, Yersinia intermedia* and *Plesiomonas shigelloides* in addition to aeromonads. BIBG and YSA suppressed growth of most of these bacteria except *Yersinia* spp. BIBG also permitted growth of *Ps. aeruginosa*. Altogether, BIBG provided the best selectivity, yet supported the growth of aeromonads. Colonies were 1–2 mm in diameter and the spot oxidase test could be performed on growth picked directly from the plates. Oxidase reactions could be determined by flooding the plates as long as no

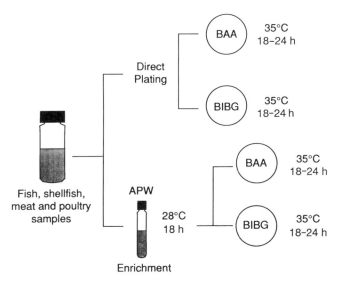

Figure 2.2   Recommended culture media and methods for isolation of *Aeromonas* spp. from food products. BAA, blood ampicillin agar (30 mg/l); BIBG, bile salts irgasan brilliant green agar; APW, alkaline peptone water, pH 8.6

xylose-fermenting colonies were present. Another advantage of BIBG is that it does not contain antibiotics so it is suitable for isolation of ampicillin-sensitive strains. The disadvantage of BIBG may be inhibition of some strains of *A. caviae* which are reported to be sensitive to brilliant green[154]. APW incubated at 28°C proved to be the better of the two enrichment broths. TSBA may underestimate aeromonads because of ampicillin sensitivity of some strains. However, inclusion of BA30 was found to be desirable since it provided easy recognition of aeromonads in foods because of its high selectivity and the ability to perform the oxidase test directly on the plate. The recommended method for isolation of aeromonads from fish, shellfish, poultry and meats is shown in Figure 2.2. The reported detection limit of this method is 100 CFU/g of food.

TSBA30 and MacConkey agar was the best combination of media for isolation of aeromonads from oysters. Sample volumes of 0.1 g/10 ml, 1 g/10 ml, 10 g/50 ml and 20g/180 ml were cultured. Modified RS broth (MRSB) and RS agar in various combinations with TSBA30 and MacConkey agar were not as effective as the combination of TSBA30 and MacConkey agar. Oysters had been frozen at −70°C for 18 months, indicating that aeromonads withstand long-term freezer storage and freeze thaw, when enrichment procedures are used for their recovery[170].

Majeed and colleagues[171] used APW and SA to isolate enterotoxigenic aeromonads from lamb purchased from a local grocery, and Hudson et al.[172] examined ready-to-eat flash foods for aeromonads using APW and SA. *Aeromonas* contamination was common.

Freitas and co-workers[173] examined pasteurized milk and white cheese using GSP supplemented with ampicillin (10 mg/l). Six of 20 milk and eight of 25 cheese samples grew aeromonads and *A. caviae* was the most common isolate. *A. hydrophila, A. sobria, A. schubertii* and atypical strains were also found.

Where verification of presumptive aeromonads is desirable, colonies are picked to nutrient agar plates and characterized for Gram reaction, oxidase reaction, resistance to O/129, fermentative metabolism, sensitivity to 6% NaCl. Alternatively, colonies may be picked to Kaper's multi-test media. Colonies producing alkaline top, acid butt, motility, indol and negative hydrogen sulphide reactions may be considered presumptive aeromonads.

At least a few colonies from each sample should be completely identified using conventional biochemical reactions or rapid commercial methods.

2.9 SPECIAL TOPICS

2.9.1 L-FORMS

McIntosh and Austin[174] recovered L-forms in low numbers (0.1–1% of recoverable organisms) from naturally and artificially infected fish using a nutrient-rich media, notably 1% brain heart infusion, supplemented with 0.5% yeast extract, 10% sucrose for osmotic stability and 10% horse serum as a detoxicant. Colonies appeared within 14 days incubation at 25°C, and were approximately 0.3–1.0 mm in diameter, brownish in colour, and buried into the agar surface. Colonies from fish tissues exhibited the typical 'fried egg' appearance characteristic of other bacterial L-forms; antibiotic-induced L-forms of *A. salmonicida* and *A. hydrophila* produced colonies 0.3–1.0 mm in diameter with dense dark centres, a distinct globose appearance and diffuse edge. In liquid culture, L-form cells ranged from 0.3–6.0 μm in diameter. Cultures reverted to the parental type upon subculture unless 20 μg/ml of benzylpenicillin was added to the culture medium. L-forms of *A. salmonicida* survived long-term storage in 10% glycerol at −70°C.

It is widely accepted that cultural enumeration methods underestimate cell populations, and the preponderance of data suggest that non-culturable but viable forms occur among other bacteria, and play an important role in the environmental survival of pathogens. Whether by production of L-forms, non-culturable but viable (dormant) forms, or by survival of small numbers of viable cells below the detection limits of culture methods, survival mechanisms remain to be established.

2.10   REFERENCES

1   Abeyta C Jr, Stelma GN. Isolation and identification of motile *Aeromonas* species. In *Bacteriological Analytical Manual, Supplement (9/87)*, 6th edn. Arlington, VA: Association of Official Analytical Chemists, 1987; 30.01–30.10.

2   Garcia-Lopez ML, Otero A, Garcia-Fernandez MC, Santos JA. [Incidence, behavior and control of *Aeromonas hydrophila* in meat and dairy products]. *Microbiologia* 1993; **9**: 49–56.

3   Palumbo S, Abeyta C, Stelma G. *Aeromonas hydrophila* group. In *Compendium of Methods for the Microbiological Examination of Foods*, Third edn. Vanderzant C, Splittstoesser DF (eds). Washington, DC: American Public Health Association, 1992; 497–515.

4   Stelma GNJ. *Aeromonas hydrophila*. In *Foodborne Bacterial Pathogens*. Doyle MP (ed). New York: Marcel Dekker, 1989; 1–19.

5   Farmer JJ III, Arduino MJ, Hickman-Brenner FW. The Genera *Aeromonas* and *Plesiomonas*. In *The Prokaryotes*, 2nd edn. Ballows A, Trüper HG, Dworkin M, Harder, W, Schleifer K-H (eds) New York: Springer-Verlag, 1991; 3012–45.

6   Joseph SW, Carnahan A. The isolation, identification, and systematics of the motile *Aeromonas* species. *Ann Rev Fish Dis* 1994; **4**: 315–43.

7   Joseph SW, Janda M, Carnahan A. Isolation, enumeration and identification of *Aeromonas* sp. *J Food Safety* 1988; **9**: 23–35.

8   Ramboarina C, Scheftel JM, Monteil H. [Mobile species of the genus *Aeromonas*: difficulties of identification and pathogenicity]. *Ann Biol Clin (Paris)* 1993; **51**: 91–100.

9   Nord CE, Sjoberg L, Wadström T, Wretlind B. Characterization of three *Aeromonas* and nine *Pseudomonas* species by extracellular enzymes and haemolysins. *Med Microbiol Immunol (Berl)* 1975; **161**: 79–87.

10   Namdari H, Cabelli VJ. The suicide phenomenon in motile aeromonads. *Appl Environ Microbiol* 1989; **55**: 543–7.

11   Namdari H, Cabelli VJ. Glucose-mediated catabolite repression of the tricarboxylic acid cycle as an explanation for increased acetic acid production in suicidal *Aeromonas* strains. *J Bacteriol* 1990; **172**: 4721–4.

12   Namdari H, Bottone EJ. Suicide phenomenon in mesophilic aeromonads as a basis for species identification. *J Clin Microbiol* 1989; **27**: 788–9.

13   Piersimoni C, Morbiducci V, De Sio G, Scalise G. Rapid presumptive phenospecies identification of mesophilic aeromonads by testing for suicidal activity. *Eur J Clin Microbiol Infect Dis* 1990; **9**: 608–10.

14   Want SV, Millership SE. Effects of incorporating ampicillin, bile salts and carbohydrates in media on the recognition and selection of *Aeromonas* spp. from faeces. *J Med Microbiol* 1990; **32**: 49–54.

15   Havelaar AH, During M, Versteegh JF. Ampicillin-dextrin agar medium for the enumeration of *Aeromonas* species in water by membrane filtration. *J Appl Bacteriol* 1987; **62**: 279–87.

16   Altwegg M, von Graevenitz A, Zollinger-Iten J. Medium and temperature dependence of decarboxylase reactions in *Aeromonas* spp. *Curr Microbiol* 1987; **15**: 1–4.

17   Esteve C, Amaro C, Biosca EG, Garay E. Biochemical and toxigenic properties of *Vibrio furnissii* isolated from a European eel farm. *Aquaculture* 1995; **132**: 81–90.

18   Lee JV, Shread P, Furniss AL, Bryant TN. Taxonomy and description of *Vibrio fluvialis* sp. nov. (synonym group F vibrios, group EF6). *J Appl Bacteriol* 1981; **50**: 73–94.

19   Janda JM, Motyl MR. Cephalothin susceptibility as a potential marker for the *Aeromonas sobria* group. *J Clin Microbiol* 1985; **22**: 854–5.

20 Carnahan A, Hammontree L, Bourgeois L, Joseph SW. Pyrazinamidase activity as a phenotypic marker for several *Aeromonas* spp. isolated from clinical specimens. *J Clin Microbiol* 1990; **28**: 391–2.

21 Figura N, Guglielmetti P. Differentiation of motile and mesophilic *Aeromonas* strains into species by testing for a CAMP-like factor. *J Clin Microbiol* 1987; **25**: 1341–2.

22 Kaper J, Seidler RJ, Lockman H, Colwell RR. Medium for the presumptive identification of *Aeromonas hydrophila* and *Enterobacteriaceae. Appl Environ Microbiol* 1979; **38**: 1023–6.

23 Toranzo AE, Santo Y, Nieto TP, Barja JL. Evaluation of different assay systems for identification of environmental *Aeromonas* strains. *Appl Environ Microbiol* 1986; **51**: 652–6.

24 De Ryck R, Struelens MJ, Surruys E. Rapid biochemical screening for *Salmonella, Shigella, Yersinia,* and *Aeromonas* isolates from stool specimens. *J Clin Microbiol* 1994; **32**: 1583–5.

25 Pathak SP, Bhattacherjee JW, Kalra N, Chandra S. Seasonal distribution of *Aeromonas hyrophila* in river water and isolation from river fish. *J Appl Bacteriol* 1988; **65**: 347–52.

26 Slade PJ, Falan MA, Al-Ghady AMR. Isolation of *Aeromonas hydrophila* from bottled waters and domestic water supplies in Saudi Arabia. *J Food Prot* 1986; **49**: 471–6.

27 Pessoa GV, Da Silva EA. [A new medium for the rapid presumptive identification of enterobacteriae, *Aeromonas* and vibrios (author's transl)]. *Ann Microbiol (Paris)* 1974; **125A**: 341–7.

28 Kalina GP, Somova AG, Grafova TI, Podosinnikova LS. [Use of the A-2 elective medium for the differentiation of *Aeromonas* from related microorganisms]. *Zh Mikrobiol Epidemiol Immunobiol* 1979; **7**: 62–6.

29 Versteegh JF, Havelaar AH, Hoekstra AC, Visser A. Complexing of copper in drinking water samples to enhance recovery of *Aeromonas* and other bacteria. *J Appl Bacteriol* 1989; **67**: 561–6.

30 Huys G, Kersters I, Vancanneyt M, Coopman R, Janssen P, Kersters K. Diversity of *Aeromonas* sp. in Flemish drinking water production plants as determined by gas-liquid chromatographic analysis of cellular fatty acid methyl esters (FAMEs). *J Appl Bacteriol* 1995; **78**: 445–55.

31 Bravo Farinas L, Monte Boada RJ, Zorrilla Aguila C, Padilla Ramirez M. [Application of the Moore swab method to the isolation of *Aeromonas* spp. from residual waters]. *Rev Cubana Med Trop* 1989; **41**: 413–18.

32 Holmberg SD, Schell WL, Fanning GR, Wachsmuth IK, Hickman-Brenner FW, Blake PA, Brenner DJ, Farmer JJ III. *Aeromonas* intestinal infections in the United States. *Ann Intern Med* 1986; **105**: 683–9.

33 Spira WM, Ahmen QS. Gauze filtration and enrichment procedures for recovery of *Vibrio cholerae* from contaminated water. *Appl Environ Microbiol* 1981; **42**: 730–3.

34 Shotts EB, Bullock GL. Bacterial diseases of fishes: diagnostic procedures for Gram-negative pathogens. *J Fish Res Board Can* 1975; **32**: 1243–7.

35 Shotts EB, Bullock GL. Rapid diagnostic approaches in the identification of Gram-negative bacterial diseases of fish. *Fish Pathol* 1976; **10**: 187–90.

36 Gracey M, Burke V, Robinson J. *Aeromonas*-associated gastroenteritis. *Lancet* 1982; **ii**: 1304–6.

37 Tsai WC, Chu SH, Lee SY, Lee PF. The effect of different brands of MacConkey and xylose lysine deoxycholate agar on the isolation of *Aeromonas hydrophila. Chung Hua Min Kuo Wei Sheng Wu Chi Mien I Hsueh Tsa Chih* 1987; **20**: 257–61.

38   Carnahan AM, Chakraborty T, Fanning GR, Verma D, Ali A, Janda JM, Joseph SW. *Aeromonas trota* sp. nov., an ampicillin-susceptible species isolated from clinical specimens. *J Clin Microbiol* 1991; **29**: 1206–10.
39   Janda JM, Dixon A, Raucher B, Clark RB, Bottone EJ. Value of blood agar for primary plating and clinical implication of simultaneous isolation of *Aeromonas hydrophila* and *Aeromonas caviae* from a patient with gastroenteritis. *J Clin Microbiol* 1984; **20**: 1221–2.
40   Richardson CJ, Robinson JO, Wagener LB, Burke V. In-vitro susceptibility of *Aeromonas* spp. to antimicrobial agents. *J Antimicrob Chemother* 1982; **9**: 267–74.
41   Chang BJ, Bolton SM. Plasmids and resistance to antimicrobial agents in *Aeromonas sobria* and *Aeromonas hydrophila* clinical isolates. *Antimicrob Agents Chemother* 1987; **31**: 1281–2.
42   Fainstein V, Weaver S, Bodey GP. In vitro susceptibilities of *Aeromonas hydrophila* against new antibiotics. *Antimicrob Agents Chemother* 1982; **22**: 513–14.
43   Altorfer R, Altwegg M, Zollinger-Iten J, von Graevenitz A. Growth of *Aeromonas* spp. on cefsulodin-Irgasan-novobiocin agar selective for *Yersinia enterocolitica. J. Clin Microbiol* 1985; **22**: 478–80.
44   Robinson J, Beaman J, Wagener L, Burke V. Comparison of direct plating with the use of enrichment culture for isolation of *Aeromonas* spp. from feces. *J Med Microbiol* 1986; **22**: 315–17.
45   Robinson J, Burke V, Worthy PJ, Beaman J, Wagener L. Media for isolation of *Aeromonas* spp. from faeces. *J Med Microbiol* 1984; **18**: 405–11.
46   Millership SE, Curnow SR, Chattopadhyay B. Faecal carriage rate of *Aeromonas hydrophila. J Clin Pathol* 1983; **36**: 920–3.
47   Moulsdale MT. Isolation of *Aeromonas* from faeces [letter]. *Lancet* 1983; **i**: 351.
48   Moyer NP, Geiss HK, Marinescu M, Rigby A, Robinson J, Altwegg M. Media and methods for isolation of aeromonads from fecal specimens. A multilaboratory study. *Experientia* 1991; **47**: 409–12.
49   Shread P, Donovan TJ, Lee JV. A survey of the incidence of *Aeromonas* in human faeces. *Soc Gen Microbiol Q* 1981; **8**: 184.
50   von Graevenitz A, Bucher C. Evaluation of differential and selective media for isolation of *Aeromonas* and *Plesiomonas* spp. from human feces. *J Clin Microbiol* 1983; **17**: 16–21.
51   Khardori N, Fainstein V. *Aeromonas* and *Plesiomonas* as etiological agents. *Annu Rev Microbiol* 1988; **42**: 395–419.
52   Millership SE, Chattopadhyay B. Methods for the isolation of *Aeromonas hydrophila* and *Plesiomonas shigelloides* from faeces. *J Hyg (Lond)* 1984; **92**: 145–52.
52a Price EH, Hunt GH. *Aeromonas* in hospital – methods of isolation [letter]. *J Hosp Infect* 1986; **8**: 309–11.
53   Stern NJ, Drazek ES, Joseph SW. Low incidence of *Aeromonas* sp. in livestock feces. *J Food Protect* 1987; **50**: 66.
54   Gray SJ. *Aeromonas hydrophila* in livestock: incidence, biochemical characteristics and antibiotic susceptibility. *J Hyg (Lond)* 1984; **92**: 365–75.
55   Effendi I, Austin B. Survival of the fish pathogen *Aeromonas salmonicida* in seawater. *FEMS Microbiol Lett* 1991; **68**: 103–6.
56   Monfort P, Baleux B. Dynamics of *Aeromonas hydrophila, Aeromonas sobria,* and *Aeromonas caviae* in a sewage treatment pond. *Appl Environ Microbiol* 1990; **56**: 1999–2006.
57   Meeks MV. The genus *Aeromonas*: methods for identification. *Am J Med Technol* 1963; **29**: 361–78.
58   McCoy RH, Seidler RJ. Potential pathogens in the environment: isolation, enumeration, and identification of seven genera of intestinal bacteria associated

with small green pet turtles. *Appl Microbiol* 1973; **25**: 534–8.

59 Seidler RJ, Allen DA, Lockman H, Colwell RR, Joseph SW, Daily OP. Isolation, enumeration, and characterization of *Aeromonas* from polluted waters encountered in diving operations. *Appl Environ Microbiol* 1980; **39**: 1010–18.

60 Schubert RH. [The occurrence of *Aeromonas* in surface waters]. *Arch Hyg Bakteriol* 1967; **150**: 688–708.

61 Schubert R. Ecology of aeromonads and isolation from environmental samples. *Experientia* 1987; **43**: 351–4.

62 Shotts EB Jr, Rimler R. Medium for the isolation of *Aeromonas hydrophila*. *Appl Microbiol* 1973; **26**: 550–3.

63 Davis, JW, Sizemore RK. Nonselectivity of Rimler–Shotts medium for *Aeromonas hydrophila* in estuarine environments. *Appl Environ Microbiol* 1981; **42**: 544–5.

64 Neilson AH. The occurrence of aeromonads in activated sludge: isolation of *Aeromonas sobria* and its possible confusion with *Escherichia coli*. *J Appl Bacteriol* 1978; **44**: 259–64.

65 Rippey SR, Cabelli VJ. Membrane filter procedure for enumeration of *Aeromonas hydrophila* in fresh waters. *Appl Environ Microbiol* 1979; **38**: 108–13.

66 Kielwein G. Ein Nährboden zur selektiven züchtung von pseudomonaden und aeromonaden. *Arch Lebensmittelhygiene* 1969; **20**: 131–3.

67 Schubert R. [The detection of aeromonads of the 'hydrophila–punctata-group' within the hygienic control of drinking water (author's transl)]. *Zentralbl Bakteriol [Orig B]* 1976; **161**: 482–97.

68 Rogol M, Sechter I, Grinberg L, Gerichter CB. Pril-xylose-ampicillin agar, a new selective medium for the isolation of *Aeromonas hydrophila*. *J Med Microbiol* 1979; **12**: 229–31.

69 Havelaar AH, Vonk M. The preparation of ampicillin dextrin agar for the enumeration of *Aeromonas* in water. *Lett Appl Microbiol* 1988; **7**: 169–71.

70 Huguet JM, Ribas F. SGAP-10C agar for the isolation and quantification of *Aeromonas* from water. *J Appl Bacteriol* 1991; **70**: 81–8.

71 Monfort P, Baleux B. [Evaluation of two culture media for the isolation and enumeration of motile *Aeromonas* in different kinds of water]. *C R Acad Sci III* 1988; **307**: 523–7.

72 Monfort P, Baleux B. Haemolysin occurrence among *Aeromonas hydrophila, Aeromonas caviae* and *Aeromonas sobria* strains isolated from different aquatic ecosystems. *Res Microbiol* 1991; **142**: 95–102.

73 Acros ML, de Vicente A, Moringo MA, Romero P, Borrego JJ. Evaluation of several selective media for recovery of *Aeromonas hydrophila* from polluted waters. *Appl Environ Microbiol* 1988; **54**: 2786–92.

74 Knøchel S. The suitability of four media for enumeration of *Aeromonas* spp. from environmental samples. *Lett Appl Microbiol* 1989; **9**: 67–9.

75 Poffé R, Op de Beeck E. Enumeration of *Aeromonas hydrophila* from domestic wastewater treatment plants and surface waters. *J Appl Bacteriol* 1991; **71**: 366–70.

76 Ribas F, Araujo R, Frias J, Huguet JM, Ribas FR, Lucena F. Comparison of different media for the identification and quantification of *Aeromonas* spp. in water. *Antonie Van Leeuwenhoek* 1991; **59**: 25–8.

77 Bernagozzi M, Bianucci F, Scerre E, Sacchetti R. Assessment of some selective media for the recovery of *Aeromonas hydrophila* from surface water. *Zentralbl Hyg* 1994; **195**: 121–34.

78 Hazen TC, Fliermans CB, Hirsch RP, Esch GW. Prevalance and distribution of

Aeromonas hydrophila in the United States. *Appl Environ Microbiol* 1978; **36**: 731–8.

79   Abeyta C, Weagant SD, Kaysner CA, Wekell MM, Stott RF, Krane MH, Peeler JP. *Aeromonas hydrophila* in shellfish growing waters: incidence and media evaluation. *J Food Protect* 1989; **52**: 7–12.

80   Venkateswaran K, Nakano H, Hashimoto H. A pad pre-enrichment technique for the isolation of members of the *Vibrionaceae*. *Microbiol Lett* 1989; **42**: 7–12.

81   Nakano H, Hashimoto H, Sasaki M, Kameyama T, Takahashi T, Kawakami H. Distribution and properties of motile *Aeromonas* and *Plesiomonas* in aquatic environment. *Jpn J Food Microbiol* 1986; **3**: 101–8.

82   Abeyta C, Kaysner CA, Wekell MM, Stott RF. Incidence of motile aeromonads from United States west coast shellfish growing estuaries. *J Food Protect* 1990; **53**: 849–55.

82a Venkateswaran K, Nakano H, Kiiyukia C, Kawakami H, Hashimoto H. Significance of sinking particles in the distribution of motile *Aeromonas* during the winter season. *Microbios* 1991; **68**: 73–85.

83   Abbey SD, Etang BB. Incidence and biotyping of *Aeromonas* species from the environment. *Microbios* 1988; **56**: 149–55.

84   Hood, MA, Baker RM, Singleton RL. Effect of processing and storing oyster meats on concentrations of indicator bacteria, vibrios and *Aeromonas hydrophila*. *J Food Prot* 1984; **47**: 598–601.

85   Nishikawa Y, Kishi T. Isolation and characterization of motile *Aeromonas* from human, food and environmental specimens. *Epidemiol Infect* 1988; **101**: 213–23.

86   Kaper, JB, Lockman H, Colwell RR, Joseph SW. *Aeromonas hydrophila*: ecology and toxigenicity of isolates from an estuary. *J Appl Bacteriol* 1981; **50**: 359–77.

87   Alonso JL, Garay E. Two membrane filter media (mADA/0129 and mSA/0129 agars) for enumeration of motile *Aeromonas* in seawater. *Zentralbl Hyg Umweltmed* 1989; **189**: 14–19.

88   Araujo RM, Arribas RM, Lucena F, Pares R. Relation between *Aeromonas* and faecal coliforms in fresh waters. *J Appl Bacteriol* 1989; **67**: 213–7.

89   Nakano H, Kameyama T, Venkateswaran K, Kawakami H, Hashimoto H. Distribution and characterization of hemolytic, and enteropathogenic motile *Aeromonas* in aquatic environment. *Microbiol Immunol* 1990; **34**: 447–58.

90   Paniagua C, Rivero O, Anguita J, Naharro G. Pathogenicity factors and virulence for rainbow trout *(Salmo gairdneri)* of motile *Aeromonas* spp. isolated from a river. *J Clin Microbiol* 1990; **28**: 350–5.

91   Mateos D, Anguita J, Rivero O, Naharro G, Paniagua C. Comparative study of virulence and virulence factors of *Aeromonas hydrophila* strains isolated from water and sediments of a river. *Zentralbl Hyg* 1992; **193**: 114–22.

92   Nazer H, Price E, Hunt G, Patel U, Walker-Smith J. Isolation of *Aeromonas* spp. from canal water. *Indian J Pediatr* 1990; **57**: 115–18.

93   Chowdhury MA, Yamanaka H, Miyoshi S, Shinoda S. Ecology of mesophilic *Aeromonas* spp. in aquatic environments of a temperate region and relationship with some biotic and abiotic environmental parameters. *Zentralbl Hyg Umweltmed* 1990; **190**: 233–56.

94   van der Kooij D. Nutritional requirements of aeromonads and their multiplication in drinking water. *Experientia* 1991; **47**: 444–6.

95   van der Kooij D, Hijnen WA. Nutritional versatility and growth kinetics of an *Aeromonas hydrophila* strain isolated from drinking water. *Appl Environ Microbiol* 1988; **54**: 2842–51.

96   Moyer NP. Clinical significance of *Aeromonas* species isolated from patients with diarrhoea. *J Clin Microbiol* 1987; **25**: 2044–8.

97   Burke V, Robinson J, Gracey M, Peterson D, Meyer N, Haley V. Isolation of *Aeromonas* spp. from an unchlorinated domestic water supply. *Appl Environ Microbiol* 1984; **48**: 367–70.

98   Burke V, Robinson J, Gracey M, Peterson D, Partridge K. Isolation of *Aeromonas hydrophila* from a metropolitan water supply: seasonal correlation with clinical isolates. *Appl Environ Microbiol* 1984; **48**: 361–6.

99   Millership SE, Chattopadhyay B. *Aeromonas hydrophila* in chlorinated water supplies. *J Hosp Infect* 1985; **6**: 75–80.

100 Cunliffe DA, Adcock P. Isolation of *Aeromonas* spp. from water by using anaerobic incubation. *Appl Environ Microbiol* 1989; **55**: 2138–40.

101 Havelaar AH, Versteegh JF, During M. The presence of *Aeromonas* in drinking water supplies in The Netherlands. *Zentralbl Hyg Umweltmed* 1990; **190**: 236–56.

102 Moyer NP, Luccini GM, Holcomb LA, Hall NH, Altwegg M. Application of ribotyping for differentiating aeromonads isolated from clinical and environmental sources. *Appl Environ Microbiol* 1992; **58**: 1940–4.

103 Holmes P, Sartory DP. An evaluation of media for the membrane filtration enumeration of *Aeromonas* from drinking water. *Lett Appl Microbiol* 1993; **17**: 58–60.

104 Mascher F, Reinthaler FF, Stunzner D, Lamberger B. *Aeromonas* species in a municipal water supply of a central European city: biotyping of strains and detection of toxins. *Zentralbl Bakteriol Mikrobiol Hyg [B]* 1988; **186**: 333–7.

105 Mariottini M, Mingoia M. [The typing of environmental strains of *Aeromonas* spp.]. *Ann Ig* 1992; **4**: 235–8.

106 Ghanem EH, Mussa ME, Eraki HM. *Aeromonas*-associated gastroenteritis in Egypt. *Zentralbl Mikrobiol* 1993; **148**: 441–7.

107 Bhat P, Shanthakumari S, Rajan D. The characterization and significance of *Plesiomonas shigelloides* and *Aeromonas hydrophila* isolated from an epidemic of diarrhoea. *Indian J Med Res* 1974; **62**: 1051–60.

108 Krovacek K, Peterz M, Faris A, Mansson I. Enterotoxigenicity and drug sensitivity of *Aeromonas hyrophila* isolated from well water in Sweden: a case study. *Int J Food Microbiol* 1989; **8**: 149–54.

109 Gonzalez C, Gutierrez C, Grande T. Bacterial flora in bottled uncarbonated mineral drinking water. *Can J Microbiol* 1987; **33**: 1120–5.

110 Hunter PR. The microbiology of bottled natural mineral waters. *J Appl Bacteriol* 1993; **74**: 345–52.

111 Warburton DW, Dodds KL, Burke R, Johnston MA, Laffey PJ. A review of the microbiological quality of bottled water sold in Canada between 1981 and 1989. *Can J Microbiol* 1992; **38**: 12–19.

112 Warburton DW, McCormick JK, Bowen B. Survival and recovery of *Aeromonas hydrophila* in water: development of methodology for testing bottled water in Canada. *Can J Microbiol* 1993; **40**: 145–8.

113 Peterkin PI, Warburton DW. MFLP-58, Enumeration of *Aeromonas hydrophila* by the hydrophobic grid-membrane filter (HGMFF). In *Compendium of analytical methods*. Montreal, Quebec: Ployscience Publications, Inc, 1988.

114 Warburton DW, Peterkin PI. Enumeration of *Aeromonas hydrophila* in prepackaged ice and water in sealed containers by the hydrophobic grid-membrane filter (HGMF) technique. MFLP-58B. In *Compendium of Analytical Methods*. Montreal: Polyscience Publications, Inc, 1993.

115 Clark JA, Vlassoff LT. Relationships among pollution indicator bacteria isolated from raw water and distribution systems by the presence–absence (P–A) test. *Health Lab Sci* 1973; **10**: 163–72.

116 Grabow WOK, DuPrez M. Comparison of m-Endo LES, MacConkey, and

Teepol media for membrane filter counting of total coliform bacteria in water. *Appl Environ Microbiol* 1979; **38**: 351–8.

117 Lupo L, Strickland E, Dufour A, Cabelli V. The effect of oxidase positive bacteria on total coliform density estimates. *Health Lab Sci* 1977; **14**: 117–21.

118 Clark JA. The influence of increasing numbers of nonindicator organisms upon the detection of indicator organisms by the membrane filter and presence–absence tests. *Can J Microbiol* 1980; **26**: 827–32.

119 Klontz GW, Anderson DP. Fluorescent antibody studies of isolates of *Aeromonas salmonicida*. *Bull Off Int Epizoot* 1968; **69**: 1149–57.

120 Shotts EB. Selective isolation methods for fish pathogens. *Soc Appl Bacteriol Symp Ser* 1991; **20**: 75S–80S.

121 Böhm, KH, Führmann H, Schlotfeldt HJ, Korting W. *Aeromonas salmonicida* from salmonids and cyprinids – serological and cultural identification. *Zentralbl Veterinarmed [B]* 1986; **33**: 777–83.

122 Boulanger Y, Lallier R, Cousineau G. Isolation of enterotoxigenic *Aeromonas* from fish. *Can J Microbiol* 1977; **23**: 1161–4.

123 Bragg RR. Health status of salmonids in river systems in Natal. III. Isolation and identification of bacteria. *Onderstepoort J Vet Res* 1991; **58**: 67–70.

124 Elliot DC, Shotts EB. Aetiology of an ulcerative disease in goldfish *Carassius auratus* (L): microbiological examination of diseased fish from seven locations. *J Fish Dis* 1980; **3**: 133–43.

125 Evelyn TPT. An aberrant strain of the bacterial fish pathogen *Aeromonas salmonicida* isolated from a marine host, the sablefish (*Anoplopoma fimbria*), and from two species of cultured Pacific salmon. *J Fisheries Res Bd Can* 1971; **28**: 1629–34.

126 McCarthy DH. Fish furunculosis caused by *Aeromonas salmonicida* var. *achromogenes*. *J Wildl Dis* 1975; **11**: 489–93.

127 Paterson WD, Douey D, Desautels D. Relationship between selected strains of typical and atypical *Aeromonas salmonicida, Aeromonas hydrophila*, and *Haemophilus piscium*. *Can J Microbiol* 1980; **26**: 588–98.

128 Duff DCB. Dissociation of *Bacillus salmonicida*, with special reference to the appearance of a G form of culture. *J Bacteriol* 1937; **34**: 49–67.

129 Ishiguro EE, Ainsworth T, Trust TJ, Kay WW. Congo red agar, a differential medium for *Aeromonas salmonicida*, detects the presence of the cell surface protein array involved in virulence. *J Bacteriol* 1985; **164**: 1233–7.

130 Udey LR. A differential medium for distinguishing Alr+ from Alr– phenotypes in *Aeromonas salmonicida*. In *13th Annual Conference and Workshop and 7th Eastern Fish Health Workshop*. Baltimore, MD, 1982.

131 Wilson A, Horne MT. Detection of A-protein in *Aeromonas salmonicida* and some effects of temperature in A-layer assembly. *Aquaculture* 1986; **56**: 23–7.

132 Cipriano RC, Bertolini J. Selection for virulence in the fish pathogen *Aeromonas salmonicida*, using Coomassie Brilliant Blue agar. *J Wildl Dis* 1988; **24**: 672–8.

133 Markwardt NM, Gocha YM, Klontz GW. A new application for coomassie brilliant blue agar: detection of *Aeromonas salmonicida* in clinical samples. *Dis Aquat Org* 1989; **6**: 231–3.

134 Cipriano RC, Ford LA. Comparison of dilution counts with standard culture methods for the detection of *Aeromonas salmonicida* from clinical specimens. *Biomed Lett* 1993; **48**: 145–50.

135 Teska JD, Cipriano RC. Nonselective nature of Coomassie brilliant blue agar for the presumptive identification of *Aeromonas salmonicida* in clinical specimens. *Dis Aquat Org* 1993; **16**: 239–42.

136 Cipriano RC, Ford LA, Teska JD, Hale LE. Detection of *Aeromonas*

salmonicida in the mucus of salmonid fishes. *J Aquat Animal Health* 1992; **4**: 114–18.

137   Ishiguro EE, Kay WW, Ainsworth T, Chamberlain JB, Austen RA, Buckley JT, Trust TJ. Loss of virulence during culture of *Aeromonas salmonicida* at high temperature. *J Bacteriol* 1981; **148**: 333–40.

138   Johnson CM, Tatner MF, Horne MT. Autoaggregation and extracellular A-layer protein in *Aeromonas salmonicida*. *Aquaculture* 1985; **46**: 163–6.

139   Chapman PF, Cipriano RC, Teska JD. Isolation and phenotypic characterization of an oxidase-negative *Aeromonas salmonicida* causing furunculosis in coho salmon *(Oncorhynchus kisutch)*. *J Wildl Dis* 1991; **27**: 61–7.

140   Altwegg M. *Aeromonas* and *Plesiomonas*: isolation procedures for pathological specimens. *Experientia* 1987; **43**: 354–5.

141   Acosta B, Perez de Leon A, Gobernado M, Camanas A. [*Aeromonas* spp. isolated from human feces. Species and pathogenicity factors]. *Enferm Infecc Microbiol Clin* 1991; **9**: 329–34.

142   Altwegg M, Lüthy-Hottenstein J. Behaviour of *Aeromonas* species in Cary–Blair transport medium at various temperatures [letter; comment]. *Eur J Clin Microbiol Infect Dis* 1992; **11**: 79–80.

143   Siitonen A, Mattila H. Effect of transport medium on recovery of *Aeromonas* species in intestinal infections [letter] [see comments]. *Eur J Clin Microbiol Infect Dis* 1990; **9**: 368–70.

144   von Graevenitz A, Zinterhofer L. The detection of *Aeromonas hydrophila* in stool specimens. *Health Lab Sci* 1970; **7**: 124–7.

145   Schubert RHW. Über den nachweis von *Plesiomonas shigelloides* Habs und Schubert, 1962 und ein elektivmedium, den inositol-brilliantgrün-gallesalz-agar. *E. Rodenwaldt Arch* 1977; **4**: 97–103.

146   Ashdown LR, Koehler JM. The spectrum of *Aeromonas*-associated diarrhoea in tropical Queensland, Australia. *Southeast Asian J Trop Med Public Health* 1993; **24**: 347–53.

147   Kelly MT, Stroh EM, Jessop J. Comparison of blood agar, ampicillin blood agar, ampicillin blood agar, MacConkey-ampicillin-Tween agar, and modified cefsulodin-Irgasan-novobiocin agar for isolation of *Aeromonas* spp. from stool specimens. *J Clin Microbiol* 1988; **26**: 1738–40.

148   Altwegg M, Jöhl M. Isolation frequency of *Aeromonas* species in relation to patient age. *Eur J Clin Microbiol* 1987; **6**: 55–6.

149   Mishra S, Nair GB, Bhadra RK, Sikder SN, Pal SC. Comparison of selective media for primary isolation of *Aeromonas* species from human and animal feces. *J Clin Microbiol* 1987; **25**: 2040–3.

150   Ashiru JL, Salau T, Rotilu IO. Incidence of *Aeromonas* species in diarrhoeic stool in University College Hospital Ibadan, Nigeria. *Comp Immun Microbiol Infect Dis* 1993; **16**: 51–4.

151   Hunt GH, Price EH, Patel U, Messenger L, Stow P, Salter P. Isolation of *Aeromonas* sp from faecal specimens. *J Clin Pathol* 1987; **40**: 1382–4.

152   Kay BA, Guerrero CE, Sack RB. Media for the isolation of *Aeromonas hydrophila*. *J Clin Microbiol* 1985; **22**: 888–90.

153   Figura N. A comparison of various media in the detection of *Aeromonas* spp. from stool samples. *Boll Ist Sieroter Milan* 1985; **64**: 167–9.

154   Desmond E, Janda JM. Growth of *Aeromonas* species on enteric agars. *J Clin Microbiol* 1986; **23**: 1065–7.

155   Wilcox MH, Cook AM, Eley A, Spencer RC. *Aeromonas* spp. as a potential cause of diarrhoea in children. *J Clin Pathol* 1992; **45**: 959–63.

156   Pascual A, Moreno MD, Ortega MC, Carranza R. Comparación de dos medios de cultivo para el aislamiento de *Aeromonas* spp. en heces [Letter]. *Enferm*

Infect Microbiol Clin 1993; **11**: 515.

157 Gobat PF, Jemmi T. Distribution of mesophilic *Aeromonas* species in raw and ready-to-eat fish and meat products in Switzerland. *Int J Food Microbiol* 1993; **20**: 117–20.

158 Palumbo SA, Maxino F, Williams AC, Buchanan RL, Thayer DW. Starch-ampicillin agar for the quantitative detection of *Aeromonas hydrophila*. *Appl Environ Microbiol* 1985; **50**: 1027–30.

159 Palumbo SA, Bencivengo MM, Del Corral F, Williams AC, Buchanan RL. Characterization of the *Aeromonas hydrophila* group isolated from retail foods of animal origin. *J Clin Microbiol* 1989; **27**: 854–9.

160 Nishikawa Y, Kishi T. A modification of bile salts brilliant green agar for isolation of motile *Aeromonas* from foods and environmental specimens. *Epidemiol Infect* 1987; **98**: 331–6.

161 Farber JM, Warburton DW, Gour L, Milling M. Microbiological quality of foods packaged under modified atmospheres. *Food Microbiol* 1990; **7**: 327–34.

162 Knøchel S, Jeppesen C. Distribution and characteristics of *Aeromonas* in food and drinking water in Denmark. *Int J Food Microbiol* 1990; **10**: 317–22.

163 Palumbo SA, Buchanan RL. Factors affecting growth or survival of *Aeromonas hydrophila* in foods. *J Food Safety* 1988; **9**: 37–51.

164 Ibrahim A, MacRae IC. Incidence of *Aeromonas* and *Listeria* spp. in red meat and milk samples in Brisbane, Australia. *Int J Food Microbiol* 1991; **12**: 263–9.

165 Pin C, Marin ML, Garcia ML, Tormo J, Selgas MD, Casas C. Incidence of motile *Aeromonas* spp. in foods. *Microbiologia* 1994; **10**: 257–62.

166 Pin C, Marin ML, Selgas MD, Garcia ML, Tormo J, Casas C. Virulence factors in clinical and food isolates of *Aeromonas* species. *Folia Microbiol (Praha)* 1994; **39**: 331–6.

167 Ciufecu C, Nacescu N, Israil A, Cedru C. Isolation of motile aeromonads from foods of animal origin. *Arch Roum Pathol Exp Microbiol* 1990; **49**: 119–29.

168 Fricker CR, Tompsett S. *Aeromonas* spp. in foods: a significant cause of food poisoning? *Int J Food Microbiol* 1989; **9**: 17–23.

169 Gobat PF, Jemmi T. Comparison of seven selective media for the isolation of mesophilic *Aeromonas* species in fish and meat. *Int J Food Microbiol* 1995; **24**: 375–84.

170 Abeyta C, Kaysner CA, Wekell MM, Sullivan JJ, Stelma GN. Recovery of *Aeromonas hyrophila* from oysters implicated in an outbreak of foodborne illness. *J Food Protect* 1986; **49**: 643–6.

171 Majeed K, Egan A, MacRae IC. Enterotoxigenic aeromonads on retail lamb meat and offal. *J Appl Bacteriol* 1989; **67**: 165–70.

172 Hudson JA, Mott SJ, Delacy KM, Edridge AL. Incidence and coincidence of *Listeria* spp., motile aeromonads and *Yersinia enterocolitica* on ready-to-eat fleshfoods. *Int J Food Microbiol* 1992; **16**: 99–108.

173 Freitas AC, Nunes, MP, Milhomem AM, Ricciardi ID. Occurrence and characterization of *Aeromonas* species in pasteurized milk and white cheese in Rio De Janeiro, Brazil. *J Food Protect* 1993; **56**: 62–5.

174 McIntosh D, Austin B. Comparison of methods for the induction, propagation and recovery of L-phase variants of *Aeromonas* spp. *J Diarrh Dis Res* 1988; **6**: 131–6.

175 Misra SK, Bhadra RK, Pal SC, Nair GB. Growth of *Aeromonas* spp. on Butzler *Campylobacter* selective agar and evaluation of the agar for the primary isolation of *Aeromonas* spp. from clinical specimens. *J Clin Microbiol* 1989; **27**: 346–7.

176 de la Morena ML, Van R, Singh K, Brian M, Murray ME, Pickering LK. Diarrhoea associated with *Aeromonas* species in children in day care centers. *J Infect Dis* 1993; **168**: 215–18.

3 Identification

SALLY E. MILLERSHIP
North Essex Health Authority, Witham, Essex, England

3.1 INTRODUCTION

The identification of bacterial colonies as *Aeromonas* spp. in primary cultures is not generally difficult. Occasionally problems may be encountered when using an oxidase test to screen colonies on primary plates, when to the casual observer the mesophilic species may be mistaken for pseudomonads. However, careful observation of colonial forms, colour and odour will readily distinguish aeromonads from other commonly encountered mesophiles. The psychrophilic members of the genus, now considered a subspecies of hybridization group 3 (=HG3), have very different ecology and growth characteristics and would not normally be confused with other organisms.

For many purposes it may be sufficient to report aeromonads as *Aeromonas* spp., *A. hydrophila* group or *A. hydrophila* complex; however, there are arguments for further identification wherever possible[1]. There is some evidence that species differ in their pathogenicity[2-4], and that the distribution of human and environmental strains differ[5]. However, little is known about the disease syndromes or pathogenic potential associated with the different species. Certainly for research studies these are compelling reasons for fully identifying isolates. In clinical laboratories, where it could be argued that routine species identification makes little difference to the individual patient, even a simple phenotyping scheme may make a considerable contribution to the understanding of the epidemiology of these organisms.

The history of the nomenclature of species within the genus has been well reviewed (Chapter 1). The tendency to use the same or similar names at different times for species defined in different ways can lead to confusion when identifications are reported, if care is not taken to use a consistent scheme. Thus *A. salmonicida* once referred to the psychrophilic strains only, but now includes all strains of HG3, some of which are mesophiles biochemically similar to *A. hydrophila*. The currently recognized phenotypes and genotypes[6-12] are summarized in Table 1.2.

The aeromonads were first described nearly 60 years ago, and many tables of biochemical characters include results from different studies where test methods are not always clearly described or in which varying methods have been used. In the tables that follow, standard test methods are assumed[14,15] unless otherwise stated. A variety of incubation temperatures from 25°C to

The Genus Aeromonas. Edited by B. Austin, M. Altwegg, P.J. Gosling and S. Joseph
© 1996 John Wiley & Sons Ltd

37°C[16-18] have been used, which may affect results (Dr A. Carnahan, personal communication). However, very little has been published on these differences with the exception of gas production from glucose[19], where some strains may be positive at an incubation temperature of 22°C but not at 37°C.

There is a need for simple identification schemes based on readily available tests, and some progress towards this has been made[18,20], but those proposed have not yet been widely used and evaluated. They were intended for use in clinical laboratories; it is not clear how valuable they would be for the identification of food, water and other environmental isolates. In addition, more species almost certainly remain to be discovered and, where accurate identification is critical, it would be wise to include molecular as well as conventional methods.

3.2   IDENTIFICATION TO GENUS LEVEL FROM PRIMARY CULTURES

3.2.1   MESOPHILIC STRAINS

Aeromonads grow readily on standard media. After overnight incubation at 37°C on horse blood agar, typical colonies are 1–3 mm in diameter, circular, convex, with a smooth edge, white to grey and translucent. Older colonies develop a greenish hue. Mesophilic strains are not usually pigment producers, except for isolates of the non-motile species *A. media*, not so far recovered from clinical samples. Motile strains producing brown pigment have been described from environmental samples[21] and occasionally from human specimens (unpublished observations). The pigment is best observed after two to three days at room temperature on nutrient agar.

Several types of haemolysis may occur around single colonies. These include a broad zone of β-haemolysis, a double zone of partial haemolysis, or a narrow zone of α-haemolysis. The presence of β-haemolysis varies with bacterial and red cell species, but is generally less common amongst strains of phenotype *A. caviae* than *A. hydrophila* or the group formerly known as *A. sobria*, whether horse or sheep blood is used[4,18]. On first isolation, almost all strains show evidence of β-haemolysis where colonies are confluent, especially if incubation is continued for 48 h. After repeated subculture, colonial variants, especially those with reduced colony size and reduced or absent zones of haemolysis, readily appear. Colonies have a peculiar sulphurous smell similar to that of the vibrios, which becomes very strong on incubation for two to three days at room temperature. Some strains show a marked tendency to become mucoid, but never to the same degree as *Klebsiella*, and although originally described as non-capsulate[22], later studies have shown that many strains do have a capsule[19].

In pure culture on MacConkey agar strains of phenotypes *A. hydrophila* and the *A. sobria* group usually appear as non-lactose fermenters. However, up to 40%[4] of isolates of phenotype *A. caviae* may be lactose-fermenting. In mixed culture with lactose-fermenting organisms, non-lactose-fermenting colonies of *Aeromonas* strains may be indistinguishable from those of the fermenters, probably from acid production by surrounding colonies[23]. Aeromonads grow on more inhibitory 'enteric' media, such as deoxycholate citrate agar, although colonies may be 1 mm or less. They do not usually grow on thiosulphate citrate bile salts agar (TCBS).

Several characters assist a presumptive identification. These include an oxidase test on media without carbohydrates. Oxidase tests are less reliable on media with a fermentable carbohydrate, as false negative results have been observed in the presence of acid[24]. Mixed cultures on filter membranes can be placed on a pad soaked with oxidase reagent and suspicious colonies subcultured promptly[25]. Aeromonads grow well on blood and enteric media in anaerobic conditions, which distinguishes them from pseudomonads, and appear as fermenters in an O/F (oxidation/fermentation) test. On indicator media where lactose has been replaced by xylose, such as deoxycholate citrate agar (XDCA)[26], they appear as non-fermenters, unlike most of the Enterobacteriaceae. Also unlike many of the Enterobacteriaceae, the aeromonads, with the exception of *A. trota*[6] and occasional strains of other species[18,23], are resistant to ampicillin. This may be incorporated into primary isolation media at a concentration of 10, 20 or 30 mg/l[23].

Confirmation of the identity of presumptive aeromonads may be performed using either conventional biochemical tests or kits. There are some problems with the latter, which will be discussed below. The current definition of *Aeromonas* ssp.[27] includes the following characters of use in identification: Members of the genus are Gram-negative, oxidase- and catalase-positive rods, which are facultatively anaerobic and ferment carbohydrates, sometimes with the production of gas. Nitrate is reduced to nitrite. Unlike many vibrios, they do not require salt for growth, and they are resistant to vibriostatic agent 0/129 (2,4-diamino-6,7-diisopropylpteridine). Several mesophilic and one psychrophilic phenotype can be distinguished. Psychrophilic strains are usually non-motile and unable to grow above 30°C. Mesophilic strains usually grow well at 37°C, and some up to 43°C. They are usually but not invariably motile.

A table of characters using conventional biochemical tests and antibiotic sensitivities for the distinction of mesophilic *Aeromonas* ssp. from other oxidase- and catalase-positive facultatively anaerobic Gram-negative species likely to be encountered in clinical or environmental samples is shown in Table 3.1[28-34]. Various methods for testing sensitivity to 0/129 have been described[36-37], the simplest of which is to place impregnated filter paper discs on a previously inoculated culture plate[37]. Suitable discs are sometimes available commercially. Authors give few details of the interpretation of the

Table 3.1   Characters for the differentiation of *Vibrio*, *Aeromonas* and *Plesiomonas*

Character	V. cholerae and V. mimicus	V. vulnificus	V. parahaemolyticus	V. alginolyticus	V. harveyi	V. cincinnatiensis	V. anguillarum	V. damsela	V. hollisae	V. natriegens	V. fluvialis and V. furnissii	Aeromonas	Plesiomonas
Nitrate reduction	+	+	+	+	+	+	+	+	+	+	+	+	+
Resistance to 0/129													
10 μg	-	d	+	+	d	+	d	-	-	+	+	+	+
150 μg	-	-	-	d	d	-	-	-	-	-	-	+	-
Sensitivity to:													
ampicillin 10 μg	-	-	+	+	+	-	+	+	D	-	+	-	D
growth in 0% NaCl	+	-	-	-	-	-	d	-	-	-	d	+	+
growth in 6% NaCl	+	+	+	+	+	+	?	?	+	+	+	-	-
ONPG	+	+	-	-	d	+	+	-	-	d	+	+	+

ONPG, *O*-nitro-phenyl-D-galactopyranoside.
+, 85% or more of strains positive; -, 15% or less strains positive; d, 16–84% of strains positive; D, varies with data source.

results, but in general any zone of inhibition of growth at all is regarded as indicating sensitivity. Initially, the test was regarded as specific for vibrios, but an increasing number of resistant strains of *Vibrio cholerae* have been described[38–40].

It is important to note that *Aeromonas* spp., especially those of phenotype *A. caviae*[41], may be misidentified as *V. fluvialis*[32] or *V. furnissii*[33], if the tests for sensitivity to 0/129 and growth in the presence of sodium chloride are omitted. The range of biochemical tests included in most commercial identification kits, although satisfactory for distinguishing the aeromonads from *Plesiomonas*[42] and many other vibrio species, do not include characters which will reliably distinguish those of phenotype *A. caviae* from *V. fluvialis* and *V. furnissii*.

In addition, mesophilic strains are usually described as universally positive for fermentation of glucose, maltose, mannitol, trehalose, fructose, galactose and negative for fermentation of xylose, dulcitol, inositol, adonitol, malonate, mucate, sorbose, erythritol and raffinose[27] following the work of Eddy and Carpenter[43,44], Ewing and colleagues[45], and Popoff and Véron[22]. Aeromonads are DNAse-positive[22], although incubation at 25°C may be necessary to demonstrate this in *A. veronii*[7], but do not produce urease, nor hydrogen suphide in conventional tests[19,43]. A number of other characters, considered universally positive or negative in early studies, have now been described as diffentiating phenotypes within the genus and are therefore more properly discussed below.

3.3.2 PSYCHROPHILIC STRAINS

After 18–24 h at 20–22°C on nutrient agar, colonies of the phenotype *A. salmonicida* are pinpoint. After four days, colonies are circular, convex, entire, friable and 1–2 mm in diameter. Pigment-producing strains are faintly yellow after five days, deepening to a brown colour if the culture is retained for a week[46].

On blood agar, colonies are similar to those on nutrient agar, but the majority of strains show β-haemolysis after two to four days at 20–22°C[46–48]. On rabbit blood agar with yeast extract, pigment appeared after 24 h, becoming an intense brown at 48 h[46]. Pigment is highly dependent upon culture medium and conditions[46].

The cultural characters and usual source of isolation of *A. salmonicida* are such that distinction from other oxidase- and catalase-positive, nitrate-reducing, facultatively anaerobic Gram-negative rods (i.e. vibrios, aeromonads and *Plesiomonas*) presents few problems. On microscopy, cultures of *A. salmonicida* grown on nutrient agar show short rods with a marked tendency to pleomorphism and coccoid forms[46]. Motile strains[49] are very rare, and the temperature maximum for growth in nutrient broth is quoted at 34.5°C[46], although adaptation to higher temperatures may be induced in the laboratory[47].

Popoff[48] examined 93 strains from around the world. Tests were incubated at 26°C, and were considered positive if a reaction occurred two to four days (four days for the methyl red test and Voges–Proskauer reactions) and negative if there was no reaction at 15 days (21 days for Moeller's decarboxylases). At least 90% of strains examined were positive for arginine and lysine decarboxylase by Moeller's method, reduction of nitrate, DNAse production, methyl red (MR) reaction, and fermentation of arabinose, glucose, mannitol, maltose, salicin and glycerol. All strains were negative for the Voges–Proskauer reaction (VP), ornithine decarboxylase (Moeller), tryptophan deaminase, urease, indole, utilization of citrate (Simmons), malonate, mucate, tartrate, production of hydrogen sulphide, and fermentation of xylose, adonitol, rhamnose, sorbitol, dulcitol, lactose (although all showed β-galactosidase activity), inositol, raffinose and cellobiose. Gas production from glucose occurred in 71% of strains only. These results confirmed those of several previous studies including those of Eddy and Carpenter[44], Ewing and colleagues[45], Griffin and co-authors[46], Smith[50] and Schubert[51]. A later study[52] of 130 strains of *A. salmonicida* subsp. *salmonicida* strains from around the world incubating at 20°C for up to two weeks also confirmed and extended these findings.

3.3   IDENTIFICATION TO SPECIES LEVEL

3.3.1   MESOPHILIC STRAINS

3.3.1.1   Phenotypic methods, pheno- and genotypes

Since the division of mesophilic strains into three groups, later known as *A. hydrophila*, *A. caviae* and *A. sobria*[27], by Popoff and Véron in 1976[22], several studies of over 100 strains with a majority of clinical isolates have shown three major clusters identified by biochemical testing[16,19,53,54]. Depending on the tests used, additional groups could be identified either within the clusters or as further small clusters. It was also recognized that the *A. hydrophila*, *A. caviae* and *A. sobria* phenotypes might be related to differences in disease presentation[54] and elaboration of toxins[4,55]. However, early DNA hybridization studies[56] soon showed that these phena included more than one genospecies. Further hybridization studies (see Chapter 1) placed the previously designated species in appropriate hybridization groups and identified several new groups. Since 1983, six new species have been proposed[7-11,21], although some hybridization groups are still unnamed because there are too few known representatives to undertake a proper study. Several attempts have been made to delineate a biochemical testing scheme which will accurately identify to the genospecies level[16-18,53]. This apparently confusing situation, where large studies have shown that most strains are in three phena, yet comprising 14 DNA hybridization groups and 10 designated species, can be resolved. Over 85% of isolates from human sources are strains of HGs 1 (*A. hydrophila*), 4 (*A. caviae*) or 8 (*A. veronii* bv sobria)[1], and most taxonomic studies have included a majority of human isolates. There have been far fewer studies examining non-clinical isolates[57-59]. Where fish or environmental isolates have been examined, although a relationship to the *A. caviae*, *A. hydrophila* and *A. sobria* complexes could be found, there was no clear division into the three major clusters seen with human isolates. Strains from each of the complexes clustered in more than one phenon. Alternatively subgroups within the same phenon were isolates from two different complexes. So far, no study has examined strains of non-clinical isolates in relation to all the currently known hybridization groups, but those performed strongly suggest that the distribution of hybridization groups amongst non-human isolates includes a higher proportion of strains from HGs other than 1, 4 and 8. The identification tables below require validation for use with environmental and fish isolates. The tables omit at least one important differential character: brown pigment production is a feature of strains of *A. media*[21,58]. Incubation is also at 35–37°C, which may not be appropriate for strains adapted to environmental conditions.

Strains of some HGs are not known from clinical sources, and others are phenotypically similar. It is difficult to distinguish HGs 1, 2 and 3, (the *hydrophila* complex[20]), those of 4 and 5 (the *caviae* complex[20]) and those of 8/10 (bv sobria), 9, 12 and 13. This last might be termed the *sobria* complex. These were the isolates identified as *A. sobria* from clinical samples, although the species correctly termed *A. sobria* is hybridization group 7, which has never been recovered from human sources. Thus for clinical laboratories, a simple approach is division of the mesophilic aeromonads into one of these three complexes formerly known as the phenotypes *A. hydrophila*, *A. caviae* and *A. sobria*. A further group must be added for strains of *A. veronii* bv veronii[7]. However, these are uncommon isolates. They are phenotypically distinct from other groups, and likely to be differentiated by any scheme for the three major complexes. Strains could then be referred to a reference laboratory for full identification. Even this simple division is useful as there is a relationship to pathogenicity as discussed above.

Identification tables for *A. hydrophila*, *A. caviae* and *A. sobria* in standard texts[60,61] have much in common with the tables suggested below, and would separate the *A. hydrophila*, *A. caviae* and *A. sobria* complexes well. *A. veronii* bv veronii would also be differentiated, but there are insufficient characters to reliably discriminate all the newer species within the *A. sobria* complex. However, for publication of clinical cases associated with *Aeromonas* disease, identification to genospecies level is important. Schemes for identification of species within the three complexes have been described. Some authors would encourage the separation of the validated species within the *A. sobria* complex[18] in clinical laboratories, but others have disputed the value of this as isolates of the more recently described species are rare[61]. Only one or two characters separate some species and it remains to be seen how well such schemes work in general use. Identification to the level of all genospecies for which phenotypic differentiation has been described requires a battery of 16 or more tests. These are not all readily available and might be best carried out in reference laboratories.

Table 3.2 is based on the work of Abbott and co-authors[20]. They examined 133 mesophilic strains of known hybridization groups. They point out that 77% of strains were from human sources, and there were few representatives of the rarer hybridization groups. The table has the merit that the characters in the first nine rows (Table 3.2A) are readily available in most laboratories, thus enabling a preliminary identification into *A. hydrophila* and *A. caviae* complexes, and some division of the *A. sobria* complex. Although earlier studies used lower incubation temperatures, recent publications most often test at 36°C as is the case here. In order to compare results with Table 3.4 '+' means >70% positive. Tests were incubated for seven days, except for antibiotic susceptibility and pyrazinamidase which were read at 24 h and 48 h, respectively.

The additional tests were given for identifying HGs within the *A. hydrophila* and *A. caviae* complexes, and for the differentiation of HGs 12

Table 3.2 Differentiation of the hybridization groups of mesophilic aeromonads. Reproduced by kind permission of the American Society for Microbiology

(A) Differentiation of the principal complexes or phenotypes

HG	A. hydrophila complex			A. caviae complex			A. sobria complex						A. sobria complex
	1 (hydrophila)	2	3 (salmonicida)	4 (caviae)	5 (media)	6 (eucrenophila)	7 (sobria)	8 (veronii bv sobria)	9 (jandaei)	10 (veronii bv veronii)	11	12 (schubertii)	13 (trota)
Lysine decarboxylase	+(100)	v(50)	v(67)	-(0)	-(0)	-(0)	+(100)	+(100)	+(100)	+(100)	v(50)	+(82)	+(100)
Ornithine decarboxylase	-(0)	-(0)	-(0)	-(0)	-(0)	-(0)	-(0)	-(0)	-(0)	+(100)	+(100)	-(0)	-(0)
Arginine dihydrolase	+(100)	+(88)	v(67)	+(100)	+(91)	+(86)	-(0)	+(100)	+(100)	-(0)	-(0)	+(91)	+(100)
Aesculin hydrolysis	+(95)	+(100)	+(100)	+(93)	+(91)	+(100)	-(0)	-(0)	-(0)	+(100)	v(50)	-(0)	-(8)
Gas from glucose	+(91)	+(75)	v(50)	-(0)	-(0)	+(100)	v(50)	+(89)	+(100)	+(80)	v(50)	-(0)	v(69)
Voges–Proskauer	+(91)	+(75)	v(67)	-(0)	-(0)	-(0)	-(0)	+(94)	+(91)	+(100)	-(0)	-(18)	-(0)
Acid from													
Arabinose	+(86)	+(100)	+(100)	+(100)	+(100)	+(86)	-(0)	-(17)	-(0)	-(0)	-(0)	-(0)	-(0)
Mannitol	+(95)	+(100)	+(100)	+(100)	+(100)	+(100)	+(100)	+(100)	+(100)	+(100)	+(100)	-(0)	v(69)
Sucrose	+(100)	+(100)	+(100)	+(100)	+(100)	+(71)	+(100)	+(100)	-(0)	+(100)	+(100)	-(0)	-(23)

Table 3.2 (continued)

(B) Differentiation within the complexes

HG	A. hydrophila complex			A. caviae complex			A. sobria complex						A. sobria complex
	1 (hydrophila)	2	3 (salmonicida)	4 (caviae)	5 (media)	6 (eucrenophila)	7 (sobria)	8 (veronii bv sobria)	9 (jandaei)	10 (veronii bv veronii)	11	12 (schubertii)	13 (trota)

Differentiation of the *hydrophila* complex

Acid from

	1	2	3
D-rhamnose	-(27)	+(75)	-(0)
D-sorbitol	-(0)	-(0)	v(67)
Salicin	+(95)	-(0)	+(100)
Lactose	-(23)	-(12)	+(100)
Gluconate oxidation	v(68)	-(0)	-(0)
Elastase	+(73)	-(0)	v(67)
Phenylpyruvic acid	v(64)	-(12)	v(67)
citrate utilization			

Differentiation of the *caviae* complex

	4	5	6
	+(93)	+(73)	-(29)
	+(100)	+(82)	-(0)

continues overleaf

Table 3.2 (continued)

(B) Differentiation within the complexes

HG	1 (hydrophila)	2	3 (salmonicida)	4 (caviae)	5 (media)	6 (eucrenophila)	7 (sobria)	8 (veronii bv sobria)	9 (jandaei)	10 (veronii bv veronii)	11	12 (schubertii)	13 (trota)
	A. hydrophila complex			A. caviae complex			A. sobria complex					A. sobria complex	
Acid from: glycerol				+(73)	+(91)	-(0)						Differentiation of *schubertii* and *trota*	
D-mannose				-(27)	+(100)	+(100)						-(0)	v(62)
H₂S (GCF)*				-(0)	-(0)	+(71)							
Haemolysis (5% SBA)				-(0)	v(45)	+(86)							
Pyrazinamidase				+(80)	-(18)	+(100)							
Corn-oil lipase												+(100)	-(0)
Acid from cellobiose												-(0)	+(100)
Cephalothin susceptibility*												+(73)	-(0)
Ampicillin susceptibility*												-(0)	+(100)

(90) = 90% positive; *Bauer-Kirby method; +, 70% or more of strains positive; v, 30–69% of strains positive; –, less than 30% of strains positive

and 13 are also included (Table 3.2B). These require some more unusual tests such as corn-oil lipase, elastase[62], pyrazinamidase[63] and H_2S production by the method of Véron and Gasser[64]. The elastase test, as originally described, requires a fine suspension of elastin particles in the agar, which is almost impossible to achieve. However, provided the agar is continuously

Table 3.3 Characters for the differentiation of the common phenotypes of mesophilic aeromonads isolated from humans. Reeproduced by kind permission of the American Society for Microbiology

	phenotype						
	hydro-phila	caviae	veronii bv sobria	jandaei	schu-bertii	trota	veronii bv veronii
Lysine decarboxylase	+	-	+	+	+	+	+
Ornithine decarboxylase	-	-	-	-	-	-	+
Aesculin hydrolysis	+	+	-	-	-	-	+
Gas from glucose	+	-	+	+	-	+[a]	+
Voges–Proskauer	+	-	+	+	v[a]	-	+
Acid from:							
Arabinose	v[a]	+	-	-	-	-	-
Mannitol	+	+	+	+	-	+[a]	+
Sucrose	+	+	+	-	-	-	+
H_2S (GCF)*	+	-	+	+	-	+[a]	+
Haemolysis (5% SBA)	+	v	+	+	+	v	+
Pyrazinamidase	+	+	-	-	-	-	
Cephalothin susceptibility*	-	-	+	-	+	-	+
Ampicillin susceptibility*	-	-	-	-	-	+	-
Carbenicillin susceptibility*	-	-	-	-	-	+	-
Colistin MIC <4 μg/ml	v	+	+	-	+	+	+
Hydrolysis of arbutin	+	+	+	-	-	v	+
Indole	+	+	+	+	-	+	
CAMP-like factor	+	-	+	v	-	-	+
Acid from:							
Salicin	v	+	-	-	-	-	+
Cellobiose	+	+	v	-	-	+	+

*Bauer–Kirby method; +, >70% positive; −, less than 30% positive; v, 30–69% of strains positive. [a] Differs from equivalent result in Table 3.2.

mixed while pouring, sufficient dispersion of larger particles can be obtained to make reading of the test feasible. Then, the zone of clearing around colonies is quite clear cut. The method for detection of H_2S production is critical[18] and, in the author's experience, even the recommended GCF medium[64] can produce intermediate results which are difficult to interpret. It should be noted that antibiotic susceptibilities were determined by the Kirby–Bauer method. The determination of β-haemolysin, although not stated, probably used 5% sheep blood agar. However, differences between horse and sheep blood agar have been noted for strains in some hybridization groups[18], and thus the table would need adaptation for laboratories where horse blood is usually used.

Careful examination of the table will show that the additional tests clearly differentiate HGs 12 and 13. However, differentiation of the HGs within the *A. hydrophila* and *A. caviae* complexes is not so clear cut and, indeed, others[12] have questioned whether it is possible at present. The table also ignores the division of HG 5 into 5A (phenotype *A. caviae*) and 5B (phenotype *A. media*).

Another scheme has been proposed by Carnahan and co-workers[18] and is shown in Table 3.3. Although it has many similarities to Table 3.2, the results cannot easily be combined. This study was a numerical analysis based on phenotype with control strains only of known hybridization groups. Thus for strains designated *A. hydrophila* and *A. caviae* more than one hybridization group may be represented. Again, the results were based on testing a majority of human isolates and the scheme is intended for use in clinical laboratories with the common species isolated from humans. Thus some hybridization groups (6,7) are ignored and HG 11 could not be differentiated. Unusually, these authors have chosen to have '+' mean 70% of strains positive, rather than 90% as is more usual. Incubation at 36°C for 72 h is considered sufficient (Dr Carnahan, personal communication). Comments regarding the more unusual tests in Table 3.2 also apply to this table. In addition, the authors found that, along with others[16,67], the CAMP test did not work as originally described and chose to use only the aerobic test. Detection of a CAMP-like factor was originally proposed to distinguish *A. hydrophila*, *A. caviae* and *A. sobria*[66]. The aeromonads were streaked at right angles to an inoculum of a β-lysin-producing *Staphylococcus aureus* strain on 5% sheep blood agar. Tests were done both aerobically and anaerobically, incubating at 37°C overnight. *A. hydrophila* produced the factor in air or anaerobically, *A. sobria* in air only and *A. caviae* not at all. Although this scheme would allow discrimination of the newer phenotypes of the *A. sobria* complex, as in Table 3.2B, this requires testing of a number of unusual characters.

The study also examined the value of alternative test methods for some characters. It showed that both bile aesculin and aesculin agars were equivalent for the detection of aesculin hydrolysis, and that gas production from glucose could be tested both in broth and on a TSI (Triple Sugar Iron) agar slant. The authors also demonstrated that characters from a widely used commercial kit system, API 20E, included in the identification table produced similar results to conventional tests. The only exception was ornithine decarboxylase, where, although only a small minority of strains were positive in either system, this was more often the case in the API 20E system (API Systems, La Balme Les Grottes, France).

A third large study[16] attempted discrimination of genospecies using a majority of tests from API 20E, API 50E and API ATB 32GN systems. Isolates were also identified by DNA hybridization or multilocus enzyme electrophoresis. Only HGs 1, 2, 3, 4, 5A, 5B and 8 were represented in the study, with a considerable majority of HG 4. Incubation was at 29°C with conventional and special tests read at 48 h and kit tests at 18–24 h. Ten characters, gas from glucose (conventional method), CAMP, haemolysin and lysine decarboxylase (special methods) and acid from arbutin, salicin, cellobiose, lactose and growth on salicin (API systems) could be used to differentiate phenotypes *A. caviae*, *A. hydrophila* and *A. sobria*. All these characters have been described elsewhere[4,16,20,22] as useful for differentiation of these groups, but care should be taken when substituting kit tests in conventional schemes. Studies of the comparability of kit and conventional tests should be carried out, as other kits do not always give similar results[12].

Four characters were useful for distinguishing hybridization groups within the *A. hydrophila* and *A. caviae* complexes. As growth and acid from D-sorbitol gave similar results, the latter is not strictly necessary and thus one kit only, API ATB 32GN, could be used. Amongst the *A. hydrophila* complex, HG 1 is distinguished from HGs 2 and 3 by growth on DL-lactate, and group 3 from HGs 1 and 2 by growth on D-sorbitol. The *A. caviae* complex is separated by the growth of HGs 4 and 5B on DL-lactate and the growth of HG4 on citrate. However, it should be noted that some HGs are separated by one character only. Since the original publication in 1990, no further attempts to use kits have been reported, and it remains to be confirmed whether the suggested characters are indeed suitable.

Following on from this work, Kämpfer and Altwegg[67] examined 176 *Aeromonas* strains from all known genospecies for 329 characters in microtitre plates. Carbon source utilization, hydrolysis of 42 chromogenic substrates and sugar fermentation in O/F media were examined. Again three clusters including HGs 7, 8, 9, 12 and 13, a second group with HGs 1, 2, 3 and 11 and a third with HGs 4, 5A, 5B and 6 representing *A. sobria*, *A. hydrophila* and *A. caviae* complexes were found. There was some cluster overlap between phena within these groups, showing their overall similarity. The *A. hydrophila* complex was difficult to separate phenotypically and, although a better separation of the *A. caviae* complex was achieved, only three characters were useful in identification. The authors suggested that multilocus enzyme electrophoresis would be required for complete differentiation of all hybridization groups. Although a table of 30 discriminatory tests was given for the 13 genospecies, the use of microtitre plates means that it would be difficult to apply in some laboratories.

3.3.1.2   Short character sets, kits and other rapid tests

Information from Table 3.4 was used to construct a dichotomous key[18] as a rapid identification system for all the currently recognized clinical species. Three characters (the Voges–Proskauer reaction and acid from arabinose and sucrose) can be read from an API 20E strip, and the remaining three tests (gas from glucose, aesculin hydrolysis in an agar formulation and resistance to cephalothin by the Kirby–Bauer method) require conventional test methods. The authors pointed out that all isolates must first be presumptively identified as aeromonads, including resistance to 150 µg of 0/129 and that the methods for aesculin and antibiotic susceptibility testing must be strictly controlled. However, ornithine decarboxylase or pyrazinamidase activity could be substituted for cephalothin sensitivity, and examination of the table will show that other substitutions are possible. The dichotomous key was validated with 60 coded strains previously identified phenotypically in another laboratory[18], and on a further 51 clinical strains[12]. In these studies, 97% and 94% of strains respectively were correctly identified. One of the problems with such a scheme is that atypical results in a single test, particularly at the top of the hierarchy, will lead to major errors in identification, and it remains to be established how well the scheme will work in general use. A Japanese group has already proposed some modifications[68].

Other short character sets have been suggested[4,69–71], but these considered only *A. caviae*, *A. hydrophila* and *A. sobria* and again, where the source of isolates is stated there has been a preponderance of human isolates. It is possible that such schemes could be used to differentiate the major complexes. However, the reactions of the newer phenotypes have not been examined and may give rise to confusion. Namdari and Bottone[69] described the 'suicide phenomenon', when isolates are inoculated into nutrient broth with 0.5% glucose and incubated for 24 h at 30°C; suicidal strains spontaneously pellet whereas non-suicidal strains show a turbid culture. *A. caviae* strains are usually suicidal, whereas *A. hydrophila* and *sobria* are not. Together with gas from glucose and aesculin hydrolysis, these three phenotypes could be differentiated reliably. Others have confirmed these findings[70,71]. The test was adapted by Wilcox and co-authors[70] to require only two tubes (including aesculin hydrolysis) by adding bromocresol purple and inverted tubes to the glucose medium to test for gas production as well as suicidal tendency. They found a 93% agreement with an eight-test conventional identification scheme.

Although researchers have been using selected tests in commercial kits to undertake taxonomic studies for some years, until recently many kit databases for Gram-negative aerobes and facultative anaerobes either did not include the aeromonads or at best designated them merely as '*A. hydrophila*' or '*Aeromonas*' spp. However, at least 14 commercial systems (most available in the USA and some elsewhere) now have *Aeromonas* species in the database, although experience so far suggests that performance is variable[12,72,73]. Interestingly, only 52% of *Aeromonas* isolates were correctly identified to genus level with API 20E systems in comparison to 100% using BBL[R] Crystal™ ID Enteric/Nonfermenter ID Kit (Becton Dickinson Microbiology Systems, Cockeysville, MD, USA)[12]. However, additional tests were required to correctly speciate the aeromonads, although the database includes *A. caviae*, *A. hydrophila* and the biovars of *A. veronii*. The equivalent API kit, API NE, is possibly rather better but still had only an 88% correlation with a conventional test scheme in identifying *A. caviae*, *A. hydrophila* and *A. sobria*[70].

Results with the carbon substrate profile system in Biolog GN Microplates (Biolog Inc. Hayward, CA, USA)[74] were more promising in so far as nine substrates, some already suggested by previous workers, could be used to discriminate these three phenotypes. The database now includes 12 named species, the biovars of *A. veronii* and DNA HGs 2/3 and 11. So far, little has been published on the use of this kit: however, it has the potential for rapid and accurate identification of all the common, and some uncommon, hybridization groups.

3.3.1.3   Molecular, chemotaxonomic and other 'special' methods

Almost every available technique has been examined as a taxonomic tool for *Aeromonas* species. Although unlikely to be used in most laboratories at present, these methods may be useful for reference laboratories in the identification of atypical strains. In the future, these methods may also provide an alternative to conventional biochemical characters for automated identification systems. The majority of studies have divided strains into *A. hydrophila*, *A. caviae* and *A. sobria* rather than into hybridization groups. In 1978, Shaw and Hodder[75] showed that the core lipopolysaccharides of what was effectively the three phenotypes *A. hydrophila*, *A. caviae* and *A. sobria* were different and could be used as an aid to classification. A decade later, Canonica and Pisano[76] showed that fatty acid profiles of these groups were also different using gas-liquid chromatography (GLC). Following this, other workers have confirmed the usefulness of fatty acids[77–79] and other cellular components such as quinones[79] as chemotaxonomic markers of *Aeromonas* pheno- and geno-species.

In 1985, Picard and Goullet showed that enzyme electrophoresis could discriminate between strains of *A. hydrophila*, *A. caviae* and *A. sobria*[80]. A large study of multilocus enzyme electrophoresis including 153 *Aeromonas* strains at 11 genetic loci showed a very good, although not perfect, correlation with DNA hybridization group amongst the 12 hybridization groups examined[81]. The authors suggested that, although the method could be used to discriminate all the hybridization groups, its major potential was in differentiating strains within the *A. hydrophila* and *A. caviae* complexes, which would require examination of only two and three separate enzymes, respectively.

Immunoblotting, where cellular proteins are first separated by electrophoresis, transferred to a nitrocellulose membrane and then visualized with labelled antisera, has been used to separate *A. hydrophila*, *A. caviae* and *A. sobria*[82]. A common protein band and outer membrane patterns specific to each group were identified. However, other studies could not identify a common genus band, nor patterns specific to species[83].

Pyrolysis mass spectrometry (Py MS) has also been used to examine 31 strains of *Aeromonas* spp.[84]. Essentially this technique heats bacterial cells until they vaporize, and examines the type and relative frequencies of different molecular species produced by mass spectroscopy. Thus it is different from all the other techniques described which study proteins, DNA, RNA or other macromolecules. There was some correlation with the *A. caviae* and *A. sobria* groups, but the *A. hydrophila* group comprised several groups. The authors noted that although clustering did not correspond well with conventional identification as *A. hydrophila*, *A. caviae* and *A. sobria*, there was a better relationship with pathogenicity. It is possible that Py MS is more sensitive than the authors realized, and was identifying different genotypes of varying virulence.

So far, only a few studies have examined the use of DNA probes and the polymerase chain reaction (PCR), although in the future such methods might be particularly useful for rapid identification of strains in environmental samples where there is heavy contamination or cells have been damaged by heat or chemicals. Specific probes have been described for *A. trota*, *A. schubertii*, *A. jandaei*, *A. hydrophila* and *A. veronii*[85–88] and it is likely that probes will soon be identified for all the known phenotypes if not hybridization groups. Work has centred on their molecular properties rather than their use in detection and identification, and therefore data on sensitivity and specificity are incomplete. Interestingly there is also a probe for aerolysin which appears to be genus-specific. This reacts even with strains of *A. caviae* which do not appear to produce aerolysin[85].

3.3.2   PSYCHROPHILIC STRAINS

It is perhaps a relief to the ordinary microbiologist that the taxonomy of these organisms is less complex than that of the mesophiles. However, this may reflect less investigation. Early studies identified several phenotypes which have been named as subspecies *salmonicida*, *achromogenes* and *masoucida*. Whether this division is strictly justified is controversial[27,89]. While there is no doubt that *A. salmonicida* subsp. *salmonicida* is a clearly differentiated homogeneous group, some authors have suggested on the basis of DNA hybridization studies that the non-pigmented subspecies *achromogenes* and *masoucida* should be combined[90,91]. Perhaps the greatest importance for those identifying wild strains is the recognition that members of the psychrophilic group of *A. salmonicida* have several variants.

Table 3.4 is a summary of all the characters that have been published as useful for the differentiation of the subspecies of *A. salmonicida*. A fourth subspecies has been described, *A. salmonicida* subsp. *smithia*[89], as a result of a numerical study of *Aeromonas* isolates from fish including 58 strains of *A. salmonicida*. This study used standard methods[41,93], and strains were incubated at 25°C for seven days. Clusters representing the three previously reported subspecies of *A. salmonicida* were identified; however, four characters (indole, aesculin, mannitol and utilization of arabinose) had reactions differing from those previously published[27]. A second, more recent, study of strains of subspecies *salmonicida*[52], although supporting earlier descriptions of *A. salmonicida* subsp. *salmonicida*, introduced further discrepancies in the description of what were otherwise described as strains closely resembling the type strain. There may be several reasons for these discrepancies. In early studies, only limited numbers of strains were available for examination, and even in later studies only a few of the rarer non-pigmented groups were examined[46,52,92]. Study methodologies have varied, especially in time and incubation temperature, and sometimes few details were stated at all.

Unfortunately also, as shown in Table 3.4, the new subspecies is somewhat non-reactive. In practice, it would be difficult to make a confident identification of such an isolate. Given the confusion surrounding the typical

Table 3.4. Characters differentiating subspecies of *A. salmonicida*

	salmonicida	Subspecies achromogenes	masoucida	smithia
Brown pigment production	+	-	-	-
Indole production	-	D	v	-
Aesculin hydrolysis	D	-	D	-
Arginine decarboxylase	D	v	+	-
β-glucosidase	+	-	-	-
Growth in 5% NaCl	v	+	+	-
Degradation of elastin	+	v	+	-
Degradation of Tween 40	+	-	+	-
Voges–Proskauer reaction	-	-	+	-
Utilization of L-arabinose	D	-	D	NK
Gas from glucose	+	-	+	NK
Acid from mannitol	+	D	+	NK

+, Usually positive; −, usually negative; v, variable; D, character differs between sources; NK, not known.

characters of the subspecies of *A. salmonicida*, and the relative rarity of the non-pigmenting strains, for many purposes until the position is clearer it may be sufficient to divide isolates into pigment-producing, i.e. *A. salmonicida* subsp. *salmonicida*, and non-pigment-producing strains. However, it should be noted that pigment production is highly dependent upon culture conditions and the constituents of the medium[46]. Pigmentation does not occur anaerobically or at an acid pH and is reduced at temperatures above the growth optimum. The medium must contain L-phenylalanine, L-tyrosine or a number of related compounds.

3.4 CONCLUSIONS

Identification of bacterial colonies as members of the genus *Aeromonas* presents few difficulties. However, the taxonomic complexities within the genus, especially amongst mesophiles, mean that further identification to species level, while not technically difficult, presents considerable interpretative problems. This is compounded by an explosion in research interest in this group of organisms, and it is almost certain that further species and biovars will be described. A number of schemes for identifying phenotypes and genotypes have been proposed but their relatively recent introduction means that so far there have been few reports of their wider use outside the authors' laboratories. Laboratories reporting species identification must decide whether they wish to separate mesophiles into *A. hydrophila*, *A. caviae* and *A. sobria* complexes only, include the newly recognized phenotypes as well or attempt identification to genotype level wherever possible.

3.5 REFERENCES

1   Janda JM. Recent advances in the study of the taxonomy, pathogenicity and infectious syndromes associated with the genus *Aeromonas*. *Clin Microbiol Rev* 1991; **4**: 397–410.
2   Dryden M, Munro R. *Aeromonas* septicaemia: relationship of species and clinical features. *Pathology* 1989; **21**: 111–14.
3   Janda JM, Brenden R. Importance of *Aeromonas sobria* in *Aeromonas* bacteremia. *J Infect Dis* 1987; **155**: 589–91.
4   Barer MR, Millership SE, Tabaqchali S. Relationship of toxin production to species in the genus *Aeromonas*. *J Med Microbiol* 1986; **22**: 303–9.
5   Millership SE, Barer MR, Tabaqchali S. Toxin production by *Aeromonas* spp. from different sources. *J Med Microbiol* 1986; **22**: 311–14.
6   Carnahan AM, Chakraborty T, Fanning GR, Verma D, Ali A, Janda JM, Joseph SW. *Aeromonas trota* sp. nov., an ampicillin susceptible species isolated from clinical specimens. *J Clin Microbiol* 1991; **29**: 1206–10.
7   Hickman-Brenner FW, MacDonald KL, Steigerwalt AG, Fanning GR, Brenner

DJ, Farmer III JJ. *Aeromonas veronii*, a new ornithine decarboxylase-positive species that may cause diarrhoea. *J Clin Microbiol* 1987; **25**: 900–6.

8   Hickman-Brenner FW, Fanning GR, Arduino MJ, Brenner DJ, Farmer JJ III. *Aeromonas schubertii*, a new mannitol-negative species found in human clinical specimens. *J Clin Microbiol* 1988; **26**: 1561–4.

9   Schubert RW, Hegazi M. *Aeromonas eucrenophila* species nova *Aeromonas caviae* a later and illegitimate synonym of *Aeromonas punctata*. Zentralbl Bacteriol Mikrobiol Hyg A 1988; **268**: 34–9.

10 Carnahan A, Fanning GR, Joseph SW. *Aeromonas jandaei* (formerly genospecies DNA group 9 *A. sobria*), a new sucrose negative species isolated from clinical specimens. *J Clin Microbiol* 1991; **29**: 560–4.

11 Carnahan AM, Chakraborty T, Fanning GR, Verma D, Ali A, Janda JM, Joseph SW. *Aeromonas trota* sp nov., an ampicillin susceptible species isolated from clinical specimens. *J Clin Microbiol* 1991; **29**: 1206–10.

12 Joseph SW, Carnahan A. The isolation, identification and systematics of the motile *Aeromonas* species. *Annu Rev Fish Dis* 1994; **4**: 315–43.

13 Martinez-Murcia AJ, Esteve C, Garay E, Collins MD. *Aeromonas allosaccharophila* sp. nov., a new mesophilic member of the genus *Aeromonas*. *FEMS Microbiol Lett* 1992; **91**: 199–206.

14 Edwards PR, Ewing HW. *Identification of Enterobacteriaceae*, 3rd edn. Minneapolis: Burgess Publishing, 1972.

15 McFaddin JF. *Biochemical Tests for Identification of Medical Bacteria*, 2nd edn. Balimore: Williams and Wilkins, 1980.

16 Altwegg M, Steigerwalt AG, Altwegg-Bissig R, Luthy-Hottenstein J, Brenner DJ. Biochemical identification of *Aeromonas* genospecies isolated from humans. *J Clin Microbiol* 1990; **28**: 258–64.

17 Arduino MF, Hickman-Brenner FW, Farmer JJ III. Phenotypic analysis of 132 *Aeromonas* strains representing 12 DNA hybridization groups. *J Diarr Dis Res* 1988; **6**: 137.

18 Carnahan AM, Behram S, Joseph SW. Aerokey II: A flexible key for identifying clinical *Aeromonas* species. *J Clin Microbiol* 1991; **29**: 2843–9.

19 Kuijper EJ, Steigerwalt AG, Schoenmakers BSCIM, Peeters MF, Zanen HC, Brenner DJ. Phenotypic characterization and DNA relatedness in human fecal isolates of *Aeromonas* spp. *J Clin Microbiol* 1989; **27**: 132–8.

20 Abbott SL, Cheung WK, Kroske-Bystrom S, Malekzadeh T, Janda JM. Identification of *Aeromonas* strains to the genospecies in the clinical laboratory. *J Clin Microbiol* 1992; **30**: 1262–6.

21 Allen DA, Austin B, Colwell RR. *Aeromonas media*, a new species isolated from river water. *Int J Syst Bacteriol* 1983; **33**: 599–604.

22 Popoff M, Véron M. A taxonomic study of the *Aeromonas hydrophila–Aeromonas punctata* group. *J Gen Microbiol* 1976; **94**: 11–22.

23 Want SV, Millership SE. Effects of incorporating ampicillin, bile salts and carbohydrates in media on the recognition and selection of *Aeromonas* spp. from faeces. *J Med Microbiol* 1990; **32**: 49–54.

24 Hunt LK, Overman TL, Otero R. Role of pH in oxidase variability of *Aeromonas hydrophila*. *J Clin Microbiol* 1981; **13**: 1054–9.

25 Millership SE, Chattopadhyay B. *Aeromonas hydrophila* in chlorinated water supplies. *J Hosp Infect* 1985; **6**: 75–80.

26 Shread P, Donovan TJ, Lee JV. A survey of the incidence of *Aeromonas* in human faeces. *Soc Gen Microbiol Q* 1981; **8**: 1.

27 Popoff M. Genus III. *Aeromonas* Kluyver and Van Niel 1936. In Krieg NR, Holt JG (eds). *Bergey's Manual of Systematic Bacteriology*, Vol 1. London: Williams and Wilkins, 1984; 545–8.

28 Bryant T, Lee JV, West PA, Colwell RR. Numerical classification of species of *Vibrio* and related genera. *J Appl Bacteriol* 1986; **61**: 437.

29 Bryant T, Lee JV, West PA, Colwell RR. A probability matrix for the identification of species of *Vibrio* and related genera. *J Appl Bacteriol* 1986; **61**: 469.

30 Brayton RR, Bode RB, Colwell RR, MacDonald MT, Hall HL, Grimes DJ, West PA, Bryant T. *Vibrio cincinnatiensis* sp. nov., a new human pathogen. *J Clin Microbiol* 1986; **23**: 104.

31 Kelly MT, Hickman-Brenner FW, Farmer JJ. *Vibrio*. In Balows A, Hausler WJ, Herrmann KL, Isenberg HD, Shadomy HJ (eds) *Manual of Clinical Microbiology*, 5th edn. Washington, DC: American Society of Microbiology, 1991: 384–95.

32 Lee JV, Shread D, Furniss AL, Bryant T. Taxonomy and description of *Vibrio fluvialis* sp. nov. (synonym group F. vibrios, group EF6). *J Appl Bacteriol* 1981; **50**: 73.

33 Brenner DJ, Hickman-Brenner FW, Lee JV, Steigerwalt AG, Fanning GR, Hollis DG, Farmer JJ III, Weaver RE, Joseph SW, Seidler RJ. *Vibrio furnissi* (formerly anaerogenic biogroup of *Vibrio fluvialis*), a new species isolated from human feces and the environment. *J Clin Microbiol* 1983; **18**: 816.

34 Reinhardt JF, George WL. Comparative in vitro activities of selected antimicrobial agents against *Aeromonas* species and *Plesiomonas shigelloides*. *Antimicrob Agents Chemother* 1985; **27**: 643–5.

35 Davis GHG, Park RWA. A taxonomic study of certain bacteria currently classified as *Vibrio* species. *J Gen Microbiol* 1962; **27**: 101.

36 Bain N, Shewan JM. Identification of *Aeromonas, Vibrio* and related organisms. In Gibbs B, Shapton DA (eds) *Identification Methods for Microbiologists, Part B*. London: Academic Press, 1968: 79.

37 Furniss AL, Lee JV, Donovan TJ. *The Vibrios*. Public Health Laboratory Service Monograph Series no. 11. London: HMSO, 1978.

38 Sundaram SP, Murthy KV. Occurrence of 2,4-diamino-6,7-diisopropylpteridine (0/129) resistance in human isolates of *V. cholerae*. *FEMS Microbiol Lett.* 1983; **19**: 115–7.

39 Kudoh YS, Matsushita S, Yamada S, Tsuno M, Ohta K, Sakai S, Ohashi M. Enterotoxin producibility and some biological features of 01 and non-01 *Vibrio cholerae* isolates. In Kuwahara S, Zinnaka Y(eds) *Proceedings of Symposium on Cholera, Gifu, 1980*. Tokyo: Toho University, 1981: 214–24.

40 Huq A, Alam M, Parveen S, Colwell RR. Occurrence of resistance to vibriostatic compound 0/129 in *Vibrio cholerae* 01 isolated from clinical and environmental samples in Bangladesh. *J Clin Microbiol* 1992; **30**: 219–21.

41 Barrow GI, Feltham RKA. *Cowan and Steel's Manual for the Identification of Medical Bacteria*. 3rd Edn. Cambridge: Cambridge University Press, 1993: 126–7.

42 Schubert RHW. *Plesiomonas*. In Krieg NR, Holt JG (eds) *Bergey's Manual of Systematic Bacteriology*, Vol. 1. London: Williams and Wilkins, 1984: 548–50.

43 Eddy BP. Further studies on *Aeromonas*. 1. Additional strains and supplementary biochemical tests. *J Appl Bacteriol* 1962; **25**: 137–46.

44 Eddy BP, Carpenter KP. Further studies on *Aeromonas*. II. Taxonomy of *Aeromonas* and C27 strains. *J Appl Bacteriol* 1964; **27**: 96–109.

45 Ewing WH, Hugh R, Johnson JG. *Studies in the* Aeromonas *group*. Atlanta, GA: US Dept Health Education and Welfare, CDC: 1961.

46 Griffin PJ, Snieszko SF, Friddle SB. A more comprehensive description of *Bacterium salmonicida*. *Trans Am Fish Soc* 1952; **82**: 129–38.

47 McIntosh D, Austin B. Atypical characteristics of the salmonid pathogen *Aeromonas salmonicida*. *J Gen Microbiol* 1991; **137**: 1341–3.

48   Popoff M. Etude sur les *Aeromonas salmonicida*. I. Caracteres biochemiques et antigeniques. *Rech Velee* 1969; **3**: 49–57.

49   Austin B, Austin DA. Expression of motility in strains of the non-motile species *Aeromonas media*. *FEMS Microbiol Lett* 1990; **68**: 123–4.

50   Smith IW. The classification of *Bacterium salmonicida*. *J Gen Microbiol* 1963; **33**: 263–71.

51   Schubert RHW. The taxonomy and nomenclature of the genus *Aeromonas* Kluvyer and van Niel 1936. Part I. Suggestions on the taxonomy and nomenclature of the aerogenic *Aeromonas* species. *Int J Syst Bacteriol* 1968; **18**: 1–7.

52   Dalsgaard I, Neilsen B, Larsen JL. Characterization of *Aeromonas salmonicida* subsp *salmonicida*: a comparative study of strains of different geographic origin. *J Appl Bacteriol* 1994; **77**: 21–30.

53   Carnahan AM, Joseph SW. Systematic assessment of geographically and clinically diverse aeromonads. *Syst Appl Microbiol* 1993; **16**: 72–84.

54   Janda JM, Reitano M, Bottone EJ. Biotyping of *Aeromonas* isolates as a correlate to delineating a species-associated disease spectrum. *J Clin Microbiol* 1984; **19**: 44–7.

55   Turnbull PCB, Lee JV, Miliotis MD, Van de Walle S, Koornhof HJ, Jeffery L, Bryant TN. Enterotoxin production in relation to taxonomic grouping and source of isolation of *Aeromonas* species. *J Clin Microbiol* 1984; **19**: 175–80.

56   Popoff M, Coynault C, Kiredijian M, Lemeline M. Polynucleotide sequence relatedness among motile *Aeromonas* species. *Curr Microbiol* 1981; **5**: 109–14.

57   Renauld F, Freney J, Boeufgras JM, Monget D, Sedaillan A, Fleurette J. Carbon substrate assimilation patterns of clinical and environmental strains of *Aeromonas hydrophila, Aeromonas sobria* and *Aeromonas caviae* observed with a micromethod. Zentralbl Bakteriol Hyg A 1988; **269**: 323–30.

58   Austin B, McIntosh D, Austin B. Taxonomy of fish-associated *Aeromonas* spp., with the description of *Aeromonas salmonicida* subsp. *smithia* subsp.nov. *Syst Appl Microbiol* 1989; **11**: 277–90.

59   Okpolwasili GC. *Aeromonas hydrophila*: variability of biochemical characteristics of environmental isolates. *J Basic Microbiol* 1991; **31**: 169–76.

60   Von Graevenitz A, Altwegg M. *Aeromonas* and *Plesiomonas*. In Balows A, Hausler WJ, Hermann KL, Isenberg HG, Shadomy HJ (eds) *Manual of Clinical Microbiology*, 5th edn. Washington DC: American Society for Microbiology, 1991: 396–410.

61   Wakabongo M, Bortey E, Meier FA, Dalton HP. Rapid identification of motile *Aeromonas*. Diagn Microbiol Infect Dis. 1992; **15**: 511–15.

62   Scharmann W. Vorkommen van Elastase bei *Pseudomonas* und *Aeromonas*. *Zentralbl Bakteriol Parasitol Infekt Hyg (Abt 1)* 1972; **A220**: 435–42.

63   Carnahan A, Hammontree L, Bourgeois L, Joseph SW. Pyrazinamidase activity as a phenotypic marker for several *Aeromonas* spp. isolated from clinical laboratories. *J Clin Microbiol* 1990; **28**: 391–2.

64   Véron M, Gasser F. Sur la detection de l'hydrogene sulfure produit par certaines enterobacteriacees dans les milieux dits de diagnostic rapide. *Ann Inst Pasteur* 1963; **105**: 524–34.

65   Ben-Ruwin RA, Rogers S, Walter K, Park RWA. A comparison between biochemical tests and the 'CAMP' reaction for identification of *Aeromonas* spp. *Lett Appl Microbiol* 1990; **11**: 244–6.

66   Figuar N, Guglielmetti P. Differentiation of motile and mesophilic *Aeromonas* strains by testing for a CAMP-like factor. *J Clin Microbiol* 1987; **25**: 1341–2.

67   Kämpfer P, Altwegg M. Numerical classification and identification of *Aeromonas* genospecies. *J Appl Bacteriol* 1992; **72**: 341–51.

68 Furuwatari C, Kawakami Y, Akahane T, Midaka E, Okimura Y, Nakayama J, Furihata K, Katsuyama T. Proposal for an Aeroscheme (modified Aerokey II) for the identification of clinical *Aeromonas* species. *Med Sci Res* 1994; **22**: 617–19.

69 Namdari H, Bottone EJ. Suicide phenomenon in mesophilic aeromonads as a basis for species identification. *J Clin Microbiol* 1989; **27**: 788–9.

70 Wilcox MH, Cook AM, Thickett KJ, Eley A, Spencer RC. Phenotypic methods for speciating *Aeromonas* isolates. *J Clin Pathol* 1992; **45**: 1079–83.

71 Piersimoni C, Morbiducci V, De Sio G, Scalise G. Rapid presumptive phenospecies identification of mesophilic aeromonads by testing for suicidal activity. *Eur J Clin Microbiol Infect Dis* 1990; **9**: 608–10.

72 Carnahan AM. Update of *Aeromonas* identification. *Clin Microbiol Newslett* 1991; **13**: 169–72.

73 Sedlacek I, Jaksl V, Prepechalova H. Identification of aeromonads from water sources (in Czech). *Epidemiol Mikrobiol Immunol* 1994; **43**: 61–6.

73 Carnahan AM, Joseph SW, Janda JM. Species identification of *Aeromonas* strains based on carbon substrate profiles. *J Clin Microbiol* 1989; **27**: 2128–9.

75 Shaw DH, Hodder HJ. Lipopolysaccharides of the motile aeromonads; core oligosaccharide analysis as an aid to taxonomic classification. *Can J Microbiol* 1978; **24**: 864–8.

76 Canonica FP, Pisano MA. Gas–liquid chromatographic analysis of fatty acid methyl esters of *Aeromonas hydrophila*, *Aeromonas sobria* and *Aeromonas caviae*. *J Clin Microbiol* 1988; **26**: 681–5.

77 Hansen W, Freney J, Labbe M, Renaud F, Yourcissowsky E, Fleurette J. Gas–liquid chromatographic analysis of cellular fatty acid methyl esters in *Aeromonas* species. *Int J Med Microbiol* 1991; **275**: 1–10.

78 Huys G, Kersters I, Vancanneyt M, Coopman R, Janssen P, Kersters K. Diversity of *Aeromonas* species as determined by gas–liquid chromatographic analysis of cellular fatty acid methyl esters (FAME). *J Appl Bacteriol* 1995; **78**: 445–55.

79 Kampfer P, Blasczyk K, Auling G. Characterisation of *Aeromonas* genomic species by using quinone, polyamine and fatty acid components. *Can J Microbiol* 1994; **40**: 844–50.

80 Picard B, Goullet Ph. Comparative electrophoretic profiles of esterases, and of glutamate, lactate and malate dehydrogenases, from *Aeromonas hydrophila*, *A caviae* and *A. sobria*. *J. Gen Microbiol* 1985; **131**: 3385–91.

81 Altwegg M, Reeves MW, Altwegg-Bissig R, Brenner DJ. Multilocus enzyme analysis of the genus *Aeromonas* and its use for species identification. *Zentralbl Bakteriol* 1991; **275**: 28–45.

82 Maruvada R, Das P, Ghosh AN, Pal SC, Nair GB. Electrophoretic mobility and immunoblot analysis of the outer membrane proteins of *Aeromonas hydrophila*, *A sobria* and *A. caviae*. *Microbios* 1992; **71**: 105–13.

83 Mulla R, Millership S. Typing of *Aeromonas* spp. by numerical analysis of immonoblotted SDS–PAGE gels. *J Med Microbiol* 1993; **39**: 325–33.

84 Magee JT, Randle EA, Gray SJ, Jackson SK. Pyrolysis mass spectrometry and numerical taxonomy of *Aeromonas* spp. *Antonie Van Leeuwenhoek* 1993; **64**: 315–23.

85 Husslein V, Chakraborty T, Carnahan A, Joseph SW. Molecular studies on the aerolysin gene of *Aeromonas* species and discovery of a species-specific probe for *Aeromonas trota* species nova. *Clin Infect Dis* 1992; **14**: 1061–8.

86 Ash C, Martinez-Murcia AJ, Collins MD. Molecular identification of *Aeromonas sobria* by using a polymerase chain reaction probe test. *Med Microbiol Lett* 1993; **2**: 80–6.

87 Ash C, Martinez-Murcia AJ, Collins MD. Identification of *Aeromonas schubertii* and *Aeromonas jandaei* by using a polymerase chain reaction probe test. *FEMS*

Microbiol Lett 1993; **108**: 151–6.

88   Dorsch M, Ashbolt NJ, Cox PT, Goodman AE. Rapid identification of *Aeromonas* species using 16S rDNA targeted oligonucleotide primers: a molecular approach based on screening of environmental isolates. *J Appl Bacteriol* 1994; **77**: 722–6.

89   Austin DA, MacIntosh D, Austin B. Taxonomy of fish-associated *Aeromonas* spp., with the description of *Aeromonas salmonicida* subsp. *smithia* subsp. nov. *Sys Appl Microbiol* 1989; **11**: 277–90.

90   McCarthy DH, Roberts RJ. Furunculosis of fish—the present state of knowledge. *Adv Aquat Microbiol* 1980; **2**: 293–340.

91   Belland RJ, Trust TJ. DNA: DNA reassociation analysis of *Aeromonas salmonicida. J Gen Microbiol* 1988; **134**: 307–15.

92   Schubert RHW. The taxonomy and nomenclature of the genus *Aeromonas* Kluvyer and van Niel 1936. Part II. Suggestions on the taxonomy and nomenclature of the anaerogenic aeromonads. *Int J Syst Bacteriol* 1967; **17**: 273–9.

93   Gerhardt P, Murray RGE, Costilow RN, Nester EW, Wood WA, Krieg NR, Phillips GB. *Manual of Methods for General Bacteriology*. Washington DC: American Society for Microbiology, 1981.

4 Subtyping Methods for *Aeromonas* Species

MARTIN ALTWEGG
University of Zürich, Zürich, Switzerland

Considering the problems to biochemically identify *Aeromonas* strains to the level of genomic species as determined by DNA–DNA hybridization, typing methods should be evaluated not only for their ability to discriminate among strains for epidemiological purposes but also for their potential as taxonomic tools. In addition to the confusing taxonomy, the lack of strains that are linked to well investigated outbreaks of disease has caused significant problems in validating subtyping methods. As a consequence, most studies have focused on the ability of methods to discriminate among independently isolated strains, providing an idea of the sensitivity of a given method. Very little is known about the stability of patterns generated and their usefulness for the investigation of suspected outbreaks in the future. In the following, some of the typing methods that have been applied to *Aeromonas* strains are reviewed.

4.1 PHENOTYPIC METHODS

4.1.1 BIOTYPING

Since specific biochemical reactions are usually not sufficient to assign unknown *Aeromonas* strains to the different genomic species[1], it is not astonishing that biotyping is of little or no value for epidemiological investigations, despite its ease of use. In one attempt to compare strains recovered from human diarrhoeal stools ($n = 187$) and from drinking water ($n = 263$), the majority (84%) belonged to only 10 biotypes[2]. Compared with other subtyping methods that have shown a considerable heterogeneity among strains of similar origin (see below), biotyping is certainly not sufficiently sensitive.

4.1.2 PHAGE TYPING

Lytic bacteriophages (or phages) (i.e. viruses capable of infecting and lysing bacterial cells) have successfully been applied for epidemiological investigations of a variety of bacteria. This method is usually available only at

The Genus Aeromonas. Edited by B. Austin, M. Altwegg, P.J. Gosling and S. Joseph
© 1996 John Wiley & Sons Ltd

reference centres because it is technically demanding and requires the maintainance of viable phages and control strains[3]. Depending on the species, there may be considerable numbers of strains that are not typable.

For the genus *Aeromonas*, Demarta and Peduzzi[4] have used a total of 25 different phages to compare 481 strains mainly isolated from water samples. In addition, 21 strains of clinical origin (stool, urine, sputum, blood) were included. Both groups shared some lysotypes, suggesting that at least some human infections were of aquatic origin. This has also been shown for wound infections, which are often associated with previous contact with contaminated water, soil or marine products during recreational or occupational activities[5]. The importance of the aquatic environment as a reservoir for strains infecting humans was confirmed in a follow-up study with 95 phages and an increased number of clinical isolates[6]. About 80% of the strains were typable. In a more recent study of strains mainly isolated from faecal specimens of patients with diarrhoea, 78 of the 96 strains (81%) were lysed by at least one of the 25 phages, and belonged to 73 different phage types[7]. This enormous diversity, however, may be somewhat misleading in that the collection of strains analysed was a selection representing different biotypes. Direct comparison as well as cluster analysis of these strains did not reveal strong associations of susceptibility to certain phages with either phenotypic *A. hydrophila*, *A. caviae* or *A. sobria*, nor with any of the DNA hybridization groups[8]. In comparison with other typing methods (see below), phage typing may provide further discrimination among strains of identical enzyme or ribotypes (Table 4.1). However, strains belonging to one phage type may also be separated from each other by other methods[7].

In Japan, *Aeromonas* phages were isolated from river water and river mud[9] and 13 phage types were defined among 594 *Aeromonas* strains. However, only 129 of these 594 strains (21.7%) which had been isolated from essentially the same sources as the phages, were typable. In contrast, 51% of

Table 4.1. Discrimination of DNA hybridization group 4 strains (*A. caviae*) of given enzyme types (ETs) by restriction enzyme analysis, ribotyping and phage typing

Strain	ET	rDNA pattern		Restriction pattern		Susceptible to phages
		*Sma*I	*Pst*I	*Sma*I	*Pst*I	
30	58	A	A	A	A	CDMNPQSTVWXY
253	58	B	B	B	B	PSVW
192	58	B	B	C	C	PUW
61	64	C	C	D	D	PSW
256	64	C	C	E	E	PSUVWX
1	49	D	D	F	F	CMQUY
157	49	E	E	G	G	BCIKLMNOPQRSTUVWXYZ
51	49	D	D	H	H	PWXY

strains isolated from patients with diarrhoea in four hospitals in Tokyo could be assigned to 28 phage types[10].

4.1.3   SEROTYPING

Serological typing has proven extremely useful in various groups of organisms. In *Salmonella*, for example, more than 2000 serotypes have been described based on somatic O- and flagellar H-antigens. This not only allowed investigation of outbreaks but also aided in the identification of primary habitats and modes of transmission, as shown very nicely for *S. enteritidis*, which may be vertically transmitted through contaminated eggs[11]. In some instances, e.g. *Escherichia coli* 0157:H7, serotyping provides an efficient means for identifying clones with special virulence features which, for this reason, are of special importance in clinical microbiology[12]. Except for very important pathogens, antisera are not commercially available and, therefore, serotyping is usually limited to reference laboratories.

Preliminary work on the serological characterization of *Aeromonas* strains included investigations with *Vibrio jamaicensis* and eight related strains that could be divided into three groups[13] and the description of 12 somatic and nine flagellar antigens in *A. hydrophila*[14]. Two decades later, 12 O antigen groups were found among 195 strains isolated from fish[15]. The first useful O serogrouping scheme for mesophilic *Aeromonas* strains defined 44 serogroups among 307 strains of *A. hydrophila* and *A. caviae*[16]. All antisera, although prepared with S form cultures, contained some R agglutinins and, therefore, had to be absorbed with R organisms prior to use. When 1513 motile mesophilic *Aeromonas* strains of clinical and environmental origin (isolated in four continents) were examined using antisera for these 44 serogroups and for an unpublished serogroup O45, about 40% of the strains were typable[17]. Among the other strains, 52 provisional serogroups were identified which increased the typability rate to 66% for strains from the United Kingdom, and to 43–68% for strains from other geographical areas. There were some differences in the geographical distribution of serogroups. In India, only 12 of 118 human and environmental isolates were not typable using the Sakazaki and Shimada scheme extended to contain 64 serogroups[18].

The relative frequencies with which the various serotypes were found in clinical and environmental isolates are listed in Table 4.2. Strains of serotypes 11, 16, 34 and AX1 seem to be the most prevalent. This is compatible with the findings that O:11 is the most important serotype among highly virulent strains[19] and is more often isolated from blood than from faecal specimens[20] whereas serotype O:34 is the most frequently found type among moderately to weakly virulent strains[21] which are more often isolated from faecal specimens than from blood[20].

Independently of the system of Sakazaki and Shimada, a serotyping scheme of *Aeromonas* species with 30 O groups and 2 K antigens was

Table 4.2. Prevalence (%) of the most frequent serotypes in clinical and environmental specimens. Values are given only for the serotypes with a frequency >2% in the relevant studies

O-serogroup	Sakazaki and Shimada[16]		Thomas et al.[17,a]	Misra et al.[18]		Merino et al.[20]	
	Hum.	Env.	Hum.+Env.	Hum.	Env.	Hum. (blood)	Hum. (faecal)
	(n=185)	(n=105)	(n=1255)	(n=68)	(n=50)	(n=68)	(n=432)
2	2.2	2.9	<	2.9	<	nd	nd
3	7.0	8.6	2.4	<	6.0	nd	nd
8	2.2	<	<	<	<	nd	nd
11	10.8	2.9	3.7	8.8	12.0	70.6	22.7
12	7.6	2.9	<	<	<	nd	nd
13	<	<	<	4.4	<	nd	nd
14	5.9	2.9	<	<	<	nd	nd
15	<	<	<	4.4	<	nd	nd
16	14.1	7.6	2.1	16.1	20.0	nd	nd
18	3.2	3.8	<	<	<	nd	nd
19	<	<	<	7.4	<	nd	nd
21	<	3.8	<	<	<	nd	nd
22	3.2	<	<	<	<	nd	nd
23	<	3.8	<	<	10.0	nd	nd
24	2.2	<	<	<	<	nd	nd
25	<	2.9	<	<	<	nd	nd
28	<	2.9	<	<	<	nd	nd
29	<	5.7	<	<	<	nd	nd
34	2.2	11.4	2.1	26.4	20.0	2.9	34.3
35	<	2.9	<	<	<	nd	nd
38	<	2.9	<	<	<	nd	nd
39	<	3.8	<	<	<	nd	nd
40	2.2	<	<	<	<	nd	nd
43	2.7	<	<	<	<	nd	nd
44	<	<	<	5.9	<	nd	nd
50	nd	nd	nd	7.4	<	nd	nd
AX1	nd	nd	7.3	nd	nd	nd	nd
AX2	nd	nd	2.6	nd	nd	nd	nd
AX3	nd	nd	2.1	nd	nd	nd	nd
Other	29.7	11.4	73.2	7.4	16.0	nd	nd
UK				8.8	12.0	26.5	43.0
R	4.9	15.2	4.6	<	<	nd	nd

a, strains from the UK only; <, serotype frequency 2% or less, nd, serotype not looked for; R, rough strains; UK, serotype unknown; Hum, human isolate; Env, environmental isolate.

described by Guinée and Jansen[22]. They used passive haemagglutination to type 306 strains isolated from surface and drinking water, food samples and faeces of patients with diarrhoea. Only 14% of 155 *A. hydrophila* strains were not typable, whereas the percentage of untypable strains was 46% for

A. sobria and 68% for *A. caviae*. These numbers reflect that 25 strains of *A. hydrophila* but only four and one of *A. sobria* and *A. caviae*, respectively, had been used for antiserum production. Although some of the serotypes of the two O-serogrouping systems are identical, one single system would be clearly desirable for future use[23]. A third serotyping system for mesophilic *Aeromonas* spp. on the basis of lipopolysaccharide antigens has been described by Fricker[24]. This system uses 16 antisera which allowed 63 of 137 (46%) strains isolated from human faeces to be typed. However, this system has not been adopted by any other group nor is anything known about its relation to the other serotyping systems.

4.1.4 RESISTOTYPING

Determining the susceptibility of clinically significant isolates to various antimicrobial agents is routinely done in diagnostic laboratories. Although an unusual antibiogram may be the first indication for the emergence of a new strain, care should be taken when interpreting such data because susceptibility patterns change rapidly (e.g. through loss or uptake of plasmids encoding resistance genes) especially under the selective pressure of the entire hospital environment[3].

The *in vitro* susceptibilities of *Aeromonas* strains have been quite constant with respect to time and location of isolation: most strains are resistant to penicillin and ampicillin but susceptible to most other drugs[25,26]. This homogeneity precludes the use of susceptibility patterns for epidemiological investigations. Conversely, susceptibility to cephalothin was significantly associated with phenotypic *A. sobria* as compared with *A. caviae* and *A. hydrophila* and thus may provide a taxonomic tool[27]. In addition, *A. jandaei* has been described as being resistant to ampicillin and cephalothin (which is consistent with the above findings because this species has previously been included in the phenotypic *A. sobria* group) and to colistin which separates *A. jandaei* from all other phenotypic species except *A. hydrophila*[28].

4.1.5 PROTEIN ANALYSIS

Separation of whole cell protein (or fractions thereof, e.g. outer membrane proteins; OMPs) according to their size by SDS–polyacrylamide gel electrophoresis (SDS–PAGE) has been effective in epidemiological investigations of various organisms[29]. The patterns produced are usually very complex and, therefore, comparisons among many different strains can be difficult. Although SDS–PAGE coupled with computer-aided analysis of patterns has proven useful for taxonomic purposes, the significance of small differences is uncertain and may be linked to technical factors such as growth conditions or protein extraction methods[3].

A method for typing *Aeromonas* species by SDS–PAGE has been described by Stephenson and colleagues[30]. Strains grown overnight were radiolabelled with ^{35}S-methionine followed by analysis of soluble proteins. Protein profiles were then compared by eye using the Dice coefficient[31]. All strains were typable and showed some bands which were relatively constant for all isolates, but reproducibility was quite low: similarity coefficients among repeatedly tested strains varied between 80 and 100%. Consequently, strains with >80% similarity were considered indistinguishable whereas strains with <70% similarity were considered to be definitely different. Such a variability in patterns precludes the use of this technique for reliable separation of individual strains. These data were confirmed in a recent study also using radiolabelled total protein profiles but in combination with computerized analysis[32]. On the other hand, strains isolated from human faeces and those isolated from taps and drains in a ward kitchen formed two well defined clusters which were not closely related to each other[30]. These findings question the validity of the assumption that human gastrointestinal *Aeromonas* infections are usually acquired from drinking water. In a follow-up study, it was shown that protein patterns obtained by radiolabelling/autoradiography were similar to those visualized by silver staining[33].

A high variability among independently isolated *Aeromonas* strains was also shown by SDS–PAGE of OMPs[34]. Almost every single isolate of HG 8 (*A. veronii* biogroup sobria and *A. veronii* biogroup veronii) and of HG 1 (*A. hydrophila* sensu stricto) had a unique OMP profile whereas the 16 strains of HG 4 (*A. caviae*) were separated into five different OMP types. In addition, this study confirmed that incubation conditions can influence protein profiles of motile *Aeromonas* strains[35]. Despite a good correlation of OMP profiles with HGs 1, 4 and 8, respectively, which allowed all isolates to be assigned to the correct HG, it is not known whether this technique may help to identify all known DNA hybridization groups known so far. It had been recognized previously that SDS–PAGE may also help to distinguish taxonomic subgroups within phenotypic *Aeromonas* species[30] but whether these subgroups correlate with the various HGs remains unknown.

The complexity of patterns produced by SDS–PAGE could drastically be reduced by immunoblotting with a polyclonal antiserum raised against whole cells of *A. sobria*[36]. Again, all strains were typable, each producing 15–20 well separated bands. Thirty types were defined among 103 isolates, but there was no correlation with biochemical phenotypes. Attempts to correlate immunoblot patterns with serotypes were not successful because only 42% of the strains included in that study could be serotyped.

4.1.6   ENZYME ANALYSIS

Instead of looking at structural proteins to characterize strains, enzyme activities can be investigated in two different ways:

1. By detecting the presence or absence of a particular enzyme using chromogenic substrates.
2. By analysing electrophoretic mobilities of such enzymes using non-denaturing gel electrophoresis.

4.1.6.1 Detection of enzyme activities

Differences in enzyme activities have proven very useful in the discrimination of related organisms which are difficult to separate by conventional biochemical reactions[37]. Most studies with *Aeromonas* strains have made use of the commercially available API ZYM system (API System SA, La Balme-les Grottes, Montalieu-Vercieu, France) or an extended system thereof provided by the same manufacturer. The API ZYM system allows rapid detection of 19 preformed enzymes including esterases, aminopeptidases, proteases, phosphatases and glycosidases. Waltman and co-authors[38] characterized 48 *Aeromonas* strains and found no significant differences in enzyme profiles of strains isolated from various sources. However, phenotypic *A. sobria* differed from phenotypic *A. hydrophila* by a lack of myristate lipase and β-glucosidase and a significant decrease in the presence of alkaline phosphatase, butyrate esterase and β-galactosidase. Most of these differences were *not* confirmed in a study that analysed biochemical and enzymatic properties of 127 *Aeromonas* strains mainly isolated in the United States, except that β-glucosidase was absent in all *A. sobria* strains but present in 60% of *A. caviae* and 79% of *A. hydrophila* strains[39]. These differences between the two studies can probably be attributed to different incubation conditions[40]. However, there were some significant differences between the three phenotypic species when exoenzymatic properties were investigated (e.g. production of elastase and haemolysin)[39]. A battery of 89 substrates was used in a study with 16 *Aeromonas* species including seven *A. hydrophila*, five *A. sobria* and four *A. caviae*[41]. Some of the enzymes examined seemed useful in the identification of the three phenotypic species but, as also mentioned by the authors, a larger number of strains needs to be analysed. It remains unclear whether or not the determination of enzyme activities is useful for the separation of the various DNA hybridization groups rather than only the phenotypic species and/or for epidemiological investigations to discriminate among strains of a given genomic species.

4.1.6.2 Enzyme electrophoresis

Much more information can be obtained by, rather than only detecting the presence or absence of an enzymatic activity, separating cellular extracts by non-denaturing gel electrophoresis followed by staining for particular enzyme activities, which again involves chromogenic substrates for visualizing bands (Fig. 4.1). Enzymes may differ in their mobilities, representing changes in their amino acid sequences and, therefore, in the genes coding for these enzymes. It has been estimated that about 80% of the allelic variants can be detected electrophoretically[42]. In the first study involving 64 *Aeromonas* strains belonging to all three major phenotypic complexes, esterases and glutamate, lactate and malate dehydrogenases were analysed by gel electrophoresis and by thin-layer isoelectrofocusing[43]. The eight species-specific zymotypes that were defined on the basis of the isoelectric points of malate dehydrogenase and the mobility of lactate dehydrogenase correlated well with the previously established DNA hybridization groups[44], whereas the other enzymes, especially the major esterase, were thought to provide useful epidemiological markers. Esterases were then used to investigate *A. hydrophila* infections in a hospital[45]. Several arguments were found to support the view that the hospital water system may have been the source of these infections: (1) motile *Aeromonas* were found in concentrations of 1–10 CFU/ml in most water samples collected at various points of the water system; (2) the same zymotype was found in two patients and in two water samples on one ward; and (3) prophylactic measures taken to avoid oral contamination of immunocompromised patients with tap water significantly reduced the number of cases of septicaemia. On the other hand, there was a great diversity of patterns found in strains from patients (eight zymotypes among 15 isolates) as well as in strains from water samples (37 zymotypes among 126 isolates).

A similar diversity was also found in 153 mainly faecal *Aeromonas* strains that were analysed for 11 different enzymes, all of which were polymorphic[46]. In all, 122 distinct allele combinations (zymotypes or enzyme types) were found among these strains. Some strains not distinguishable by enzyme typing were shown to be different by phage typing and by restriction enzyme analysis and ribotyping[7] (Table 4.3). When these enzyme data were analysed using pairwise weighted distance coefficients (which represent the genetic

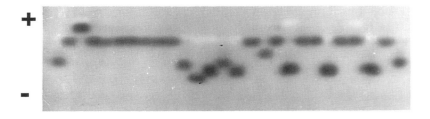

Figure 4.1. Starch gel electrophoresis and staining for adenylate kinase activity of cellular extracts from various *Aeromonas* strains. Migration was from cathode to anode

Table 4.3. Further discrimination of *Aeromonas* strains of identical phage type by enzyme electrophoresis, restriction enzyme analysis and ribotyping

Strain	Phage type	Enzyme type	rDNA pattern *Pst*I	Restriction pattern *Pst*I
61	1	64	A	A
179	1	27	B	B
138	2	53	C	C
167	2	99	D	D
6	3	16	E	E
275	3	65	F	F

relatedness between isolates) and UPGMA cluster analysis[42], an excellent, though not perfect, correlation was found with DNA–DNA hybridization data[46]. Despite the genetic complexity of the genus *Aeromonas*, the use of multilocus enzyme electrophoresis might serve as the sole method for species determination. However, its major advantage will be in conjunction with phenotypic tests. For the *A. hydrophila* complex, just two enzymes are sufficient to separate the three DNA hybridization groups (see also Chapter 1). The situation is somewhat more complex for the *A. caviae* group where three enzymes allowed the identification of hybridization groups 4, 5A, 5B and 6. A combination of four enzymes was sufficient to separate reliably all those DNA hybridization groups that usually occur in clinical specimens[47].

A subdivision within *A. hydrophila*, *A. caviae* and *A. sobria*, respectively, by multilocus enzyme electrophoresis was also observed by Tonolla and colleagues[48], who analysed 128 strains of clinical and environmental origin. The distribution of reference strains included in that study suggested that these subgroups most probably represent some of the DNA hybridization groups. This would be in accordance with a previous study[46]. In addition, clinical isolates, with few exceptions, were shown to be genetically distinct from those collected in the environment. The hypothesis that most cases of *Aeromonas*-related diarrhoea are acquired from water could therefore not be confirmed.

4.2   GENOTYPIC METHODS

4.2.1   PLASMID ANALYSIS

Plasmid analysis was among the earliest DNA-based techniques applied to epidemiological studies, and its value for tracing routes of infections has been documented for a variety of organisms[49]. The method is relatively simple and does not require very sophisticated equipment. Strains to be investigated are lysed to prepare a plasmid extract and are then subjected to agarose gel electrophoresis and ethidium bromide staining. The reproducibility of plasmid profiles may be confounded by the fact that a single plasmid can be present in different molecular forms (closed circle, open circle, linear), which all migrate differently in agarose gels. This problem can be eliminated by digesting the plasmid DNA with restriction endonucleases prior to electrophoresis. However, plasmid analysis is considered of little epidmiological value in the genus *Aeromonas* because the plasmid carriage rate has been found to be very low, with only 20 of 75 strains analysed harbouring at least one extrachromosomal element[50]. In addition, it should be borne in mind that plasmids can be acquired or lost relatively easily, as shown for many Gram-positive and Gram-negative bacterial species. Consequently, epidemiologically related isolates can exhibit different plasmid profiles[51].

4.2.2   RESTRICTION ENZYME ANALYSIS AND RIBOTYPING

Chromosomal DNA provides a better target for nucleic acid-based typing techniques than plasmid DNA. It is present in every single cell, and can be isolated without much effort in sufficient quantities for further characterization. High-molecular-weight chromosomal DNA is usually digested ('cut') with restriction enzymes, followed by separating the resulting fragments through agarose gel electrophoresis and visualizing them with ethidium bromide staining for relatively small fragments produced when frequently cutting restriction enzymes are used. Larger fragments produced with rarely cutting enzymes are better separated by pulsed-field gel electrophoresis (PFGE). Patterns produced by restriction enzyme analysis (REA) may be very complex, making a comparison of large numbers of strains inconvenient or even impossible. This problem can be circumvented by transferring the separated fragments to a solid support (nitrocellulose or nylon membranes) followed by hybridization with a radioactively or a non-radioactively labelled probe, e.g. with ribosomal RNA or with a plasmid containing an rRNA gene[52,53]. This procedure is called ribotyping, and reduces the number of bands to about 5–20. Reproducibility and stability of ribotyping patterns appear to be excellent, as shown by identical patterns with strains subcultured *in vitro* many times[54]. *In vivo* stability could be demonstrated in a patient from whom *A. caviae* was repeatedly isolated over a period of more than three years without any change in the ribotyping pattern[55], as well as in several pairs of *A. caviae* and *A. hydrophila* strains isolated from the same patients which were ribotyped blindly[56].

When 58 *Aeromonas* strains were ribotyped by digestion with *Pst*I a very high degree of heterogeneity was observed[7]. Few, if any, of the strains exhibited identical patterns. We therefore compared strains that had previously been characterized by enzyme electrophoresis and phage typing. Several groups of strains of identical enzyme type were clearly separable by DNA analysis (Table 4.1). In addition, strains of a single phage type could be distinguished not only by REA but also by ribotyping and enzyme electrophoresis (Table 4.3). It was also evident that REA is somewhat more sensitive than ribotyping, for example, strains 30, 192 and 253 were all different by REA yet two strains had identical ribotypes (Table 4.1, Figure 4.2). The latter two strains were also very similar in their susceptibility to *Aeromonas*-specific phages. Similar results were published by Kuijper and co-authors[34] for 23 strains from DNA HGs 1 ($n = 5$), 4 ($n = 10$), and 8 ($n = 8$). All strains had different digestion patterns. Most of these strains had also been analysed by SDS-PAGE of OMPs, giving different patterns for all four *A. hydrophila* and all seven *A. sobria* isolates. However, some *A. caviae* strains with identical OMP patterns were clearly separable by REA and also by ribotyping.

4.2.2.1 Ribotyping as an epidemiological tool

The usefulness of ribotyping for epidemiological investigations has been extensively investigated by Moyer and co-authors[55,57]. In one study, they characterized strains isolated from five patients with gastroenteritis and from the source water, treatment plant and distribution system of a small public water supply[57]. All patient strains were shown to be unique, but identical ribotypes of *A. hydrophila* and *A. sobria* were isolated from multiple sites in

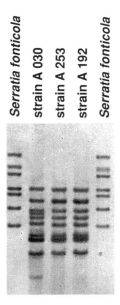

Figure 4.2. Hybridization of *Aeromonas* DNA digested with restriction enzyme *Sma*1 with radiolabelled rRNA of *Escherichia coli* after agarose gel electrophoresis and transfer to nitrocellulose membrane (ribotyping). The size (in base pairs) of fragments generated by digestion of *Serratia fonticola* DNA with *Hin*dIII is 14'596, 11'853, 10'601, 8'520, 7'892, 6'634, and 5'419, respectively (from top to bottom)

the water system. Based on these results, which indicated colonization of a well, sand filters and the softener, with the potential for sporadic contamination of distribution water, operational deficiencies were detected and could be corrected. Despite the fact that clinical and environmental strains were clearly different, the possibility of waterborne transmission of strains leading to gastrointestinal infection could not be excluded because the cultures investigated in this study were isolated from the water system two to three months after cases of gastroenteritis first appeared in that community.

Again no direct link could be documented between isolates from an untreated farm well and from four persons with presumably *Aeromonas*-associated diarrhoea[55]. However, the isolates from the three children (but not from their mother) were identical by ribotyping, suggesting a common source. In a second case cluster observed in that study, *A. sobria* and *A. caviae* were repeatedly isolated from the stools of an infant with kwashiorkor between the ages of three and 15 months. The child was placed in a foster home at three months of age, which coincided with the onset of chronic *Aeromonas*-associated diarrhoea. After treatment with trimethoprim-sulfamethoxazole, diarrhoea recurred with only *A. caviae* being isolated from the stool. Within four months, both foster parents developed acute diarrhoea and *A. caviae* was isolated from their stools as sole bacterial pathogen. The ribotypes of the *A. caviae* isolates from the child and the foster mother were indistinguishable by ribotyping whereas the isolate from the foster father was not available for typing. These data strongly suggest familial transmission of a particular *Aeromonas* strain.

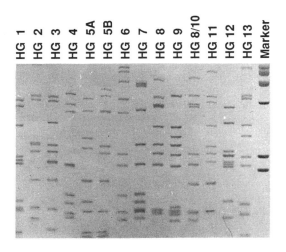

Figure 4.3. Ribotyping of representative strains of the various *Aeromona* is DNA hybridization groups using pGML1 for hybridization. The size (in kilobases) of the marker fragments is 23.1, 9.4, 6.6, 4.3, 2.3 and 2.0, respectively (from top to bottom)

4.2.2.2   Ribotyping as taxonomic tool

While ribotyping using rDNA or plasmid pKK3535 containing an entire operon coding for 16S, 5S and 23S rRNA of *E. coli* provides an excellent tool for epidemiological investigations in the genus *Aeromonas*, with major bands above a size of 3–4 kilobases (kb), no obvious correlation was found between these patterns and taxonomic grouping[56,58]. However, smaller bands of 0.8–4 kb became clearly visible when hybridization was performed with plasmid pGML1, which is a pGEM11zf(+) plasmid containing a 576 bp *Hind*III fragment representing a region of the 5' half of the 16S rRNA gene of *E. coli*[58]. When more than 150 *Aeromonas* strains of known DNA hybridization group were ribotyped using *Sma*I for digestion and pGML1 for hybridization, an excellent correlation was found between banding patterns and hybridization groups. For some groups, such as HG 8/10 (*A. veronii* biogroup sobria and *A. veronii* biogroup veronii) and HG 9 (*A. jandaei*), no reliable separation has been achieved because patterns are very similar and only few strains of some groups have been included in the database[58]. In addition, strains of the newly described species *A. allosaccharophila* and *A. encheleia* had not been included in that study at all. On the other hand, strains of HG 2 and HG 3, which are difficult to separate phenotypically could be assigned to one of the groups beyond any reasonable doubt[59,60].

4.3   OUTLOOK

We now have available a number of very sensitive methods for subtyping strains of the genus *Aeromonas*. Even so, the sources for most infections remain obscure due to the extreme diversity and ubiquitous occurrence of aeromonads, and to the fact that very few epidemiological studies have been performed because of their limited clinical significance. In addition, data on the aquatic origin of strains causing gastrointestinal disease are contradictory, and additional, prospective ecological investigations are necessary to elucidate the role of environmental reservoirs in human infections. So far, only foods have been proven to serve as vehicles for transmission of *Aeromonas*-associated diarrhoea. Although aeromonads are often found in seafood, beef, pork, chicken and vegetables[61,62], few investigations have documented an association with human illness[63–65].

Despite the excellent correlation between results obtained by some of the subtyping methods and the genomic species, DNA–DNA hybridization cannot be replaced as the method for the definition of new species because it is not clear whether a new ribotyping pattern or a new enzyme type represent a completely new hybridization group or just a variant of an existing species. On the other hand, a polyphasic approach may be very helpful for the classification of strains in this extremely heterogeneous group of organisms.

4.4 REFERENCES

1 Carnahan A, Altwegg M. This volume, Chapter 1.
2 Havelaar AH, Schets FM, van Silfhout A, Jansen WH, Wieten G, van der Kooij D. Typing of *Aeromonas* strains from patients with diarrhoea and from drinking water. *J Appl Bacteriol* 1992; **72**: 435–44.
3 Arbeit RD. Laboratory procedures for the epidemiologic analysis of microorganisms. In Murray PR, Baron EJ, Pfaller MA, Tenover FC, Yolken RH (eds). *Manual of Clinical Microbiology*, 6th edn. Washington, DC: American Society for Microbiology, 1995: 190–208.
4 Demarta A, Peduzzi R. Etude épidémiologique des *Aeromonas* par lysotypie. *Estratto dalla Rivista Italiana di Piscicoltura e Ittiopatologia* 1984; **19**: 148–55.
5 Janda JM, Abbott SL, Carnahan AM. *Aeromonas* and *Plesiomonas*. In Murray PR, Baron EJ, Pfaller MA, Tenover FC, Yolken RH (eds) *Manual of Clinical Microbiology*, 6th edn. Washington DC: American Society for Microbiology, 1995: 477–82.
6 Peduzzi R, Demarta A. Comparaison au moyen de la lysotypie entre souches d'*Aeromonas* de pisciculture et souches isolées de l'homme. *Schweiz Z Hydrol* 1986; **48**: 161–9.
7 Altwegg M, Altwegg-Bissig R, Demarta A, Peduzzi R, Reeves MW, Swaminathan B. Comparison of four typing methods for *Aeromonas* species. *J Diarrh Dis Res* 1988; **6**: 88–94.
8 Altwegg M, unpublished results.
9 Fukuyama M, Kamimura T, Itoh T, Hara M, Tabuchi K, Murata M, Kohzaki K. Studies on motile *Aeromonas* infection: 2. Development of a bacteriophage typing system for motile *Aeromonas*. *Kansenshogazu Zasshi* 1991; **65**: 813–19.
10 Fukuyama M, Kawakami K, Imagawa Y, Itoh T, Hara M, Tabuchi K. Studies on motile *Aeromonas* infection: 3. Phage typing of motile *Aeromonas* isolated from patients with diarrhoea. *Kansenshogaku Zasshi* 1992; **66**: 628–31.
11 Mishu B, Griffin PM, Tauxe RV, Cameron DN, Hutcheson RH, Schaffner W. *Salmonella enteritidis* gastroenteritis transmitted by intact chicken eggs. *Ann Intern Med* 1991; **115**: 190–4.
12 Griffin PM, Tauxe RV. The epidemiology of infections caused by *Escherichia coli* 0157:H7, other enterohemorrhagic *E. coli*, and the associated hemolytic uremic syndrome. *Epidemiol Rev* 1991; **13**: 60–98.
13 Caselitz FH, Schönn O. Serologische Studien an *Vibrio jamaicensis* und verwandten Vibrionenstämmen. *Zschr Tropenmed Paras* 1956; **7**: 50–5.
14 Ewing WH, Hugh R, Johnson G. Studies on the *Aeromonas* group. CDC Monograph, USDHEW, Communicable Disease Center, Atlanta, GA 1961.
15 Leblanc D, Mittal KR, Oliver G, Lallier R. Serogrouping of motile *Aeromonas* species isolated from healthy and moribund fish. *Appl Environ Microbiol* 1981; **42**: 56–60.
16 Sakazaki R, Shimada T. O-serogrouping scheme for mesophilic *Aeromonas* strains. *Japan J Med Sci Biol* 1984; **37**: 247–55.
17 Thomas LV, Gross RJ, Cheasty T, Rowe B. Extended serogrouping scheme for motile mesophilic *Aeromonas* species. *J Clin Microbiol* 1990; **28**: 980–4.
18 Misra SK, Shimada T, Bhadra RK, Pal SC, Nair GB. Serogroups of *Aeromonas* species from clinical and environmental sources in Calcutta, India. *J Diarrh Dis Res* 1989; **7**: 8–12.
19 Kokka RP, Janda JM, Oshiro LS, Altwegg M, Shimada T, Sakazaki R, Brenner DJ. Biochemical and genetic characterization of autoagglutinating phenotypes of

Aeromonas species associated with invasive and noninvasive disease. *J Infect Dis* 1991; **163**: 890–4.

20 Merino S, Camprubi S, Tomas JM. Incidence of *Aeromonas* spp. serotypes 0:34 and 0:11 among clinical isolates. *Med Microbiol Lett* 1993; **2**: 48–55.

21 Merino S, Camprubi S, Tomas JM. Characterization of an O-antigen bacteriophage from *Aeromonas hydrophila*. *Can J Microbiol* 1992; **38**: 235–40.

22 Guinée PAM, Jansen WH. Serotyping of *Aeromonas* species using passive haemagglutination. *Zentralbl Bakteriol Hyg A* 1987; **265**: 305–17.

23 Shimada T, Kosako Y. Comparison of two O-serogrouping systems for mesophilic *Aeromonas* spp. *J Clin Microbiol* 1991; **29**: 197–9.

24 Fricker CR. Serotyping of mesophilic *Aeromonas* spp. on the basis of lipopolysaccharide antigens. *Lett Appl Microbiol* 1987; **4**: 113–6.

25 Fainstein V, Weaver S, Bodey GP. In vitro susceptibilities of *Aeromonas hydrophila* against new antibiotics. *Antimicrob Agents Chemother* 1982; **22**: 513–14.

26 Burgos A, Quindos G, Martinez R, Rojo P, Cisterna R. In vitro susceptibility of *Aeromonas caviae*, *Aeromonas hydrophila* and *Aeromonas sobria* to fifteen antimicrobial agents. *Eur J Clin Microbiol Infect Dis* 1990; **9**: 413–17.

27 Janda JM, Motyl MR. Cephalothin susceptibility as a potential marker for the *Aeromonas sobria* group. *J Clin Microbiol* 1985; **22**: 854–5.

28 Carnahan A, Fanning GR, Joseph SW. *Aeromonas jandaei* (formerly genospecies DNA group 9 *A. sobria*), a new sucrose-negative species isolated from clinical specimens. *J Clin Microbiol* 1991; **29**: 560–4.

29 Mulligan ME, Peterson LR, Kwok RYY, Clabots CR, Gerding DN. Immunoblots and plasmid fingerprints compared with serotyping and polyacrylamide gel electrophoresis for typing *Clostridium difficile*. *J Clin Microbiol* 1988; **26**: 41–6.

30 Stephenson JR, Millership SE, Tabaqchali S. Typing of *Aeromonas* species by polyacrylamide-gel electrophoresis of radiolabelled cell proteins. *J Med Microbiol* 1987; **24**: 113–18.

31 Dice LR. Measures of the amount of ecologic association between species. *Ecology* 1945; **26**: 297–302.

32 Arzese A, Pipan C, Piersimoni C, Scalise G, Morbiducci V, Botta GA. Characterization of mesophilic *Aeromonas* from clinical specimens by computerized analysis of SDS–PAGE protein profiles and by enzymatic activity. *Microbiologica* 1993; **16**: 333–42.

33 Millership SE, Want SV. Typing of *Aeromonas* species by protein fingerprinting: comparison of radiolabelling and silver staining for visualizing proteins. *J Med Microbiol* 1989; **29**: 29–32.

34 Kuijper EJ, van Alphen L, Leenders E, Zanen HC. Typing of *Aeromonas* strains by DNA restriction endonuclease analysis and polyacrylamide gel electrophoresis of cell envelopes. *J Clin Microbiol* 1989; **27**: 1280–5.

35 Statner B, Jones MJ, George WL. Effect of incubation temperature on growth and soluble protein of motile *Aeromonas* strains. *J Clin Microbiol* 1988; **26**: 392–3.

36 Mulla R, Millership S. Typing of *Aeromonas* spp. by numerical analysis of immunoblotted SGS–PAGE gels. *J Med Microbiol* 1993; **39**: 325–33.

37 Gruner E, von Graevenitz A, Altwegg M. The API ZYM system: a tabulated review from 1977 to date. *J Microbiol Methods* 1992; **16**: 101–18.

38 Waltman II WD, Shotts EB, Hsu T-C. Enzymatic characterization of *Aeromonas hydrophila* complex by the API ZYM system. *J Clin Microbiol* 1982; **16**: 692–6.

39 Janda JM. Biochemical and exoenzymatic properties of *Aeromonas* species. *Diag Microbiol Infect Dis* 1985; **3**: 223–32.

40 Gray SJ. Characterization of *Aeromonas* sp. in the API ZYM system. *Med Lab Sci* 1987; **44**: 287–9.

41 Carnahan AM, O'Brien M, Joseph SW, Colwell RR. Enzymatic characterization of three *Aeromonas* species using API peptidase API 'oxidase', and API esterase test kits. *Diagn Microbiol Infect Dis* 1988; **10**: 195–203.

42 Selander RK, Caugant DA, Ochman H, Musser JM, Gilmour MN, Whittam TS. Methods of multilocus enzyme electrophoresis for bacterial population genetics and systematics. *Appl Environ Microbiol* 1986; **51**: 873–84.

43 Picard B, Goullet P. Comparative electrophoretic profiles of esterases, and of glutamate, lactate and malate dehydrogenases, from *Aeromonas hydrophila, A caviae* and *A. sobria. J Gen Microbiol* 1985; **131**: 3385–91.

44 Popoff MY, Coynault C, Kiredjian M, Lemelin M. Polynucleotide sequence relatedness among motile *Aeromonas* species. *Curr Microbiol* 1981; **5**: 109–14.

45 Picard B, Goullet P. Epidemiological complexity of hospital *Aeromonas* infections revealed by electrophoretic typing of esterases. *Epidemiol Infect* 1987; **98**: 5–14.

46 Altwegg M, Reeves MW, Altwegg-Bissig R, Brenner DJ. Multilocus enzyme analysis of the genus *Aeromonas* and its use for species identification. *Zentralbl Bakteriol* 1991; **275**: 28–45.

47 Altwegg M, Steigerwalt AG, Janda JM, Brenner DJ. Identification of *Aeromonas* species by isoenzyme analysis. *Abstr Annu Meet Am Soc Microbiol* 1989, New Orleans.

48 Tonolla M, Demarta A, Peduzzi R. Multilocus genetic relationships between clinical and environmental *Aeromonas* strains. *FEMS Microbiol Lett* 1991; **81**: 193–200.

49 Mayer LW. Use of plasmid profiles in epidemiologic surveillance of disease outbreaks and in tracing the transmission of antibiotic resistance. *Clin Microbiol Rev* 1988; **1**: 228–43.

50 Chang BJ, Bolton SM. Plasmids and resistance to antimicrobial agents in *Aeromonas sobria* and *Aeromonas hydrophila* isolates. *Antimicrob Agents Chemother* 1987; **31**: 1281–2.

51 Mickelsen PA, Plorde JJ, Gordon KP, Hargiss C, McClure J, Schoenknecht FD, Condie F, Tenover FC, Tompkins LS. Instability of antibiotic resistance in a strain of *Staphylococcus epidermidis* isolated from an outbreak of prosthetic valve endorcarditis. *J Infect Dis* 1985; **152**: 50–8.

52 Grimont F, Grimont PAD. Ribosomal ribonucleic acid gene restriction patterns as potential taxonomic tools. *Ann Inst Pasteur/Microbiol* 1986; **L137B**: 165–75.

53 Altwegg M, Mayer LW. Bacterial molecular epidemiology based on a non-radioactive probe complementary to ribosomal RNA. *Res Microbiol* 1989; **140**: 325–33.

54 Stull TL, LiPuma JJ, Edlind TD. A broad-spectrum probe for molecular epidemiology of bacteria: ribosomal RNA. *J Infect Dis* 1988; **157**: 280–6.

55 Moyer NP, Martinetti Lucchini G, Holcomb LA, Hall NH, Altwegg M. Application of ribotyping for differentiating aeromonads isolated from clinical and environmental sources. *Appl Environ Microbiol* 1992; **58**: 1940–4.

56 Carey PE, Eley A, Wilcox MH. Assessment of a chemiluminescent universal probe for taxonomical and epidemiological investigations of *Aeromonas* sp. isolates. *J Clin Pathol* 1994; **47**: 642–6.

57 Moyer NP, Martinetti G, Lüthy-Hottenstein J, Altwegg M. Value of rRNA gene restriction patterns of *Aeromonas* spp. for epidemiological investigations. *Curr Microbiol* 1992; **24**: 15–21.

58 Martinetti Lucchini G, Altwegg M. rRNA gene restriction patterns as taxonomic tools for the genus *Aeromonas*. *Int J Syst Bacteriol* 1992; **42**: 384–9.

59 Altwegg M. A polyphasic approach to the classification and identification of *Aeromonas* strains. *Med Microbiol Lett* 1993; **2**: 200–5.

60 Ali A, Carnahan A, Altwegg M, Lüthy-Hottenstein J, Joseph S. *Aeromonas bestiarum.* sp nov. (formerly genomospecies DNA group 2 *A. hydrophila*), a new species isolated from non-human sources. *Med Microbiol Lett* 1996; **5**: 156–65.

61 Nishikawa Y, Kishi T. Isolation and characterization of motile *Aeromonas* from human, food and environmental specimens. *Epidemiol Infect* 1988; **101**: 213–23.

62 Palumbo SA, Bencivengo MM, DelCorral F, Williams AC, Buchanan RL. Characterization of the *Aeromonas hydrophila* group isolated from retail foods of animal origin. *J Clin Microbiol* 1989; **27**: 854–8.

63 Altwegg M, Martinetti Lucchini G, Lüthy-Hottenstein J, Rohrbach M. *Aeromonas*-associated gastroenteritis after consumption of contaminated shrimp. *Eur J Clin Microbiol Infect Dis* 1991; **10**: 454–6.

64 Krovacek K, Dumontet S, Eriksson E, Baloda SB. Isolation, and virulence profiles, of *Aeromonas hydrophila* implicated in an outbreak of food poisoning in Sweden. *Microbiol Immunol* 1995; **39**: 655–61.

65 Tanaka K, Yamamoto M, Matsumoto M, Saito M, Funabashi M, Yoshimatsu S. An outbreak of food poisoning suspected due to *Aeromonas* and characteristics of the isolated strains. *Nippon Koshu Eisei Zasshi* 1992; **39**: 707–13.

5 The Ecology of Mesophilic *Aeromonas* in the Aquatic Environment

PHILIP HOLMES
Severn Trent Water Ltd, Long Eaton, Nottingham, England

LYNDA M. NICCOLLS AND DAVID P. SARTORY
Severn Trent Water Ltd, Tewkesbury, England

5.1 INTRODUCTION

In almost all aquatic environments throughout the world, mesophilic members of the genus *Aeromonas* may be found. Numbers may be high in both polluted and unpolluted habitats and even in chlorinated drinking water. Studies on the occurrence of mesophilic aeromonads and the ecological factors that affect their numbers in the aquatic environment are of increasing interest to environmental and drinking water microbiologists, as well as to public health officials and epidemiologists.

Although it is now over a century since Zimmerman first demonstrated the presence of *Aeromonas* in the aquatic environment[1], it is only during the past two decades that an increased understanding of the ecological aspects of this group of bacteria has begun to emerge. Previously the oxidase-positive, Gram-negative bacilli have been considered by the water industry to be nuisance organisms and of little clinical significance, frequently being confused with members of the Enterobacteriaceae. With the development of specific isolation media, many formulations of which contain high concentrations of inhibitors to reduce levels of competing background flora, there is now a reasonable body of data on the factors influencing the occurrence of *Aeromonas* in the aquatic environment. However, these selective agent- and nutrient-rich media may result in low recovery of aeromonads derived from nutrient-limited environments, or those subjected to chemical or physical stresses (e.g. chlorinated drinking water), and thus weight data in favour of the more rapidly growing strains of mesophilic *Aeromonas*. It has been demonstrated that many bacteria have survival mechanisms enabling them to persist under adverse conditions[2,3], and, although they may still be viable, they may be non-culturable on selective and non-selective media. There is, however, no information relating to the occurrence of this phenomenon to mesophilic aeromonads, although

The Genus Aeromonas. Edited by B. Austin, M. Altwegg, P.J. Gosling and S. Joseph
© 1996 John Wiley & Sons Ltd

Monfort and Baleux[4] demonstrated high numbers of *A. sobria* cells present in filter-sterilized brackish water 90 days after inoculation, when culturable cells could no longer be recovered. However, the metabolic activity of these cells was not tested. In general, therefore, it may be difficult to relate the results of laboratory-based studies to bacterial behaviour in the natural environment.

The taxonomy of this group was not well understood until recently. Studies on the genotypic characteristics have resulted in the description of a number of closely related genospecies. In ecological terms, however, we can still refer to the four classically described species, divided into two groups.

1.   The motile mesophilic species—*A. hydrophila, A. caviae* and *A. sobria*.
2.   The psychrophilic and non-motile *A. salmonicida*.

The mesophilic aeromonads have been implicated in diseases of poikilothermic aquatic animals (e.g. red leg disease in frogs); see Chapter 7. Infections may also occur in warm-blooded animals including humans (see Chapters 6 and 7), generally due to ingress via a wound from contaminated waters[5]. Aeromonad-associated gastrointestinal infections have recently been described, mainly from areas where *Aeromonas* are known to be present in potable water supplies, but to date there has been no firm evidence that transmission has occurred via this route.

The distribution of mesophilic species in the aquatic environment may be related to the levels of pollution in the water. It has been reported by several authors including Schubert[6], Araujo and colleagues[7] and Stecchini and Domenis[8] that *A. caviae* predominates in sewage and waters with a high degree of faecal pollution. In less polluted waters, either fresh or marine, *A. caviae* and *A. hydrophila* were almost equally distributed. *A. sobria* can be found in unpolluted waters, brackish waters and sewage effluents. Large numbers of aeromonads, especially *A. caviae*, could therefore be considered to be indicative of nutrient-rich conditions of water. Typical levels of *Aeromonas* in different aquatic environments are presented in Table 5.1. The numbers of mesophilic *Aeromonas* species in some aquatic environments have been reported to show a seasonal variation, with populations greatest in the warmer months of the year for both temperate freshwater lakes[9] and chlorinated drinking water[10].

The aquatic environments for which the ecology of aeromonads has been investigated can be conveniently divided into four water types: freshwater, seawater, sewage and potable water. There has been substantial interest in the incidence of *Aeromonas* during water treatment and distribution recently, not only with respect to public health concerns but also with regard to their role as indicators of regrowth potential and biofilm development.

Table 5.1.   Typical numbers of mesophilic *Aeromonas* species in aquatic environments

Environment	Typical counts (CFU/ml)
Domestic sewage sludge	$>10^8$
Crude sewage	$10^6 - 10^8$
Treated sewage	$10^3 - 10^5$
Wastewater	$10^2 - 10^7$
Rivers receiving sewage discharges	$10 - 10^4$
Clean rivers, lakes, storage reservoirs	$1 - 10^2$
Seawater	$10^{-2} - 10^2$
Drinking water, at treatment plant	$10^{-2} - 10$
Drinking water, in distribution system	$10^{-2} - 10^3$
Groundwaters	<1

Data adapted from references 11, 23 and 80.

5.2   FRESHWATER ENVIRONMENTS

The freshwater environment covers a wide range of habitats, from oligotrophic groundwaters and alpine lakes to eutrophic lakes and rivers. While groundwaters support a limited population of microorganisms, surface waters, particularly rivers and lakes, provide habitats for a diverse range of flora and fauna, and experience a greater variation in physical and chemical conditions.

5.2.1   GROUNDWATERS

Groundwaters are normally poor in nutrients with the overlying rocks tending to determine the mineral content. The presence of fissures may result in some nutrient enrichment. There is a lack of data on the occurrence of *Aeromonas* in pristine boreholes, but where coliform bacteria are not detected aeromonads are also unlikely to be isolated in significant numbers. Havelaar and co-workers[11] recovered low numbers of *Aeromonas* (maximum count 35 CFU/100 ml) from deep aerobic and anaerobic groundwaters in the absence of coliforms. *A. hydrophila* predominated in these waters, although *A. caviae* made up to 30–40% of the population. In a shallow limestone borehole in central England where coliforms are isolated during the summer and autumn, species of mesophilic *Aeromonas* were recovered at counts of up to 200 CFU/100 ml with no apparent seasonal variation (Niccolls and Holmes, unpublished data).

Using gas–liquid chromotographic analysis of cellular fatty acid methyl esters (FAMEs), Huys and co-authors[12] found that aeromonads from two groundwaters in Belgium consisted solely of strains belonging to hybridization groups 2 and 3 of the *A. hydrophila* complex. The numbers present were not reported, but the groundwaters were free of coliforms, although one did have low levels of faecal streptococci.

In a survey of 64 still mineral waters in Europe, Havelaar and colleagues[13] failed to recover any *Aeromonas*. Six of these waters were inoculated with a tapwater-derived isolate of *A. hydrophila* which have been pregrown under low-nutrient conditions. The isolate only survived and multiplied when the autochthonous bacteria had been removed by pasteurization, indicating that natural mineral waters and, by inference, groundwaters do not normally contain sufficient nutrients to sustain significant populations of *Aeromonas* to compete with the natural flora.

5.2.2 SURFACE FRESHWATERS

Surface waters are far more variable than groundwaters, with a wide range of trophic state, from oligotrophic to hypereutrophic, and are more readily affected by human activities. Recent interest in *Aeromonas* distribution in surface waters stems from the importance of these waters as sources of drinking water, as recreational amenities and as fisheries, linked with an increasing awareness of the potential pathogenicity of mesophilic aermonads.

The first study relating the occurrence of mesophilic aeromonads to environmental factors was that of Schubert[6], who reported that anaerogenic aeromonads (= *A. caviae*) were the dominant strains in polysaprobic rivers and waters receiving sewage effluent, whilst *A. hydrophila* was dominant in oligotrophic waters.

Hazen and colleagues[14] recovered aeromonads from a wide range of freshwater environments in the United States, from pristine alpine lakes to turbid ponds and rivers, with lotic habitats supporting larger populations than lentic. Using Rimler–Shotts medium, *A. hydrophila* was isolated from waters with temperatures of 4–45°C, with highest densities at 35°C, but could not be isolated at temperatures above 45°C. No relationship with pH was observed and *A. hydrophila* was isolated over the range pH 5.2–9.8, although laboratory studies indicated that it was unable to grow below pH 4.0 or above pH 10[14]. In lakes receiving heated waters, large populations of *A. hydrophila* (>60 000 CFU/100 ml) developed when the power plants were in operation when the upper water layer temperature reached 29.5°C[15,16]. *A. hydrophila* survived longer and in larger numbers in the hypolimnetic waters of these lakes. Rippey and Cabelli[17] demonstrated that growth of a strain of *A. hydrophila* originally isolated from a highly oligotrophic fresh water was limited when incubated at 22°C in untreated waters of various nutrient enrichments. Growth at this temperature in filtered-autoclaved water, however, was rapid with generation times ranging from 1.26 h in eutrophic waters to 12.8 h in the more oligotrophic waters.

Hazen[18] also reported a significant negative correlation between numbers of *A. hydrophila* and dissolved oxygen, while also gaining strong positive correlations with numbers of faecal coliforms and heterotrophic plate count bacteria. He was able to generate a predictive model for the abundance of *A. hydrophila* based upon six water quality parameters (temperature, dissolved oxygen, orthophosphate, total Kjeldahl nitrogen, ammonia and chlorophyll α)

The potential of using *A. hydrophila* populations for assessing trophic status of freshwater lakes was investigated by Rippey and Cabelli[19] using the Relative Trophic Index (RTI). They found that although the recoverable heterotrophic population did not correlate with the RTI, a good correlation was observed with the *A. hydrophila* portion of that population. The strongest correlations were obtained for total phosphorus, chlorophyll α and Secchi depth, but not dissolved oxygen, ammonia or orthophosphate. A strong seasonality of population numbers was evident, with early summer to mid-autumn maxima of $10^3/100$ ml declining to <10 in the winter. Further work[9] showed a highly significant correlation between thermotolerant aeromonads and the trophic state of freshwater in lakes in the north and north-eastern United States. In contrast, using the Trophic State Index developed by Walker[20], which is also based upon total phosphorus, chlorophyll α and Secchi depth, Rhodes and Kator[21] were unable to predict lake trophic status for several lakes in south-eastern United States, and did not observe any statistical correlation between mesophilic aeromonad populations and the parameters used to establish trophic status. They also reported inconsistent relationships with temperature and seasonality, with larger winter populations (10^2–$10^3/100$ ml) than were reported by Rippey and Cabelli[19].

Pathak and co-workers[22] reported *A. hydrophila* numbers in Indian rivers peaking in spring and autumn with a decline in summer, but this may in part be attributed to monsoon rains in July and August. Other studies of lowland European rivers have shown no detectable seasonal pattern[11,23]. Mesophilic *Aeromonas* numbers in the lower reaches of the river Severn, England, ranged from 8×10^2 to 3.9×10^5 CFU/100 ml (mean 6.3×10^4 CFU/100 ml) (Niccolls and Holmes, unpublished data). In two rivers in the Netherlands receiving wastewater discharges and agricultural runoff, the aeromonad populations were highly variable, with counts reaching 10^5 CFU/100 ml during late summer[11]. *A. caviae* and other unidentified anaerogenic aeromonads predominated. In these waters, therefore, mesophilic aeromonad numbers appear to be more dependent upon human activities.

These studies emphasize the difficulty of identifying single factors which impact upon aeromonad populations in freshwater. Aquatic environments are infinitely variable and the interrelationships are accordingly complex.

5.3   MARINE WATERS

The majority of studies of *Aeromonas* in saline waters have been initiated by
their importance as fish pathogens or as a potential pathogen of humans and
have, therefore, tended to focus on commercial and sport fisheries and
recreational waters. Consequently most have taken place in coastal or
estuarine waters. Mesophilic aeromonads can be classified as halotolerant,
being associated with discharges direct to the sea or via rivers and streams. In
their study, Hazen and co-authors[14] concluded that although *A. hydrophila*
was not generally considered to be a marine bacterium, it could be found
naturally in marine systems which interface with fresh water, and at all
salinities except the most extreme. In general, populations of saline waters
were higher (up to 9×10^5 cells/100 ml) than in freshwater habitats, but with
a much larger variation in numbers.

Several studies[24-26] have shown that aeromonad numbers normally exceed
those of faecal coliforms and may be found in areas free of faecal coliforms.
As with the freshwater studies, seasonal variations in abundance have been
noted, with a rise in spring and a decrease after autumn[26,27]. In the tidal
estuary of the Anacosta River, Washington DC, Seidler and co-workers[27]
recorded a peak of 3×10^4 CFU/100 ml in August, declining to a minimum by
February and rising again in May when the water temperature rose to 22.5°C.
Populations in the sediments fluctuated dramatically with change in
temperature, dropping by $10^3 - 10^4$-fold from an August maximum of $>10^5$/g.
Kaper and colleagues[26] noted a similar seasonal variation in planktonic *A.
hydrophila* populations in Cheasapeake Bay, but did not find a wide
fluctuation in sediment populations associated with the seasons. Correlations
with *A. hydrophila* numbers were demonstrated for temperature, total viable
counts and faecal coliforms and inversely for dissolved oxygen and salinity.
In a study of a coastal bay in Japan, Nakano and co-workers[28] found that *A.
caviae* was the most abundant of the mesophilic aeromonads, while *A. sobria*
predominated in brackish waters. *Aeromonas* numbers correlated with those
for faecal coliforms, but not with temperature. While *A. hydrophila* and *A.
sobria* maxima occurred during summer and autumn, *A. caviae* were seen in
high numbers during the winter (<13°C) season. Monfort and Baleux[4] also
reported *A. sobria* as the dominant aeromonad in a brackish water cove
receiving sewage effluent.

Laboratory studies of the survival of *A. hydrophila* or *A. sobria* in saline
waters by Araujo and colleagues[25], Monfort and Baleux[4], and Lowcock and
Edwards[29] have shown that when *Aeromonas* were inoculated into sterile
seawater there was at first an acute reduction in the numbers of culturable
cells, followed by an increase, the extent of which was dependent on the
concentration of organic matter present. Thereafter, there was a further
decline. Although Monfort and Baleux[4] monitored total cell numbers which
remained high during the decline in culturability, the viability of these cells

was not assessed. Evidence that species of *Aeromonas* may enter an ultramicrobacterial viable non-culturable state comes from the work of MacDonell and Hood[30] who filtered seawater through a $0.2 \mu m$ polycarbonate membrane, and then enriched the filtrate with dilute nutrient broth. After incubating this primary enrichment resuscitation culture for at least 21 days at 21°C or 35°C, aliquots were used as inocula for dilute nutrient agar plates. Some of the isolated strains which developed into normal-sized bacteria after nutrient conditioning were phenotypically identified as *Aeromonas*.

5.4 SEWAGE TREATMENT

Aeromonads are not considered to be normal inhabitants of the human gastrointestinal tract, although studies by Geldreich[31] demonstrated that approximately 1% of healthy adults carry them. This figure may increase to 20% in South East Asia where there is consumption of contaminated raw seafood coupled with poorer sanitation[32]. The presence of mesophilic *Aeromonas* in raw sewage and in treated effluents in numbers equivalent to those of faecal coliforms indicates that they are able to multiply in this environment.

As in freshwater and marine habitats there is a population seasonality of *Aeromonas* in sewage and waters receiving sewage effluents. In a study of sewage treatment ponds in southern France, Monfort and Baleux[33] found that numbers of mesophilic aeromonads rose in the summer months (mean pond temperatures of >20°C) to $10^5 - 10^6$ CFU/100 ml, and fell to 10^3 CFU/100 ml in the winter when generation times are greatly increased. This is in contrast to faecal coliforms which survive better in the colder winter temperatures than in summer when radiation and interactions with algae can have a negative effect. The *Aeromonas* populations in the ponds were dominated by *A. caviae* regardless of season, while in the outflow *A. caviae* predominated in the winter months only, and *A. sobria* was the dominant species from spring to autumn. *A. sobria* was present in very low numbers in raw sewage, and this rise in numbers indicates that *A. sobria* is adapted to warm water eutrophic habitats. The *A. hydrophila* component of the population was 0–15% in the influent waters and 4–15% in the effluents. Similar patterns were observed by Boussaid and colleagues[34] and Hassani and co-workers[35] for *Aeromonas* levels in stabilization ponds in Morocco, which peaked in October to April. They were dominated by *A. caviae*, with *A. sobria* predominating in the effluent regardless of season (representing 93–100% of the effluent aeromonad population). In these studies, however, the winter temperature did not fall below 10°C and this may have influenced effluent *Aeromonas* population patterns. This dominance by *A. sobria* under warm water conditions parallels the findings of Rhodes and Kator[21] for freshwater lakes, and Nakano and colleagues[28] for coastal marine waters.

A. caviae and *A. veronii* biotype *veronii* also predominated in the influent waters (70.3% and 16.8% of the aeromonad population respectively) and effluents (49.4% and 34.9% respectively) of 10 activated sludge plants in Italy during the summer months May to August[8]. *A. veronii* biotype sobria represented less than 5% of the effluent aeromonad population even though temperatures were 12–23°C.

Overall reduction in aeromonads by sewage treatment varies with the treatment process. Boussaid and co-workers[34] observed reductions of 93% in stabilization ponds and the data of Monfort and Baleux[33] show reductions of $10^1 - 10^2$ in summer and $10^3 - 10^4$ in winter. Reductions of *A. hydrophila* of 99.9% by an activated sludge plant and of 98% by a trickling filter plant in Belgium were observed by Poffe and Op de Beek[36]. A similar figure of 96.5% reduction of mesophilic aeromonads by activated sludge plants was reported by Stecchini and Domenis[8]. Despite these removal figures large numbers of mesophilic aeromonads are discharged into receiving waters. There has been little work on sewage sludges, but Poffe and Op de Beek[36] found that three-month old anaerobically (methane) fermented and partially dried sludge from trickling filters contained more than 10^6 *A. hydrophila* per gram of dried sludge.

Thus *A. caviae* is the dominant aeromonad in raw sewage and many effluents. Schubert[6] found that the anaerogenic aeromonads (*A. hydrophila* subsp. *anaerogenes* and *A. punctata* subsp. *caviae*) were more abundant in sewage-derived water than the aerogenic aeromonads. These anaerogenic strains correspond to *A. caviae* under the current classification. Ramteke and co-authors[37] related *A. caviae* numbers to sewage-polluted river and groundwater sources in India. This preponderance of *A. caviae* in sewage treatment and receiving waters has led to proposals for its use as an indicator of faecal pollution[37].

It has been noted that a higher percentage of *A. hydrophila* and *A. sobria* are toxigenic compared with *A. caviae*[33,38,39]. The species composition of the aeromonad population in sewage effluents may, therefore, be of particular interest where such effluents are used for irrigation of crops or are discharged into recreational waters.

5.5   POTABLE WATER TREATMENT AND DISTRIBUTION

The aim of water treatment is not to eliminate all microorganisms but to reduce the total bacterial burden and eliminate enteric pathogens. Potable water consequently can have a diverse flora. Traditionally, bacteriological monitoring of drinking water supplies has focused upon detection of *Escherichia coli* and the related coliform bacteria. Attention has recently been given to the occurrence of biofilms and associated organisms that occur in distribution systems, including the mesophilic species of *Aeromonas*.

Aeromonads may be seeded into drinking water supplies as a result of incomplete or ineffective water treatment. Large populations within distribution systems are often due to multiplication (often referred to as aftergrowth), particularly at the ends of networks supplied by treatment works abstracting from lowland surface water sources. Although bacteria including *Aeromonas*, can be isolated from the main body of water, the majority of bacteria within a distribution system are associated with biofilms on the surface of pipe walls[40]. The presence of a biofilm can itself lead to a range of problems, including taste and odour complaints and microbially enhanced corrosion. The production of extracellular products by the biofilm-associated microflora helps protect resident cells from disinfectants and provides a relatively nutrient-rich environment. Nutrient-stressed suspended bacteria also become more resistant to disinfectants due to changes reflected in cellular modifications and the synthesis of unique starvation/stress protein[3,41]. Inhibition of synthesis of these stress proteins resulted in reduction of resistance to disinfectants. It has been demonstrated that Gram-negative bacteria may survive in nutrient-poor aquatic environments by entering a quiescent 'viable non-culturable' state, whereby they become metabolically inert and more resistant to many toxic substances[2]. Although viable, these cells are typically difficult to culture, particularly on selective media. The majority of studies into the occurrence of the 'viable non-culturable' state have involved members of the Enterobacteriaceae and species of *Vibrio*; it is safe to assume that the mesophilic aeromonads will exhibit similar survival strategies.

The mesophilic aeromonads are nutritionally versatile, *A. hydrophila* being capable of utilizing a wide range of low-molecular-weight compounds (amino acids, carbohydrates and carboxylic acids), peptides and long-chain fatty acids when present in concentrations as low as $0.1\ \mu g$ of carbon per litre[42,43]. This versatility, particularly for biopolymers, makes these bacteria ideally suited for growth as components of biofilms in distribution systems.

5.5.1 *AEROMONAS* IN DRINKING WATER TREATMENT

As aeromonads can occur in large numbers in some water sources (particularly lowland rivers and reservoirs), and thus potentially enter distribution systems as a result of ineffective water treatment, an understanding of the fate of *Aeromonas* species through the treatment processes is needed. This area is only recently being addressed.

Meheus and Peeters[44] conducted a preliminary survey of the reduction of aeromonads at different stages of the treatment process of a treatment works in Belgium with cumulative reductions under summer and winter conditions being:

following flocculation/sedimentation	30 – 60%
following rapid sand filtration	70 – 90%
following granular activated carbon	80 – 90%
following hyperchlorination/direct filtration	99 – 100%
following slow sand filtration	98 – 100%

Neither the mode of cleaning nor the age of the filters appeared to influence the elimination of *Aeromonas* during slow sand filtration.

Havelaar and colleagues[11] found low numbers of mesophilic *Aeromonas* in the final waters of 20 Dutch plants treating surface and ground waters. A maximum count of 470 CFU/100 ml was recorded from a plant treating deep aerobic groundwater. High counts obtained at some works were often associated with poorly functioning filters, or those filter units which were operated intermittently to meet variable water demand.

We studied the impact of the type of material used for rapid gravity filters at a treatment works in central England, which takes water directly from the river Severn[23]. Raw water concentrations of *Aeromonas* ranged from 8×10^2 to 3.9×10^5 CFU/100 ml. Coagulation and clarification resulted in a mean reduction of 90% and aeromonads were undetectable after postclarifier chlorination (average level of free chlorine of 2.6 mg/1). Aeromonads were, however, recovered from the rapid gravity filters, indicating seeding with viable but non-culturable cells. There was a marked difference between sand-based rapid gravity filters and those employing granulated activated carbon (GAC) (Figure 5.1). *Aeromonas* were recovered on only one occasion from the sand filters as chlorine levels were maintained through the beds. However, chlorine was rapidly removed by the activated carbon filters, resulting in concentrations of less than 0.1 mg/1 in the effluent. *Aeromonas* were recovered throughout the year, with elevated numbers between July and September when water temperatures were highest (Figure 5.1). Samples taken five days a week from the final water resulted in four occasions where *Aeromonas* were recovered (maximum count of 1 CFU/100 ml).

In a study of five water treatment plants in Belgium, Huys and co-authors[12] and Kersters and colleagues[45] reported a mean reduction of 99.7% in aeromonad numbers following flocculation-decantation and chlorination. Slow sand filtration reduced aeromonad numbers by 98.9%. Rapid gravity sand filters also reduced *Aeromonas* numbers, but at one plant the sand filter effluent contained high numbers of aeromonads. Increased levels of *Aeromonas* were also obtained from the effluents of activated carbon filters. At four of the plants the treatment processes did not affect the species composition of the *Aeromonas* population. At the fifth, however, which treated surface water, there was a marked shift following slow sand filtration from dominance by *A. hydrophila* and *A. sobria* to a predominance of *A. caviae*[12].

Water treatment may, therefore, markedly reduce levels of *Aeromonas*, but these bacteria are capable of establishing significant populations in GAC-based treatment processes. Low numbers may be recovered from the final

(a)

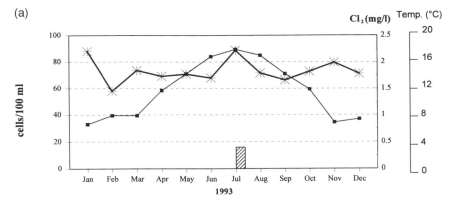

(b)

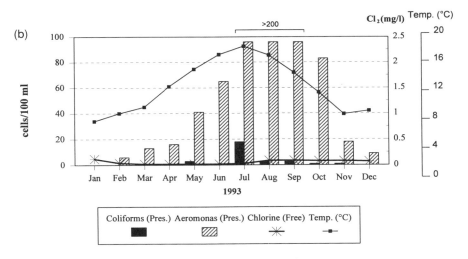

Figure 5.1. Presumptive coliforms and *Aeromonas* levels from (a) sand rapid gravity filters and (b) GAC rapid gravity filters from a treatment works abstracting from a lowland river[23]

waters of water treatment plants which meet water quality standards for the hygienic indicator organisms.

5.5.2   *AEROMONAS* IN DISTRIBUTION SYSTEMS

Aeromonads are readily isolated from municipal drinking water systems, sometimes at quite high levels[11,23,46,47]. A range of frequencies of *Aeromonas* occurrence has been reported from distribution system studies. A survey of parts of London and Essex, UK, gave isolation rates for *A. hydrophila* from chlorinated drinking water of 25% in summer and 7% in the winter, compared with rates of 82% in the summer and 75% in the winter for untreated water[48]. Knochel and Jeppesen[49] examined drinking water in Denmark and found only 28% of samples were positive, with counts of 1–40 CFU/100 ml, with *A. hydrophila* making up 97% of isolates. In contrast, Ghanem and colleagues[50] state that 90% of domestic water supplies in areas of Cairo contained *Aeromonas*, while from a survey of three distribution systems in Sweden, Krovacek and co-authors[46] reported that 85% of samples were positive for presumptive *Aeromonas* with a maximum count of 860 CFU/100 ml. *A. hydrophila* accounted for 67% of the strains isolated, the remainder being *A. sobria*. Stelzer and colleagues[47] recorded a maximum count of 240 *Aeromonas*/100 ml in a drinking water supply in Germany with an isolation frequency for *A. hydrophila* of 37% and for *A. sobria* of 57%. The higher counts were obtained from points furthest from the treatment works (>10 km).

Havelaar and co-workers[11] reported regrowth of aeromonads in 16 of 20 Dutch distribution systems examined. Geometric mean counts in the second half of 1986 varied between 1 and 440 CFU/100 ml, with a maximum count of 3300 CFU/100 ml being obtained from a river water sourced supply. Growth of *Aeromonas* generally occurred in the peripheral parts of distribution systems and was particularly associated with drinking water derived from anaerobic groundwaters containing methane. *Aeromonas* densities usually showed a seasonal pattern with peak values occurring in late summer, when temperatures were highest. There was no obvious relationship with total organic carbon (TOC) or heterotrophic plate counts. All three genospecies were recovered during the survey with either *A. hydrophila* or *A. caviae* tending to predominate, although *A. sobria* did predominate in one system, but generally this species tended to be recovered in low numbers only. LeChevallier and colleagues[51] reported that the occurrence of aeromonads in 27% of samples taken over an 18-month period for a chlorinated supply in Oregon, United States was due entirely to *A. sobria*. Regrowth problems due to blooms of single species are not uncommon in drinking water supplies, and have also been reported for coliforms[52].

Although it is accepted that species of mesophilic *Aeromonas* are commonly resident in drinking water distribution systems, there are few data on the factors affecting their occurrence. It is, however, generally reported that higher rates of isolation and larger populations occur during the warmer months and at the peripheries of distribution systems[11,47,48,51].

Burke and colleagues[10] studied the occurrence of *Aeromonas* species in the metropolitan water supply of Perth, Western Australia over a one-year period. Using stepwise multiple regression analysis a relationship was demonstrated for *Aeromonas* in chlorinated water with water temperature and residual chlorine. *Aeromonas* counts were allocated into six 'score' categories. A seasonal variation in mean *Aeromonas* counts closely paralleled mean water temperature in samples which were either unchlorinated or had free chlorine values consistently less than 0.3 mg/l. For the parts of the system where the free chlorine was generally greater than 0.3 ml/l the *Aeromonas* 'score' (*A*) could be predicted by the regression equation:

$$A = 7.29 + 0.28T - 17.6C + 0.76TC$$

where *T* = water temperature, *C* = free chlorine and *TC* reflects the interaction between temperature and free chlorine. Isolation of *Aeromonas* spp. from drinking water lacking chlorine was generally associated with water temperature greater than 14.5°C.

We studied a large supply system in central England where 22 locations were sampled over a two-year period[23,53]. A single treatment works supplies surface water-derived drinking water to two well defined distribution networks (Figure 5.2). Leg A is of short retention time and historically has had no aftergrowth problems. Several zones on Leg B have experienced aftergrowth problems; it is more complicated and has longer retention times. The age of water at the ends of the system may exceed 72 h. Using Ryan's medium[54] mesophilic *Aeromonas* populations were monitored along with those for coliforms and heterotrophic plate count bacteria, together with temperature, free chlorine and age of water (as estimated by a Severn Trent Water model). Populations of *Aeromonas* in the distal areas of the long Leg

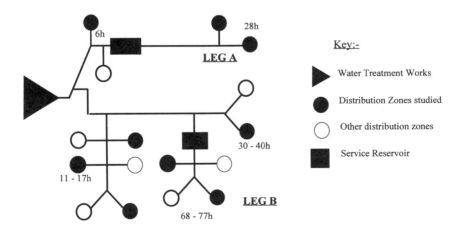

Figure 5.2. Schematic diagram of a water distribution system in central England for which a model for the occurrence of *Aeromonas* was derived[53]

B distribution network frequently exceeded 200 CFU/100 ml during the warm water seasons. The data were summarized into three-monthly seasonal means and statistically analysed using an iterative modelling process to establish category thresholds and derivation of a model using logistic regression analysis (generalized linear models[55]). No relationship between *Aeromonas* incidence and coliforms or heterotrophic plate counts was found. A model of best fit relating the probability of occurrence of *Aeromonas* to temperature, free chlorine and age of water was derived whereby:

$$\text{Logit } (p) = -5 - 1.7(\text{temperature category}) + 10.7(\text{chlorine category})$$
$$+ 4.4(\text{age of water category}) - 4.5(\text{chlorine category}$$
$$\times \text{ age of water category})$$

and the probability of *Aeromonas* being absent $(p) = \log(p/(1 - p))$,

for which the categories were:-

for water temperature: $1 = <10°C$, $2 = 10\text{–}14°C$ and $3 = >14°C$
for free chlorine: $1 = >0.1$ mg/1 and $2 = <0.1$ mg/1
for age of water: $1 = <20$ h, $2 = 21\text{–}34$ h and $3 = >35$ h.

Significant increases in risk of incidence of *Aeromonas* occur when the mean seasonal temperature exceeds 14°C and this is exacerbated where the mean free chlorine concentration falls below 0.1 mg/1, agreeing with the findings of Burke and colleagues[10]. The impact of age of water was only significant when the mean free chlorine was <0.1 mg/1.

From these studies it is apparent that water temperature and free chlorine are the principal factors so far recognized as significantly influencing the growth of *Aeromonas* in drinking water supplies. Given the nutrional versatility of these bacteria at low concentrations of organic compounds[42], however, the organic carbon content (assimilable organic carbon, AOC, or biodegradable organic carbon, BDOC) of the water may also have a significant role to play in their occurrence. This has yet to be studied within water supply distribution systems.

The mesophilic aeromonads have a wide temperature tolerance range, growing between 4°C and 42°C and surviving up to 55°C[56,57]. Hazen and colleagues[14] isolated *A. hydrophila* from aquatic habitats over a temperature range of 4–45°C. We isolated aeromonads from chlorinated drinking water during late winter/early spring when water temperatures were around 7°C[23]. Using the technique of Palumbo and colleagues[56], we tested drinking water isolates of *A. hydrophila*, *A. sobria* and *A. caviae* for growth at 4°C, 12°C and 21°C. All three strains were capable of growth at the lower temperatures, although the population maxima for *A. caviae* were significantly reduced (Figure 5.3). Majeed and co-workers[58] demonstrated that *A. hydrophila* and *A. sobria* were also able to produce enterotoxins and haemolysins at 5°C.

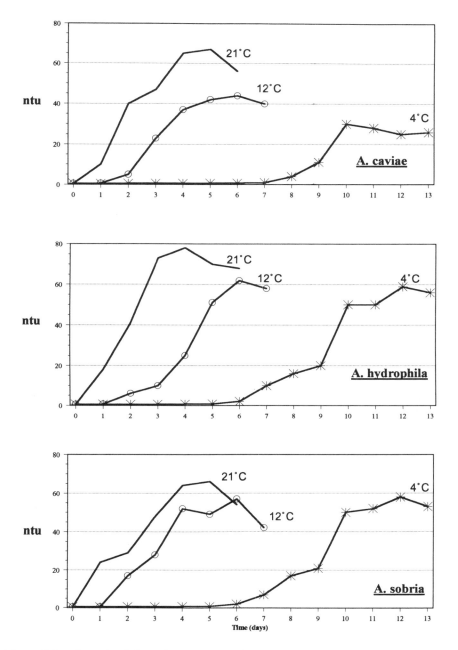

Figure 5.3.   Growth of three drinking water isolates of *Aeromonas* at 4°C, 12°C and 21°C determined by increase in turbidity in BHI broth[23]

Thus, although large populations in drinking water supplies tend to be associated with the warm water periods, mesophilic aeromonads are capable of growth within distribution systems throughout the year.

Control of *Aeromonas* in drinking water is typically achieved with increased disinfection and it appears that free cells of *Aeromonas* are relatively susceptible to the commonly employed chlorine-based disinfectants. Using the disc assay procedure of Ridgeway and Olsen[59] plus clinical and environmental strains of *A. hydrophila*, *A. sobria*, *A. caviae* and *A. veronii*, Knochel[60] found that the mesophilic aeromonads were generally more susceptible to chlorine and monochloramine than a control group consisting of Enterobacteriaceae and pseudomonads. Medema and colleagues[61] found that laboratory-grown and environmental *Aeromonas* were also susceptible to chlorine dioxide (≤ 0.2 mg/l). Despite this relative susceptibility of aeromonads to chlorine-based disinfectants it may take some considerable time and elevated chlorine concentrations (>0.2 mg/l) in the distribution system to control their numbers[62]. This is probably due to their association with biofilms. Mackerness and co-authors[63] found that *A. hydrophila* became readily established within a mixed heterotrophic bacterial biofilm and was unaffected by addition of 0.3 mg/l monochloramine. There was evidence that the biofilm-associated *A. hydrophila* would also survive 0.6 mg/l monochloramine, which was sufficient to eradicate biofilm-associated *E. coli*. Holmes and Niccolls[23] investigated biofilms from exhumed pipe lengths, employing the methods of LeChevallier and colleagues[64] and found that 30% contained aeromonads with an average population of 118 CFU/g wet weight. Following disinfection with 1 mg/l chlorine, *Aeromonas* could still be isolated from 10% of the pipe lengths, with an average population of 51 CFU/g wet weight. These data indicate that although free cells of *Aeromonas* may be relatively susceptible to disinfection, populations may survive high chlorine dosing when associated with biofilms.

5.5.3   HEALTH SIGNIFICANCE OF *AEROMONAS* IN DRINKING WATER

The health significance of detecting mesophilic aeromonads in public water supplies is not well understood as few outbreaks are documented and establishing epidemiological links is difficult. In May 1984 a sudden increase in numbers of aeromonads leaving a water treatment plant in The Hague, Netherlands was reported[65]. The strains isolated showed strong cytotoxic properties. There were also reports of increased *Aeromonas* occurrences from other parts of the Netherlands and, concomitantly, from Australia[10,39] where it was suggested that there may be a connection between cases of *Aeromonas*-associated diarrhoea and the numbers of *Aeromonas* in the drinking water. As a response, in 1985, the health authorities in the Netherlands defined 'indicative maximum values' for *Aeromonas* densities in drinking water. These values are currently 20 CFU/100 ml for water leaving the production plant and 200 CFU/100 ml for drinking water in distribution[65]; levels above these are considered to be indicative of treatment or distribution problems meriting investigation.

It is increasingly accepted that some strains of *Aeromonas* are enteropathogens possessing a range of virulence factors (enterotoxins, cytotoxins, haemolysins and invasive ability)[66] which can be expressed by both environmental and clinical strains of *Aeromonas*[67]. Enteropathogenic strains of *A. hydrophila, A. sobria* and *A. caviae* have all been isolated from cases of gastroenteritis[50,68].

Millership and co-authors[69] found that 28% of *Aeromonas* isolates from chlorinated and unchlorinated drinking water were cytotoxic, mainly *A. hydrophila*, of which 83% of isolates were positive, as was a single isolate of *A. sobria*. Conversely, none of the strains of *A. caviae*, which represented 50% of the isolates, exhibited cytotoxicity. During our studies[23], we found that 20% of *Aeromonas* isolates exhibited phenotypic characteristics associated with enterotoxicity. Of these, 75% were *A. hydrophila*, 14% were *A. sobria*, 9% were *A. caviae* and the remaining were *A. schubertii*. In contrast, Burke and colleagues[39] reported that 61% of aeromonads isolated from an unchlorinated municipal supply in Australia were enterotoxigenic, and 64% produced haemolysins. Krovacek and co-workers[46] found that in Swedish chlorinated and unchlorinated drinking water, 100% of *A. hydrophila* and 70% of *A. sobria* were haemolytic, but less than 30% of the isolates were shown to be enterotoxigenic. Kirov and colleagues[70] found that 53.6% of isolates of *A. hydrophila* hybridization group (HG) 1 and 55.9% of HG 3 isolates from water expressed two or more virulence factors.

Despite the association of virulence factors with drinking water aeromonads, there is increasing evidence that those strains isolated from the environment belong to different groups than those associated with gastroenteritis. Havelaar and co-workers[71] typed 187 *Aeromonas* isolates from human diarrhoeal stools and 263 from drinking water. There was little similarity between strains from the stools and those from drinking water. This was most pronounced for *A. caviae*, which was the dominant aeromonad in both sets of samples. For *A. hydrophila* prevalence may be related to hybridization groups. Both Kirov and colleagues[70] and Hanninen[72] found that HG 1 was associated with clinical specimens, while HG 3, and to a lesser extent HG 2, predominated in water and environmental samples. It appears that this may be reflected in the maximum growth temperature (t_{max}) of the homology groups. Hanninen and colleagues[73] have reported that those hybridization groups of *Aeromonas* associated with clinical samples (HGs 1, 4, 9/10 and 13) generally had a t_{max} of 40–44°C, while isolates from freshwater (HGs 3 and 11) had t_{max} values between 36.5°C and 37.5°C.

Claims that the drinking water supply is responsible for increased incidence of *Aeromonas*-associated gastroenteritis have been made. Ghanem and co-authors[50] considered that, since 90% of the domestic water supplies in Cairo were positive for aeromonads and 56% of isolates produced enterotoxins, the supply was a major source of *Aeromonas* infections. Investigating a case of long-term diarrhoea in an 18-month-old child, it was concluded[74] that the cause was *A. hydrophila* from a private unchlorinated well where counts ranging from 70 CFU/100 ml to 6.4×10^4 CFU/100 ml were obtained. The majority of isolates were enterotoxin producers. In studies of chlorinated and unchlorinated municipal water supplies in Western Australia, Burke and co-authors[10,79] demonstrated a relationship between aeromonad occurrence and seasonal incidence of *Aeromonas*-associated gastroenteritis, where faecal isolates closely followed the distribution pattern of *Aeromonas* spp. in the drinking water.

Although these reports indicate a possible relationship between *Aeromonas* in drinking water and increased incidence of aeromonad-related illness, the evidence is tenuous. In the one case where serotyping of faecal and water isolates was compared, the two groups were not related[75]. Following a number of cases of diarrhoea in children within a small community water supply in Iowa, where aeromonads had been isolated from water treatment and distribution samples, without any evidence of treatment problems and with maintenance of effective chlorine residuals (0.3 mg/l), ribotyping of stool and drinking water isolates showed that those from faeces were of different serogroups than those from the drinking water[75]. The isolates from the water treatment works and distribution all belonged to the same serotype. The water supply, therefore, could not be implicated as the source of the infections.

Thus, although aeromonads are frequently isolated from drinking water systems, and some strains may exhibit enterotoxic properties, further epidemiological studies are required to ascertain whether there is any relationship between cases of *Aeromonas*-associated diarrhoea and presence of these organisms in drinking water. To date there is no firm evidence that transmission occurs via this route, but in the absence of more definitive proof of their public health significance it would be advisable to control numbers of aeromonads in drinking water supplies, primarily by limiting regrowth possibilities within distribution systems.

5.5.4   *AEROMONAS* IN BOTTLED MINERAL WATERS

Bottled natural mineral waters are increasingly consumed in the developed world and in some areas can be the principal source of drinking water. There is, however, limited information on the occurrence of *Aeromonas* in these waters. Havelaar and colleagues[13] did not recover aeromonads from samples of 64 European still mineral waters, concluding that species of *Aeromonas* could not grow to any significant extent in these nutrient-poor waters, and are not able to compete with the autochthonous flora. Hunter and Burge[76] similarly failed to isolate *Aeromonas* from 58 bottles of mineral water. Although Slade and co-authors[77] isolated *A. hydrophila* from 41 of 95 (43%) bottles of mineral water in Saudi Arabia, these were from a single Lebanese brand.

Warburton and colleagues[78] failed to recover *A. hydrophila* from 410 samples of Canadian bottled waters, even after 30 days storage at room temperature. Further studies, however, showed that this organism could survive and proliferate to levels in excess of 10^5 CFU/100 ml in water stored at room temperature[79]. These waters were, presumably, relatively nutrient-rich. Based upon these findings and the opportunistic pathogenicity of *A. hydrophila*, a standard of 0 CFU/100 ml in bottled waters has been proposed for this organism[79]. In contrast, Havelaar and co-workers[13] felt that routine monitoring for *Aeromonas* in natural mineral waters could not be recommended, but that aeromonads should be investigated when examining new sources.

5.6 CONCLUSIONS

The mesophilic species of *Aeromonas* are ubiquitous in surface fresh and marine waters, but it appears that they are not normally part of the autochthonous bacterial population of groundwaters. Despite a nutritional versatility these aeromonads do not appear to compete well in low-nutrient waters and, thus, they are generally either not recovered or recovered in low numbers only. This may be a reflection on the isolation media currently employed for the recovery of the mesophilic aeromonads, which tend to be nutrient-rich and contain relatively high concentrations of selective agents. There is evidence that, like many other Gram-negative bacteria, species of *Aeromonas* also enter a 'viable but non-culturable' state. This needs to be investigated further. In nutrient-enriched waters *Aeromonas* spp. can attain very large populations, particularly at higher temperatures. Sewage effluents are also a major allochthonous source of the mesophilic aeromonads in the aquatic environment.

Species of *Aeromonas* have been shown to be associated with gastroenteritis and with wound infections acquired via contaminated water. Consequently there is increasing interest in the significance of their occurrence in drinking water supplies. *Aeromonas* can be readily isolated from drinking water distribution systems, sometimes in relatively high numbers, particularly from lowland river water-derived supplies during warm water periods. They appear to survive well when associated with biofilm development in pipes; chlorine concentration, temperature and nutrient status are the principal controlling factors so far identified. Within water distribution systems mesophilic aeromonads are able to proliferate at low temperatures (4°C).

There is increasing evidence that those species of *Aeromonas* associated with clinical infections are generally from differing hybridization groups than those isolated from drinking water and environmental sources, although the environmental strains can still express a number of virulence factors. Given the claims that water supplies may be associated with increased incidence of aeromonad-related gastroenteritis, the clinical significance of those hybridization groups isolated from water and the environment needs clarification.

5.7 ACKNOWLEDGEMENTS

This article is published with the permission of Severn Trent Water Ltd; the views expressed are those of the authors and do not necessarily reflect those of the Company.

5.8 REFERENCES

1   Zimmerman OER. Die Bakterien unsere Trink- und Nutz-wasser inbesondere des Wassers der Chemnitzer Wasserleitung. I. Reihe. *Elfter Bericht der Naturwissenschaftlichen zu Chemnitz* 1890; **1**: 38–9.
2   Roszak DB, Colwell RR. Survival strategies of bacteria in the natural environment. *Microbiol Rev* 1987; **51**: 365–79.
3   Martin A, Harakeh S. Effect of starvation on bacterial resistance to disinfectants. In McFeters GA (ed) *Drinking Water Microbiology*. New York: Springer-Verlag, 1990: 88–103.
4   Monfort P, Baleux B. Distribution and survival of motile *Aeromonas* spp. in brackish water receiving sewage treatment effluent. *Appl Environ Microbiol* 1991; **57**: 2459–67.
5   Semel JD, Trenholme G. *Aeromonas hydrophila* water-associated traumatic wound infection: A review. *J Trauma* 1990; **30**: 324–7.
6   Schubert RHW. The relationship of aerogenic to anaerogenic aeromonads of the 'hydrophila-punctata-group' in river water depending on the load of waste. *Zentralbl Bakteriol Hyg 1 Abt Orig B* 1975; **160**: 237–45.
7   Araujo RM, Arribas RM, Pares R. Distribution of *Aeromonas* species in waters with different levels of pollution. *J Appl Bacteriol* 1991; **71**: 182–6.
8   Stecchini ML, Domenis C. Incidence of *Aeromonas* species in influent and effluent of urban wastewater purification plants. *Lett Appl Microbiol* 1994; **19**: 237–9.
9   Rippey SR, Cabelli VJ. Use of thermotolerant *Aeromonas* group for the trophic classification of freshwaters. *Water Res* 1989; **23**: 1107–14.
10 Burke V, Robinson J, Gracey M, Peterson D, Partridge K. Isolation of *Aeromonas hydrophila* from a metropolitan water supply: Seasonal correlation with clinical isolates. *Appl Environ Microbiol* 1984; **48**: 361–6.
11 Havelaar AH, Versteegh JFM, During M. The presence of *Aeromonas* in drinking water supplies in the Netherlands. *Zentralbl Hyg* 1990; **190**: 236–56.
12 Huys G, Kersters I, Vancanneyt M, Coopman R, Janssen P, Kersters K. Diversity of *Aeromonas* sp. in Flemish drinking water production plants as determined by

gas–liquid chromatographic analysis of cellular fatty acid methyl esters (FAMEs). *J Appl Bacteriol* 1995; **78**: 445–55.

13 Havelaar AH, Toorop-Bouma A, Medema G. The occurrence and significance of *Aeromonas* in water with special reference to mineral water. *Riv Ital Igiene* 1990; **50**: 349–56.

14 Hazen TC, Fliermans CB, Hirsch RP, Esch GW. Prevalence and distribution of *Aeromonas hydrophila* in the United States. *Appl Environ Microbiol* 1978; **36**: 731–8.

15 Fliermans CB, Gorden RW, Hazen TC, Esch GW. *Aeromonas* distribution and survival in a thermally altered lake. *Appl Environ Microbiol* 1977; **33**: 114–22.

16 Hazen TC, Fliermans CB. Distribution of *Aeromonas hydrophila* in natural and man-made thermal effluents. *Appl Environ Microbiol* 1979; **38**: 166–8.

17 Rippey SR, Cabelli VJ. Growth of characteristics of *Aeromonas hydrophila* in limnetic waters of varying trophic state. *Arch Hydrobiol* 1985; **104**: 311–19.

18 Hazen TC. A model for the density of *Aeromonas hydrophila* in Albermarle Sound, North Carolina. *Microb Ecol* 1983; **9**: 137–53.

19 Rippey SR, Cabelli VJ. Occurrence of *Aeromonas hydrophila* in limnetic environments: relationship of the organism to trophic state. *Microb Ecol* 1980; **6**: 45–54.

20 Walker WW. Use of hypolimnetic oxygen depletion rate as a trophic state index for lakes. *Water Res Res* 1979; **15**: 463–70.

21 Rhodes MW, Kator H. Seasonal occurrence of mesophilic *Aeromonas* spp. as a function of biotype and water quality in temperate freshwater lakes. *Water Res* 1994; **28**: 2241–51.

22 Pathak SP, Bhattacherjee JW, Kalra N, Chandra S. Seasonal distribution of *Aeromonas hydrophila* in river water and isolation from river fish. *J. Appl Bacteriol* 1988; **65**: 347–52.

23 Holmes P, Niccolls LN. Aeromonads in drinking water supplies—their occurrence and significance. *J Chartered Inst Water Environ Manage* 1995; **5**: 464–9.

24 Alonso JL, Amoros I, Botella MS. Enumeration of motile *Aeromonas* in Valencia coastal waters by membrane filtration. *Water Sci Technol* 1991; **24**: 125–8.

25 Araujo RM, Pares R, Lucena F. The effect of terrestrial effluents on the incidence of *Aeromonas* spp. in coastal waters. *J Appl Bacteriol* 1990; **69**: 439–44.

26 Kaper JB, Lockman H, Colwell RR, Joseph SW. *Aeromonas hydrophila*: ecology and toxigenicity of isolates from an estuary. *J Appl Bacteriol* 1981; **50**: 359–77.

27 Seidler RJ, Allen DA, Lockman H, Colwell RR, Joseph SW, Daily OP. Isolation, enumeration and characterization of *Aeromonas* from polluted waters encountered in diving operations. *Appl Environ Microbiol* 1980; **39**: 1010–18.

28 Nakano H, Kameyama T, Venkateswaran K, Kawakami H, Hashimoto H. Distribution and characterization of hemolytic, and enteropathogenic motile *Aeromonas* in aquatic environment. *Microbiol Immunol* 1990; **34**: 447–58.

29 Lowcock D, Edwards C. Survival of genetically-marked *Aeromonas hydrophila* in water. *Lett Appl Microbiol* 1994; **19**: 121–3.

30 MacDonell MT, Hood MA. Isolation and characterisation of ultramicrobacteria from a Gulf Coast estuary. *Appl Environ Microbiol* 1982; **43**: 566–71.

31 Geldreich EE. Bacterial populations and indicator concepts in faeces, sewage, stormwater and solid wastes. In Berg G (ed) *Indicators of Viruses in Water and Food*. Ann Arbor: Ann Arbor Science, 1978: 51–99.

32 Sack DA, Chowdhury KA, Hug A, Kay BA, Sayeed S. Epidemiology of *Aeromonas* and *Plesiomonas* diarrhoea. *J Diarrh Dis Res* 1988; **6**: 107–12.

33 Monfort P, Baleux B. Dynamics of *Aeromonas hydrophila, Aeromonas sobria*

and *Aeromonas caviae* in a sewage treatment pond. *Appl Environ Microbiol* 1990; **56**: 1999–2006.

34   Boussaid A, Baleux B, Hassani L, Lesne J. *Aeromonas* species in stabilization ponds in the arid region of Marrakesh, Morocco, and relation to fecal-pollution and climatic factors. *Microb Ecol* 1991; **21**: 11–20.

35   Hassani L, Imziln B, Boussaid A, Gauthier M.J. Seasonal incidence of and antibiotic resistance among *Aeromonas* species isolated from domestic wastewater before and after treatment in stabilization ponds. *Microb Ecol* 1992; **23**: 227–37.

36   Poffe R, Op de Beeck E. Enumeration of *Aeromonas hydrophila* from domestic wastewater treatment plants and surface waters. *J Appl Bacteriol* 1991; **71**: 366–70.

37   Ramteke PW, Pathak SP, Gautam AR, Bhattacherjee JW. Association of *Aeromonas caviae* with sewage pollution. *J Environ Sci Health* 1993; **A28**: 859–70.

38   Ashbolt NJ, Ball A, Dorsch M, Turner C, Cox P, Chapman A, Kirov SM. The identification and human health significance of environmental aeromonads. *Water Sci Technol* 1995; **31**: 263–9.

39   Burke V, Robinson J, Gracey M, Peterson D, Meyer N, Haley V. Isolation of *Aeromonas* spp. from an unchlorinated domestic water supply. *Appl Environ Microbiol* 1984; **48**: 367–70.

40   Block JC. Biofilms in drinking water distribution systems. In Bott TR, Melo, L, Fletcher M, Capdeville B (eds) *Biofilms—Science and Technology*. Dordrecht: Kluwer, 1992: 469–85.

41   Berg JD, Martin A, Roberts PV. Growth of disinfectant resistant bacteria and simulation of natural aquatic environments in the chemostat. In Jolley RL (ed) *Water Chlorination: Environmental Impact and Health Effects*. Ann Arbor: Ann Arbor Science, 1981: 219–43.

42   van der Kooij D. Nutritional requirements of aeromonads and their multiplication in drinking water. *Experientia* 1991; **47**: 444–6.

43   van der Kooij D, Hijnen WAM. Nutritional versatility and growth kinetics of an *Aeromonas hydrophila* strain isolated from drinking water. *Appl Environ Microbiol* 1988; **54**: 2842–51.

44   Meheus J, Peeters P. Preventive and corrective actions to cope with *Aeromonas* growth in water treatment. *Water Supply* 1989; **7**: 10-1–10-4.

45   Kersters I, Huys G, Janssen P, Kersters K, Verstraete W. Influence of temperature and process technology on the occurrence of *Aeromonas* sp. and hygienic indicator organisms in drinking water production plants. Presented at the Fifth International *Aeromonas—Plesiomonas* Symposium, Edinburgh, Scotland, April 1995.

46   Krovacek K, Faris A, Baloda SB, Lindberg T, Peterz M, Mansson I. Isolation and virulence profiles of *Aeromonas* spp. from different municipal drinking water supplies in Sweden. *Food Microbiol* 1992; **9**: 215–22.

47   Stelzer W, Jacob J, Feuerpfeil J, Schulze E. A study of the prevalence of aeromonads in a drinking water supply. *Zentralbl Mikrobiol* 1992; **147**: 231–5.

48   Millership SE, Chattopadhyay B. *Aeromonas hydrophila* in chlorinated water supplies. *J Hosp Infect* 1985; **6**: 75–80.

49   Knochel S, Jeppesen C. Distribution and characteristics of *Aeromonas* in food and drinking water in Denmark. *In J Food Microbiol* 1990; **10**: 317–22.

50   Ghanem EH, Mussa ME, Eraki HM. *Aeromonas*-associated gastro-enteritis in Egypt. *Zentralbl Mikrobiol* 1993; **148**: 441–7.

51   LeChevallier MW, Evans TM, Seidler RJ, Daily OP, Merrell BR, Rollins DM, Joseph SW. *Aeromonas sobria* in chlorinated drinking water supplies. *Microb Ecol* 1982; **8**: 325–33.

52 Edberg SC, Patterson JE, Smith DB. Differentiation of distribution systems, source water, and clinical coliforms by DNA analysis. *J Clin Microbiol* 1994; **32**: 139–42.

53 Holmes P, Niccolls LN, Sartory DP. Factors influencing the incidence of *Aeromonas* during water treatment and distribution. Presented at the 5th International *Aeromonas–Plesiomonas* Symposium, Edinburgh, Scotland, April 1995.

54 Holmes P, Sartory DP. An evaluation of media for the membrane filtration enumeration of *Aeromonas* from drinking water. *Lett Appl Microbiol* 1993; **17**: 58–60.

55 McCullagh P, Nelder JA. *Generalised Linear Models*, (2nd edn). London: Chapman and Hall, 1989.

56 Palumbo SA, Morgan DR, Buchanan RL. Influence of temperature, NaCl and pH on the growth of *Aeromonas hydrophila*. *J Food Sci* 1985; **50**: 1417–21.

57 Rouf MA, Rigney MM. Growth temperatures and temperature characteristics of *Aeromonas*. *Appl Microbiol* 1971; **22**: 503–6.

58 Majeed KN, Egan AF, Mac Rae IC. Production of exotoxins by *Aeromonas* spp. at 5°C. *J Appl Bacteriol* 1990; **69**: 332–7.

59 Ridgeway HF, Olsen BH. Chlorine resistance patterns of bacteria from two drinking water distribution systems. *Appl Environ Microbiol* 1982; **44**: 972–87.

60 Knochel S. Chlorine resistance of motile *Aeromonas* spp. *Water Sci Technol* 1991; **24**: 327–30.

61 Medema GJ, Wondergem E, van Dijk-Looyaard AM, Havelaar AH. Effectivity of chlorine dioxide to control *Aeromonas* in drinking water distribution systems. *Water Sci Technol* 1991; **24**: 325–6.

62 Edge JC, Finch PE. Observation on bacterial aftergrowth in water supply distribution systems: implications for disinfection strategies. *J Inst Water Environ Manage* 1987; **1**: 104–10.

63 Mackerness CW, Colbourne JS, Keevil CW. Growth of *Aeromonas hydrophila* and *Escherichia coli* in a distribution system biofilm model. *Proc UK Symp Health-Related Water Microbiology*. London, IAWPRC, 1991: 131–8.

64 LeChevallier MW, Babcock TM, Lee RG. Examination and characterisation of distribution system biofilms. Appl Environ Microbiol 1987; **53**: 2714–24.

65 van der Kooij D. Properties of aeromonads and their occurrence and hygienic significance in drinking water. *Zentralbl Bakt Hyg* 1988; **187**: 1–17.

66 Kirov SM. The public health significance of *Aeromonas* spp. in foods. *Int J Food Microbiol* 1993; **20**: 179–98.

67 Cahill MM. Virulence factors in motile *Aeromonas* species. *J Appl Bacteriol* 1990; **69**: 1–16.

68 Deodhar LP, Saraswathi K, Vardukar A. *Aeromonas* spp. and their association with human diarrheal disease. *J Clin Microbiol* 1991; **29**: 853–6.

69 Millership SE, Barer MR, Tabaqchali S. Toxin production by *Aeromonas* spp. from different sources. *Med Microbiol* 1986; **22**: 311–4.

70 Kirov SM, Hudson JA, Hayward LJ, Mott SJ. Distribution of *Aeromonas hydrophila* hybridization groups and their virulence properties in Australian clinical and environmental strains. *Lett Appl Microbiol* 1994; **18**: 71–3.

71 Havelaar AH, Schets FM, van Silfhout A, Jansen WH, Wieten G, van der Kooij D. Typing of *Aeromonas* strains from patients with diarrhoea and from drinking water. *J Appl Bacteriol* 1992; **72**: 435–44.

72 Hanninen ML. Phenotypic characteristics of the three hybridization groups of *Aeromonas hydrophila* complex isolated from different sources. *J Appl Bacteriol* 1994; **76**: 455–6.

73 Hanninen ML, Salmi S, Siitonen A. Maximum growth temperature ranges of

Aeromonas spp. isolated from clinical or environmental sources. *Microb Ecol* 1995; **29**: 259–67.

74 Krovacek K, Peterz M. Enterotoxigenicity and drug sensitivity of *Aeromonas hydrophila* isolated from well water in Sweden: a case study. *Int J Food Microbiol* 1989; **8**: 149–54.

75 Moyer NP, Luccini GM, Holcomb LA, Hall NH, Altwegg M. Application of ribotyping for differentiating aeromonads isolated from clinical and environmental sources. *Appl Environ Microbiol* 1992; **58**: 1940–4.

76 Hunter PR, Burge SH. The bacteriological quality of bottled natural mineral waters. *Epidemiol Infect* 1987; **99**: 439–43.

77 Slade PJ, Falah MA, Al-Ghady AM. Isolation of *Aeromonas hydrophila* from bottled waters and domestic water supplies in Saudi Arabia. *J Food Protect* 1986; **49**: 471–76.

78 Warburton DW, Dodds KL, Burke R, Johnston MA, Laffey PJ. A review of the microbiological quality of bottled water sold in Canada between 1981 and 1989. *Can J Microbiol* 1992; **38**: 12–19.

79 Warburton DW, McCormick JK, Bowen B. Survival and recovery of *Aeromonas hydrophila* in water: development and methodology for testing bottled water in Canada. *Can J Microbiol* 1994; **40**: 145–8.

6 Human Pathogens

J. MICHAEL JANDA AND SHARON L. ABBOTT
California Department of Health Services, Berkeley, USA

6.1 INTRODUCTION

While *Aeromonas* is undoubtedly more commonly isolated from patients with gastroenteritis, reports of *Aeromonas* sepsis, wound and ocular infections, peritonitis and meningitis, as well as a number of miscellaneous infections have appeared increasingly in the literature. Unlike gastroenteritis, these infections are often reported to have fatal or serious debilitating outcomes, such as amputation. Also notable and at variance with diarrhoeal disease is the narrower spectrum of *Aeromonas* species encountered. Although *A. hydrophila* and *A. veronii* biotype sobria accounted for the majority of cases of sepsis and wounds, *A. jandaei, A. trota, A. veronii* biotype veronii and *A. schubertii* were also isolated whereas *A. caviae*, a rather common cause of milder, self-limiting cases of gastroenteritis, was rarely seen. Another striking point of difference was age, gastroenteritis is generally associated with young children; however, the overwhelming number of cases of miscellaneous infections involved adults.

Although there are exceptions, as will be noted in the following pages, *Aeromonas* sepsis generally arises secondary to gastroenteritis or wound infections and is associated with high mortality rates. As for wound, ocular and other infections the portal of entry is usually a penetrating or traumatic injury that is environmentally contaminated. It will be seen that underlying illnesses, in particular cirrhosis, or immunosuppressive states play major roles in the acquisition and outcome of these infections as well.

6.2 SEPTICAEMIA

By far the most serious life-threatening complication of systemic infection with *Aeromonas* is septicaemia. Generally considered an uncommon illness, various surveys have reported an incidence of *Aeromonas* bacteraemia ranging from a low of <0.15–0.3% to a high of 25%[1,2]. In the United States, the overall frequency of *Aeromonas* sepsis appears to be very low (<1%), while several studies from Southeast Asia have reported significantly higher rates. Lee and colleagues[3] monitored the number of *Aeromonas* bacteraemias at the Chinese University Hospital over a four-year period

The Genus Aeromonas. Edited by B. Austin, M. Altwegg, P.J. Gosling and S. Joseph
© 1996 John Wiley & Sons Ltd

(1981–1984). Aeromonads accounted for 2.3% of all bacteraemias, with little variation in the annual incidence of *Aeromonas* sepsis (2.0–3.0%) during this interval. A more recent investigation at the Prince of Wales Hospital in Hong Kong between 1984 and 1989 found a 1.8% rate, a value comparable to that found in the earlier Chinese study[4]. In Western Australia, *Aeromonas* was the single most common cause of bacteraemia in one medical institution over a 16-month period[1]. The sizeable difference (10- to 20-fold) in the relative frequency of *Aeromonas* sepsis noted in various geographical regions of the world may be attributed to a number of factors, including ethnicity, environmental exposure, climatic conditions, dietary habits, and access to adequate health care-related programmes and facilities.

Persons developing *Aeromonas* bacteraemia typically present with classic signs of sepsis associated with other Gram-negative bacillary pathogens. Common symptoms include fever (94–95%) and chills (70%)[3,5]. Patients who become septic with *Aeromonas* often exhibit signs of gastrointestinal (GI) involvement, including abdominal pain, nausea, vomiting and diarrhoea[3,6]. While previous GI infection is not surprising, the relatively high frequency with which respiratory symptoms accompany *Aeromonas* bacteraemia is. Lee and co-authors[3] found that 15% of all patients presenting with *Aeromonas* bacteraemia developed cough, dyspnoea, or rhinorrhoea associated with the septicaemic episode. In another study, three of 17 cancer patients developed pulmonary infiltrates during their bacteraemic bouts although respiratory cultures (e.g. sputum) failed to yield *Aeromonas*[5]. A similar observation was made by Picard and colleagues[7] in which seven of 15 (47%) individuals with positive blood cultures developed bilateral infiltrates, some accompanied by multiple abscesses or haemorrhaging. The significance of these observations is not known. Hypotension (18–32%) is a less common finding[3,5]. Sometimes patients present with unusual clinical manifestations including rhabdomyolysis, cerebellar haematoma and acute renal failure where aeromonads are unlikely to be suspected as the cause of septicaemia[4].

Most cases of *Aeromonas*-associated bacteraemia can be roughly categorized into one of four broad categories (Table 6.1). These categories are primarily defined on the basis of patient demographics, the host's physiological status, and accompanying aeromonad-associated syndromes. More than 90% of all cases of *Aeromonas* septicaemia fall into group 1, and these are generally associated with immunologically impaired adults that have one of several underlying medical conditions including cancer; most but not all of these are bloodstream infections arising from prior GI tract colonization/infection with subsequent invasion. Janda and Brenden[8] found that 69% of all patients with positive blood cultures for *Aeromonas* had diarrhoeal symptoms immediately preceding their septic episode. Dryden and Munro[6] found 54% of patients to simultaneously present with signs of sepsis and gastroenteritis in a survey of 13 patients with *Aeromonas* septicaemia; one of these seven patients was positive for *Aeromonas* in their

Table 6.1.   Major categories of *Aeromonas* septicaemia

Group	Age	Portal of entry	Immuno-compro-mised*	Under-lying disease	Fresh-water exposure	Associated syndromes	Mor-tality rate (%)
I	Adults	GI	Yes	Yes	Un-common	Occasional	30–50
II	Infants <2 years	GI	Yes	Yes	Un-common	Meningitis	37
	Adults >34 years						
III	All ages	Wound	No	Common	Common	Myo-necrosis	>85
IV	Adults	Unknown	No	No	Unknown	None	75

GI, gastrointestinal tract
*for a majority of cases

faeces during the bacteraemic phase. Group II comprises individuals that not only develop septicaemia but subsequently exhibit clinical manifestations of CNS disease (meningitis) via extension of the infectious process across the blood–brain barrier. Despite the fact that the blood of 25% of these individuals is culture-negative for aeromonads, the eventual progression to meningitis implies a haematogenous dissemination[9]. A third highly specialized group of patients are those that develop *Aeromonas* septicaemia in conjunction with infection of soft tissues, most commonly of the lower extremities. This group is particularly significant in light of the high mortality rate associated with it[10]. The final group of persons presenting with *Aeromonas* sepsis are those with no detectable immunological deficiency or underlying medical/physiological disorder. This group is exceedingly rare, and there are few case reports in the literature documenting such illnesses. Some investigators suspect that such a category does not truly exist and that an undetected imbalance or dysfunction exists that predisposes such people to septicaemia.

There have been four major studies published on *Aeromonas* septicaemia since 1986, and the collective results of these investigations are listed in Table 6.2. Surprisingly, despite the fact that these studies were conducted in different geographical regions of the world, their overall results are remarkably similar. *Aeromonas* bacteraemia is most often seen in older males with classic signs of sepsis. The overwhelming predominance of *Aeromonas* septicaemia in men may be a reflection of increased environmental exposure to aeromonads during occupational or recreational

Table 6.2 Major surveys of *Aeromonas*-associated septicaemia

Reference	Population	Patients (no.)	Age (years) (mean)	M/F ratio	Bacteraemia		Acquisition (%)		Clinical Presentation (%)				Underlying Disease (%)			Mortality (%)
					Mono-microbic	Poly-microbic	Commu-nity	Nosoco-mial	Sepsis	Hepato-biliary disease	Periton-itis	Cellulitis/ Myo-necrosis	Malig-nancy	Hepato-biliary Disease	Other	
Duthie et al.[4]	Hospital	40	64	1.7	65	35	68	32	45	18	10	5	35	60	5	27.5
Janda et al.[2]	Multi-centre	53	58.2	1.6	72	28	NR	NR	69	11	3	11	53	22	25	29
Dryden and Munro[6]	Hosp-ital	13	65.8	1.6	69	31	62	38	69	0	0	0	54	23	23	46
Lee et al.[3]	Hosp-ital	38	34	2.2	75	25	50	50	70	10	8	10	32	42	25	32

NR, not reported

activities, and to the higher frequency at which liver disease, a predisposing condition to *Aeromonas* infection, is seen in men. There are, however, no published validation studies to confirm this hypothesis.

In addition to classic signs of sepsis, patients bacteraemic with aeromonads can present with one of several other clinical conditions. Probably the most common of these syndromes concerns infections/obstructions of the biliary tract. These biliary tract complications usually involve inflammation of the gallbladder (cholecystitis), the ducts (cholangitis) and/or formation of stones. Duthie and colleagues[4] found that almost half of all biliary tract infections associated with *Aeromonas* sepsis involved recurrent pyogenic cholangitis. This condition, which involves acalculous formation, subsequent obstruction of the biliary tract and later bacterial contamination of bile with secondary bacteraemia, seems to be uniquely associated with people inhabiting Southeast Asian countries. Another common manifestation of *Aeromonas* sepsis is peritonitis accompanied by complaints of abdominal pain and fever with occasional references to diarrhoea and vomiting[11]. In a recent review of cases of *Aeromonas*-associated peritonitis, more than 70% of all individuals had positive blood cultures[12], the overall fatality rate for this group was 65%. Less frequently encountered presentations include pancreatitis, fever of unknown origin, GI tract complaints, and empyema. Some of these less common presentations, such as pancreatitis, may be an indirect manifestation of other underlying medical problems (alcoholism). A less common manifestation of sepsis involves cutaneous lesions of the skin. These may present as typical cases of cellulitis or as ecthyma gangrenosa lesions (10–15% of patients) secondary to unrecognized signs of overt sepsis. Most patients that develop *Aeromonas* sepsis are immunocompromised with either cancer or hepatobiliary complication (75–95% of patients) as the major underlying disease[13–15]. Malignancies commonly associated with *Aeromonas* septicaemia involve haematological neoplasias such as leukaemia and lymphoma (Table 6.3). However, a large number of other malignant processes have been linked to cases of sepsis, these include solid tumours of the GI tract, pancreas, lung, liver and bladder[2,3]. The other common theme associated with a predisposition to developing *Aeromonas* sepsis are abnormalities of the hepatobiliary tree, particularly the liver. In the ground-breaking report and review of aeromonad infections by von Graevenitz and Mensch[16], one of two bacteraemic patients with *Aeromonas* infection suffered from Laennec's cirrhosis, a condition characterized by liver enlargement with necrosis and inflammation associated with acute alcohol injury. This observation has been repeatedly verified in numerous case studies and epidemiological investigations on underlying diseases linked to *Aeromonas* sepsis. Often patients presenting with *Aeromonas* bacteraemia have no clinical/laboratory history of cirrhosis preceding their septic episode. However, a careful review of the medical history of such persons reveals ethanol abuse or a longstanding history of alcohol consumption. Other

Table 6.3. Underlying conditions associated with *Aeromonas* septicaemia

Underlying disease	Relative frequency observed
Malignancy Acute myelocytic leukaemia Acute lymphocytic leukaemia Chronic myelocytic leukaemia Erythroleukaemia Multiple myeloma Oat cell carcinoma Adenocarcinoma Solid tumours/sarcomas	40–50%
Biliary tract Choledocholithiasis Cholecystitis Cholangitis Recurrent pyogenic cholangitis Gallstones	10–15%
Liver Laennec's cirrhosis Wilson's disease Viral hepatitis Ethanol abuse	5–15%
Diabetes Diabetes mellitus	3–5%
Renal Nephrotic syndromes Chronic renal failure	3–5%
Pancreatitis	3–5%
Trauma	1–2%
Cardiac anomalies Tetralogy of Fallot Congestive heart disease Chronic heart disease Endocarditis Acute myocardial infarction	<1%
Gastrointestinal Inflammatory bowel disease Digestive haemorrhages Colonic polyps Crohn's disease Gastroenteritis	<1%
Haematological Aplastic anaemia Thalassaemia Waldenstrom's macroglobulinaemia	<1%
Respiratory Pneumonia	<1%

Based on >200 cases of *Aeromonas* septicaemia
A majority of the data derived from the following references: Cordingley and Rajanayagam[1], Picard *et al.*[7], Harris *et al.*[5], Janda and Brenden[8], Dryden and Munro[6], Ong *et al.*[62], Janda *et al.*[2], and Duthie *et al.*[4]

conditions leading to liver damage, such as hepatitis-induced cirrhosis, have been associated with sepsis, although less prominently. In addition to these medical conditions leading to an immunocompromised state, a number of other underlying diseases have been reported in association with *Aeromonas* sepsis which are not directly linked to the patient's immunological status. These include diabetes, cardiac anomalies, GI tract syndromes, respiratory problems, and trauma (Table 6.3).

Two studies have investigated *Aeromonas* septicaemia in select patient populations. Sirinavin and colleagues[17] reviewed 20 cases of sepsis occurring in paediatric patients in one hospital in Thailand over a five-year period. Almost half (45%) of these children had leukaemia or aplastic anaemia (neutrophil counts <500 cells/mm^3); the mortality rate in this group was 78% despite the fact that most children received combination chemotherapy which included an aminoglycoside. For 11 other children in the study with a variety of other underlying complications (thalassaemia, cirrhosis) and who received similar medical intervention, the observed case fatality rate was only 27%. In a second series of adult patients retrospectively analysed in the M. D. Anderson Hospital in Houston, Texas, 24 individuals were found to have had *Aeromonas* bacteraemia over a 13-year period (1970–1983)[5]. All but one of these persons (96%) were receiving treatment for haematological malignancy at the time of their septic crisis. The overall mortality rate in this

Table 6.4. Prognosis and mortality rates associated with various presentations of *Aeromonas* sepsis

Population	Clinical presentation	Immunological status	Underlying disease	Mortality rate	Prognosis
Adults, children	Sepsis with/without ecthyma gangrenosa lesions, peritonitis	Compromised	Malignancies, cirrhosis	>50%	Poor
Adults	Sepsis, abdominal pain, diarrhoea	Normal	Biliary disease	<25%	Good
Adults	Varies	Normal	Cardiac anomalies, respiratory problems, diabetes, GI syndromes	<25%	Good

Adapted and modified from Lee *et al.*[3]

study was 28%. Shock, hypotension and the development of ecthymotic lesions are all poor prognostic signs associated with *Aeromonas* septicaemia. Composite prognostic features associated with select patient populations are listed in Table 6.4.

Laboratory findings have revealed some interesting information regarding *Aeromonas*-associated septicaemia. Although 14 nomenspecies are presently recognized within the genus, only five to date (*A. hydrophila, A. caviae, A. veronii, A. jandaei, A. schubertii*) have been documented to cause sepsis in humans. Both biotypes of *A. veronii* (formerly referred to as '*A. sobria*') have been reported to cause bacteraemia, including the ornithine decarboxylase-positive *A. veronii* biotype veronii which recently mimicked *Vibrio cholerae* septicaemia in a 77-year-old man[72]. *A. hydrophila, A. caviae* and *A. veronii* cause more than 95% of all reported episodes of *Aeromonas* bacteraemia. However, there are important differences between these three species in regard to the clinical settings in which they cause disease. Almost all cases of monomicrobic sepsis caused by aeromonads involve either *A. hydrophila* or *A. veronii* (Figure 6.1). This is in sharp contrast to the association of *A. caviae* with polymicrobic sepsis. These observations were originally reported by Janda and Brenden[8] and have been subsequently confirmed by other investigators. The clinical differences noted are apparently explained by the greater invasive and pathogenic potential of the former two species, and while *A. caviae* is isolated from persons with more serious underlying conditions (malignancy), a higher fatality rate has been associated with *A. veronii* infection in some studies but not others[6,8]. Most monomicrobic

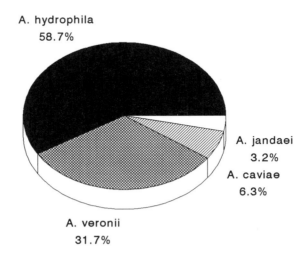

Figure 6.1. Species distribution of 63 cases of monomicrobic bacteraemia due to *Aeromonas*. Data from Dryden and Munro[6], Janda *et al.*[2], and Duthie *et al.*[4]

bacteraemias (>60%) involve one of four serogroups out of more than 90 *Aeromonas* serogroups, with 0:11, 0:16, 0:18 and 0:34 predominating[2]. Serogroups 0:11 (S layer, capsule) and 0:34 (capsule) possess additional surface appendages which may aid their dissemination. In cases of polymicrobic sepsis, aeromonads are most commonly isolated with members of the family Enterobacteriaceae (*Escherichia coli* > *Klebsiella* > *Enterobacter*) and less often with streptococci and pseudomonads. Polymicrobic sepsis is most commonly seen in persons with biliary disease[3]. Between 15 and 33% of patients with positive blood cultures for *Aeromonas* are simultaneously positive at other anatomical sites for this bacterium; these include, in decreasing order of occurrence, wounds, the biliary tract, catheters/instrumentation, faeces, and a variety of miscellaneous sources. Among immunocompetent individuals, septic patients often (>50%) have elevated white blood cell counts (WBC) and 40% have low immunoglobulin (IgM, IgG, or IgA) levels in their serum[3].

Epidemiological studies have found that a majority of the cases of *Aeromonas* septicaemia occur during the warmer months of the year, mimicking aeromonad-associated gastroenteritis[5,6]. However, at least one major study found a significant number of *Aeromonas* bacteraemias in the United States occurring between January and March, suggesting that the immune status of the host and virulence properties of the strain might be more important factors in causing disease than the infectious dose[2]. Unlike most *Vibrio* infections, a vehicle of infection for most cases of *Aeromonas* septicaemia cannot be identified. Cumulative data on 43 episodes of *Aeromonas* sepsis revealed only two instances (5%) where a history of freshwater, outdoor exposure or contact with fish could be linked to specific cases of septicaemia[5,6,8]. The failure to document these environmental exposures as sources of infection suggests that most illnesses arise through the ingestion of contaminated potable water or foods (meat, dairy produce). Although specific factors that enhance the risk of developing *Aeromonas* bacteraemia have not been adequately addressed due to frequency data, patients with catheterization or other medical instrumentation, receiving antineoplastic chemotherapy, or with low neutrophil counts (<1000 cells/mm^3) seem especially prone to developing sepsis.

Parras and colleagues[9] have recently reported on a fatal case of *A. veronii* (formerly *A. sobria*) meningitis in a 54-year-old male with alcoholic cirrhosis, and reviewed the medical literature concerning the role of aeromonads in CNS-related disease. To date, there have been only eight reported cases of *Aeromonas* meningitis in the literature since 1960[9]. These patients fall into two main categories. The first involves neonates or infants ($n = 4$) ranging in age between 13 days and 2 years. Two of these infants had underlying blood dyscrasias (beta-thalassaemia, sickle cell disease) while the other two had no obvious problems other than their age. The overall mortality rate for this group was 50%. A second population consists of adults ($n = 4$) with a mean age of 47.8 years. Two of these persons had underlying liver disorders while a head trauma was the precipitating event leading to meningitis in another. Only 50% of the adult cases had positive blood cultures for *Aeromonas* in conjunction with meningeal signs. The only adult to die from *Aeromonas*-associated meningitis was a 54-year-old male who developed pneumonia, peritonitis, myositis and ecthyma gangrenosum in addition to his CNS disease[9]. In addition to the previously mentioned case, the species identification of only one other CSF isolate has been determined according to current taxonomic standards. In that instance *A. veronii* was recovered from the CSF but not the blood of a 66-year-old male with meningitis[18].

In 1983, Heckerling and associates reported on a case of *Aeromonas*-associated myonecrosis and reviewed the literature in regards to deep muscle (soft tissue) infections with this organism[10]. They noted in their review of this subject that all reported cases of aeromonad-associated myonecrosis in which blood cultures were simultaneously positive for *Aeromonas* were uniformly fatal. This observation has basically held true over the past 12 years, with some minor exceptions. A fatal case of *Aeromonas* myonecrosis in a 50-year-old male developed after intravenous insertion of a cannula[27]. The patient had been previously diagnosed with cirrhosis and suffered from oesophageal varices and scleropathy. Within 48 h of cannula insertion to both legs (one leg tried first which failed) the patient developed bilateral oedema, subcutaneous gas formation, and bullae of the lower extremities. Within 5 h, the gangrenous process had progressed so rapidly that it proved fatal. Both legs yielded heavy pure growth of *A. hydrophila*, blood cultures were not performed but the clinical symptoms (fever, DIC (disseminated intravascular coagulation)) indicate that if taken they would have been positive. Moses and colleagues[19] reported a fatal case of *A. hydrophila* myonecrosis of the right forearm in a 36-year-old man with idiopathic aplastic anaemia. The patient had undergone bone marrow transplantation five years earlier for aplastic anaemia. Although *Aeromonas* was clearly involved in the course of the infection, mucormycosis of the brain appeared to be a significant contributing factor to the patient's death. The only reported case of *Aeromonas* myonecrosis with positive blood cultures in which the patient survived was recently published on a 41-year-old male with longstanding ethanol abuse and GI bleeding. This cirrhotic individual rapidly developed myonecrosis of both calves which involved soft tissue gas formation and necessitated triple amputation (including a unilateral upper extremity amputation). Although the patient initially survived the rapidly progressive myonecrosis condition, he died two weeks later due to hepatorenal syndrome, sepsis (unspecified) and cardiac arrest. In approximately half of these cases, a recognized source of the initial infection can be determined (warm water packs, fishbone, sewage) while in a majority of the remaining persons a medical procedure (cannula insertion, bone marrow transplantation, immunosuppressive therapy) appeared to be involved. Golik and collaborators[20] have recently summarized the literature with regard to reports of *Aeromonas* sepsis in non-immunocompromised patients. Between 1973 and 1990, approximately 24 cases of sepsis due to aeromonads have been described in individuals without pre-existing immunocompromised states (cancer, liver disease). However, in most of these reports septic individuals had other underlying illness (cardiac disease, biliary problems), were elderly (>70 years of age) implying decreased immunological competency, or sustained obvious traumas/insults which predisposed them to subsequent infection. Currently, there have been only four convincing cases of sepsis in healthy ('normal') people (Table 6.5). Kovarik and co-investigators[21] described the first case of *Aeromonas* sepsis in a healthy 30-year-old male. He initially presented with a six-week history of fever, chills, myalgia and productive cough. His initial medical workup was unremarkable. Over a 19-day period he subsequently developed signs of pulmonary and CNS disease and died of a cerebral disorder. The only pathogen isolated (at post-mortem) was *Aeromonas*, which was found in a number of organs. The second case of sepsis in a healthy individual was reported by Scott and others in 1978[22]. This was an acute presentation in a 29-year-old male who developed severe right-sided pleuritic chest pain 15 h before his death. After receiving initial medication (penicillin G, meperidine), he was sent home but returned to the emergency room 6 h prior to his death with increasing chest pain and dyspnoea. He developed progressive respiratory problems and died as a result of cardiorespiratory arrest. A third fatal case of *Aeromonas* sepsis occurred in a 65-year-old

Table 6.5. *Aeromonas* sepsis in normal individuals

Reference	Sex	Age	Culture-positive sites		Antibiotic
			Pre-mortem	Post-mortem	treatment
Kovarik *et al.*[21]	M	30	None	Lung, lymph nodes, spleen	Multiple
Scott *et al.*[22]	M	29	Blood, sputum	Lungs	Penicillin
Janda and Brenden[8]	F	65	None	Blood, lung, pleural fluid, peritoneal fluid, pericardial fluid	None
Janda and Clark[65]	F	23	Blood	NA	Cipro-floxacin, metroni-dazole

NA, not applicable

woman with no underlying problems who presented to the emergency room with abdominal pain, oedema (feet), and heart failure. Post-mortem cultures from this rapidly fatal infection yielded *A. hydrophila* from multiple sites. The only non-fatal case of *Aeromonas* septicaemia in a healthy individual stems from a report of polymicrobic sepsis involving *Aeromonas* in a 23-year-old college student. The patient initially presented with fever and GI symptoms. Blood cultures revealed *A. caviae* and an *Enterobacter* species. The implicated vehicles of infection were fish and Chinese food. Although recovery was uneventful after appropriate therapy she was readmitted to the hospital 3 months later with a diagnosis of ulcerative colitis, suggesting that although unnoticed at the time this condition may have preceded her bacteraemia. The common underlying condition for each of the three fatal cases of *Aeromonas* septicaemia was pulmonary involvement based upon clinical symptomatology or post-mortem cultures; such rare complications may have resulted in the high fatality rate observed for these normal individuals.

6.3   WOUND INFECTIONS

Of the infections caused by *Aeromonas*, only gastroenteritis is more common than wounds. Despite this, their involvement in these infections is not always readily appreciated, often to the detriment of the patient and the frustration of the physician who is faced with treating an infection that can be rapid and progressively fatal.

Aeromonas wounds fall into three categories, discussed by the order of frequency with which they occur and the severity of damage they cause. The lesser of the three evils, cellulitis is the most frequently encountered manifestation of *Aeromonas* wound infections. It is an acute inflammation of subcutaneous tissue characterized by redness and induration with indistinct borders and may arise either from trauma related to some type of injury or secondary to sepsis. Histological sections reveal heavy infiltrates of inflammatory cells; however, bacteria will rarely be discernible[23]. If blood flow to the affected areas is reduced, there is subsequent soft tissue involvement. Fortunately, these infections will generally subside with adequate chemotherapy and debridement and a complete recovery is the expected outcome. However, in the following two less commonly seen presentations, the outcome is likely to be fatal. The patients with these infections are usually diabetic, cirrhotic, agranulocytopenic or otherwise compromised. The first of these two conditions, myonecrosis or bullous lesions, is characterized by the liquefaction of muscles with blackening of the tissue which may be gangrenous with gas formation. Typically, these patients require aggressive antimicrobial therapy and debridement; those individuals who fail to respond to these measures may require amputation of the affected extremity. When there is accompanying sepsis the infection may, more often than not, be fatal. The third presentation, ecthyma gangrenosum, is a cutaneous necrotic or gangrenous pustule that occurs secondary to sepsis. Unlike cellulitis, histological sections are characterized by large numbers of bacteria with minimal inflammatory cells present[23]. Lesions have an erythematous border surrounding a vesicle which can progress to necrosis of the soft tissue within 24 h. *Pseudomonas* is usually the implicated agent; however, the increasing association of *Aeromonas* with this syndrome is such that now both organisms are considered to produce ecthyma with a frequency that is disproportionate to the number of bacteraemias they cause[23].

Most wound infections involving *Aeromonas* should be easily recognized as the vast majority follow a well defined pattern. An example that illustrates the elements of this pattern is a case report by Lineaweaver[24], although any number of case reports would also serve. In this instance, a 40-year-old man sustained multiple lacerations to his foot during a boating accident. Three days after surgery to repair the damage the patient experienced fever and

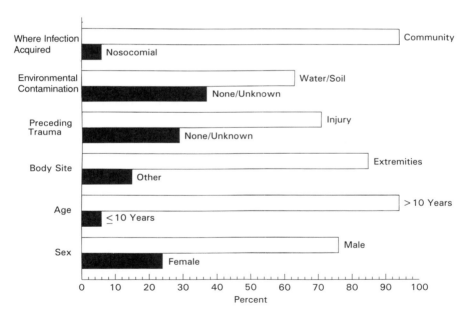

Figure 6.2. Demographics of *Aeromonas* wound infections. Data from Francis *et al.*[66], Grant and Hoddinott[67], Isaacs *et al.*[33], Joseph *et al.*[34], Lineaweaver *et al.*[24], Moses *et al.*[19], Newton and Kennedy[68], Raynor *et al.*[28], Revord *et al.*[29], Siddiqui *et al.*[27], Suthipintawongs and Wanvaree[69], Voss *et al.*[25], Warrier *et al.*[70], Wells *et al.*[35], Werner and Rutherford[32], and Vukmir[71]

chills and subsequently developed cellulitis in his left foot and calf. The patient was treated with an antibiotic, in this case cefotaxime, the involved areas were debrided of necrotic tissue and the patient recovered uneventfully from his infection. The notable aspects of this case are as follows: the infection was community acquired, it involved an adult male, there was a preceding trauma to an extremity and the wound was environmentally contaminated. The following data from an additional 16 reports and reviews published in the literature between 1980 and 1995 on 138 patients with *Aeromonas* wounds further substantiate the profile just described (Figure 6.2; see figure legend for references).

Virtually all *Aeromonas* wound infections are community acquired; of the 139 cases referred to above only seven were believed to be nosocomially acquired. Interestingly, sources for nosocomial infections are rarely, if ever, established although hospital water supplies have yielded cultures of aeromonads[25,26]. It is also frequently assumed that hospital-acquired infections arise endogenously through the GI tract even though the organism may not be isolated from the patient's faeces at the time of the wound infection[27].

Yet another notable feature of *Aeromonas* wound infections is the fact that men are about three times (76% versus 24% of patients when sex was indicated) more likely than women to develop these infections. While this could indeed simply be a matter of greater opportunity for exposure, a discussion raised previously with septicaemic patients, a more intriguing explanation has been proposed by Baddour[26]. In a recent editorial, he speculated that perhaps regulation of certain *Aeromonas* adhesins may be affected by hormonal or metabolic conditions of the host, much as is seen with OMP II regulation in *Neisseria gonorrhoeae*. In short, phenotypic variation regulating the expression of these adhesins could result in favouring the binding of *Aeromonas* to male tissue.

One more remarkable and perhaps unusual feature of *Aeromonas* wound infections concerns the age of the patient involved. Where age was given specifically and not as a range (as in the case of two reviews including 58 patients) there were only five patients under the age of 10 (6%); of the remaining patients 24 (23%) were between the age of 11 and 21 and 59 (70%) were over 21 (range 21 to 88). This finding is odd since most infections (71%), regardless of age, were preceded by penetrating trauma from an accident, and such incidents are particularly common in young children.

Most precipitating trauma were of the normal variety, i.e. nail puncture, boating or automobile accidents particularly motorcycles, fractures or wounds attained during recreational/home repair/work activities. However, an unlikely acquisition has been reported in several instances, including two cases from alligator attacks[28] and one from a piranha bite[29]. When no penetrating injury occurred prior to the development of their wounds, patients were almost invariably compromised, a finding that will be discussed in more detail later. Notably, the above three unusual reports reflect probably the most outstanding aspect with regards to the pattern of *Aeromonas* wound infections: environmental contamination. Overall, 63% of the patients in the 17 case reports reviewed were determined to have soil- or water-contaminated wounds. In a series of 28 wound infections reported by Voss and colleagues[25], 43% were water-related injuries whereas Kelly and co-authors[30] reported that 17 of 27 community-acquired wound infections involved soil or water. That environmental contamination is a significant consideration in *Aeromonas* wounds is hardly surprising because *Aeromonas* species are routinely isolated from soil and water. Indeed, the organism has been reported in virtually all freshwater sources in the United States[31]. One would particularly expect to see higher incidences of *Aeromonas* wounds in the summer months as warm temperatures lead to increased numbers of organisms in the water during this time. Though most individual case reports do not mention seasonal acquisition, Kelly and colleagues[30] reported that 76% of extraintestinal infections acquired in the community occurred in their wet season, with the highest frequency occurring in January, the middle of Australia's wet months. Frequency of isolations remained constant and lower during their dry season. Likewise, in California, where the incidence of wound infections based on reported cases was 0.7 per million population, infections peaked in the summer months[32]. While Isaacs and co-workers[33] could find no seasonal distribution in their study of 27 patients, they allowed that the temperate climate of Auckland may have accounted for the lack of seasonality.

The final piece of the pattern involves wounds on extremities versus other anatomical sites. The site of infection could be determined for 113 patients, and of those 96 (85%) involved an extremity. The scalp or facial area was the next most common site, with lacerations of the scalp accounting for most of these infections. However *Aeromonas* was isolated from a nose injury acquired in a fall[33] and a dual infection involving *A. jandaei* and *A. veronii* was reported in a 10-year-old child who was shot in the eye with a BB from an air rifle[34]. Two days later the child went swimming while in summer camp and developed a purulent discharge from the skin overlying the lacrimal sac. A hip puncture wound attained in an automobile accident, a thigh wound acquired while swimming[32] and a soft tissue infection of the throat[35] accounted for the remaining sites of infection.

Seventy-two (64%) of the patients in the case reports were healthy prior to their wound infections, and there were four patients who otherwise fit a typical *Aeromonas* wound infection pattern but who were immunocompromised. But patients who did not have a prior penetrating injury and/or no environmental contact with their wound were invariably compromised; some of these infections were nosocomially acquired. In three of the more notable cases involving immunocompromised individuals the patients were also bacteraemic and these were discussed under septic infections. Two of these patients were admitted to the hospital for GI bleeding, and one had a history of alcohol abuse while the other had cirrhosis. The third patient had a bone marrow transplant for aplastic anaemia and a concomitant mucormycosis brain infection. Patients who acquired their infections nosocomially had undergone a variety of interventions, including Caesarian section, gastrectomy and transverse colectomy.

No discussion of *Aeromonas* wounds would be complete without mention of the infections acquired by the use of medicinal leeches. For an informative (and rather fascinating) description of the application of leeches in microvascular surgery, the reader is referred to an excellent article by Abrutyn[36]. The increased use of *Hirudo medicinalis*, a medicinal leech, to relieve venous congestion after plastic surgery had led to the publication of no fewer than eight reports involving 14 patients since 1971. Mercer and colleagues[37] reported six *Aeromonas* wound infections involving leeches in a three year period for a 20% infection rate. At another institution, three of 42 patients experienced infection following medicinal leech therapy for a 7% infection rate[38]. Infections resulting from the use of medicinal leeches were predicted[39] because leeches harbour *Aeromonas* in the gut where it serves to aid in the digestion of the blood meal. In addition to the gut, *Aeromonas* has also been isolated from the anterior and posterior suckers, the mucous trail of the leech and the tank of water used to maintain them[39,40]. Because of the presence of *Aeromonas* and the risk it posed, Whitlock recommended leeches be counterindicated for use after surgery. However, because they are extremely effective medical devices, it is doubtful that their use will be discontinued. For now, it is generally accepted that the patient exposed to leech therapy be treated prophylactically with a third-generation cephalosporin, aminoglycoside, chloramphenicol, tetracycline, Bactrim or ciprofloxacin[41]. Patients involved in these infections are usually immunocompetent, although some have undergone reconstructive surgery for mastectomy or other carcinomas[37,42]. Symptoms include erythema, oedema and fever; severity ranged from purulent discharge to necrosis. Surgical debridement of necrotic tissue was frequently required.

The causative agents reported in wound infections, regardless of source or underlying condition of the patient, were invariably *A. hydrophila* or *A. veronii* (formerly *A. sobria*). Of course, it should be pointed out that a number of these reports predate the expansion of this genus to the currently recognized 14 nomenspecies, and most of the reports fail to include diagnostic biochemical methods. In the case where *A. jandaei* was isolated concurrently with *A. veronii* from the eye, it is difficult to discern the role of the *A. jandaei* strain since another known virulent *Aeromonas* was also present. The capability of *Aeromonas* strains to cause wound infections is easily understood given that they produce a number of extracellular factors, including haemolysins and, in particular, a number of different proteases, as well as amylase.

6.4 OCULAR INFECTIONS

Reports of *Aeromonas* ocular involvement, which run the gamut from asymptomatic carriage to corneal ulcers, are relatively uncommon—there were more case reports in the literature of infections caused by leech therapy than ocular infections. Wearing of contact lenses or, as with wounds, a penetrating injury to the eye with concomitant water contamination figured prominently in the infections reported. The exception was in a 55-year-old man with a history of alcohol abuse who developed *A. hydrophila* endophthalmitis believed to have been acquired endogenously by haematogenous spread, although blood and urine cultures were negative (no mention was made of stool testing)[43]. Smith[44] also reported a case of asymptomatic carriage in a 75-year-old man admitted for cataract surgery whose presurgical eye swabs yielded a pure culture of *A. hydrophila*. There was no sign of infection at the time of surgery. He was treated with 0.5% ophthalmic solution of chloramphenicol and later with oral tetracycline without any sign of infection thereafter. Likewise, Sire and colleagues[45] reported on an 86-year-old woman with persistent bilateral ectropion who developed an *A. hydrophila* bilateral blepharoconjunctivitis. The ectropion was thought to be a predisposing factor (also age?) and tapwater in contact with eyes the most likely source.

Water is also the probable source of infections in contact lens wearers who develop *Aeromonas* conjunctivitis and, in fact, has been isolated from the inside of a contact lens case of an individual with conjunctivitis who washed her lenses with tapwater[46]. Most other reported infections involving a trauma to the eye[44,46,47] preceding the infection. In one such infection the patient developed a corneal ulcer surrounded by three abscesses after being hit in the eye by a reed; the causative organism was *A. veronii*. An unusual dual infection with *A. hydrophila* and *Plesiomonas shigelloides* occurred in a small boy with endophthalmitis whose injury was caused by a fishhook that lacerated the eye. Ocular infections were generally treated with ophthalmic chloramphenicol or aminoglycoside (usually gentamicin) solutions.

6.5 RESPIRATORY TRACT INFECTIONS

Actual infections of the respiratory tract are relatively uncommon, with most isolations of *Aeromonas* at these sites representing transient colonization as opposed to significant disease[48]. When true aeromonad infections of the respiratory tract are encountered, the overwhelming number are in adult men who had either a pre-existing disease or who aspirated the organisms during near drownings. The infections were usually fatal if the organism was isolated from both a respiratory site and blood, as was the case in five of six patients, the lone survivor being a 13-year-old boy[49-52]. Of these six cases, four acquired the infection by swimming or an accident during which water was aspirated into the respiratory tract, and all were previously healthy persons. The remaining two patients were compromised, one of whom developed empyema and had a history of cirrhosis secondary to hepatitis, and the other developed a cavitating bilateral pneumonia and had chronic renal failure as an underlying condition. Both died rapidly within 24 and 36 h of admission to the hospital, respectively. The source for the empyema infection was believed to be a haematogenous spread from the GI tract. This was likely with the other patient as well since on autopsy a pseudomembranous enteritis of the small bowel with large numbers of Gram-negative bacilli was seen. In an additional two cases, *A. hydrophila* was isolated in pure culture from sputum and bronchial washings in a patient with a lung abscess and from three swabs of the epiglottis in a case of epiglottitis[53,54]. The patient with the lung abscess was a previously healthy 59-year-old farmer who presented with fever, pleuritic chest pain and blood-tinged purulent sputum. He was treated with cefuroxime and recovered; no mention was made of the possible source of the infection. The 8-year-old girl with epiglottitis had β-thalassaemia and required a tracheostomy with intravenous cefotaxime, intramuscular ceftriaxone and, after removal of the tracheostomy, oral cotrimoxazole. She recovered. The source of this infection was thought to be water or hands contaminated with faeces, since Thailand has an *Aeromonas* carriage rate in the stool of between 16 and 27%[53].

6.6 BONE AND JOINT INFECTIONS

Osteomyelitis caused by *Aeromonas* is rare, and is invariably preceded by fractures or lacerations that are water-contaminated. It has been reported in both immunocompromised and healthy persons. Karam and colleagues[55] reported two cases of osteomyelitis in normal hosts where *A. hydrophila* was diagnosed by X-ray and isolation in pure culture from bone and sinus tract specimens. The patients were treated with intravenous gentamicin and oral tetracycline, and intravenous cefamandole and oral tetracycline, respectively. In an earlier report, a healthy man with a fracture, sustained in a motorcycle accident, that was contaminated by stagnant ditch water developed septicaemia, cellulitis and ultimately osteomyelitis[56]. Despite debridement and treatment with gentamicin, the patient's leg was amputated to halt his deteriorating condition.

In 1976 Chmel and Armstrong[57] reported two cases of *Aeromonas* arthritis. Both patients had leukaemia, developed an effusion of their knee joints with *Aeromonas* sepsis and died. In both cases, *A. hydrophila* was isolated from blood and synovial fluid.

6.7 INTRA-ABDOMINAL INFECTIONS

Peritonitis caused by *Aeromonas* spp. is not a rare event, as evidenced in an article published by Munoz and colleagues[12]. In this presentation of five cases and review of an additional 29 reports, several points were apparent. Yet again, there was a predominance of adult male patients, 72% versus 28% women (mean age 56.9). Where the origin was known, seven cases were nosocomial and 22 community-acquired, and all patients but one were compromised (25 cirrhotic patients with spontaneous bacterial peritonitis, five patients on continuous ambulatory peritoneal dialysis (CAPD), and three with perforated bowel). *Aeromonas* was directly responsible for eight deaths, including the only previously healthy patient who developed disseminated disease and died. *A. hydrophila* was the predominant causative agent in 27 cases, followed by five cases of *A. veronii* biotype sobria and two cases of *A. caviae*. The *A. caviae* from this review and an additional case[58] all involved patients undergoing CAPD. Sources suggested for these infections included water, but this was rarely proved[12], except for one case of *A. caviae* in a CAPD patient who sprayed her catheter connector site from a container used for an alcohol-based disinfectant but which currently contained water used to spray houseplants. *A. caviae* was cultured from the container but not the water inside it, nor was it found in the patient's stool[59].

A rare case of *A. hydrophila* cholecystitis was reported by Parsons[60]. Laparotomy showed inflammation of the gallbladder and an abscess was found in histological sections. Bile and blood were culture-positive for *A. hydrophila*. *E. coli* which was also isolated from the blood but not the bile, was considered unlikely to be the cause of the necrotizing process seen in the gallbladder.

6.8 MISCELLANEOUS INFECTIONS

Occasionally aeromonads are isolated from or associated with unexpected disease syndromes. Haake and co-authors[61] published a case report where an '*A. hydrophila* group' organism was isolated from blood, lung, muscle tissue and purulent material from the wrist in a 74-year-old man with suppurative venous thrombophlebitis. The patient had underlying chronic lung disease and was admitted to the hospital for partial small bowel obstruction. The initial site of the thrombophlebitis, which spread from the wrist to the upper arm with myonecrosis of the underlying muscle and soft tissue, was an intravenous catheter site. The catheter was removed when the patient noted pain and swelling and the wrist was wrapped in towels heated by tap water. The source could have been water used on the towels or the organism could have spread from the GI tract to the bloodstream; in either case, prednisone used to treat the lung disease probably compounded the rapid spread.

Endocarditis represents another unusual manifestation of *Aeromonas* infection[62]. Echocardiography, performed on a patient with multiple underlying diseases who was admitted for chest pain and fever, indicated left-sided endocarditis. *A. hydrophila* was isolated from the patient's blood. Cefazolin and gentamicin resolved the fever but the patient died from hepatic failure.

An unusual association of colonic carcinoma with *Aeromonas* colitis was noted in a brief 1994 report[63]. The bloody diarrhoea suffered by a patient ultimately led to the discovery of the adenocarcinoma (by biopsy) during colonoscopy. The patient's source of infection was from beer cans chilled in contaminated river water.

Robson and colleagues[64] suggested that *A. hydrophila* enterocolitis may be associated with haemolytic–uraemic syndrome (HUS), after isolating this organism from the stool of two of 82 (2%) children with typical HUS and from whom no known shiga-like toxin-producing organisms were found.

6.9   REFERENCES

1   Cordingley FT, Rajanayagam A. *Aeromonas hydrophila* bacteremia in haematological patients. *Med J Aust* 1981; **1**: 364–5.

2   Janda JM, Guthertz LS, Kokka RP, Shimada T. *Aeromonas* species in septicemia: laboratory characteristics and clinical observations. *Clin Infect Dis* 1994; **19**: 77–83.

3   Lee L-N, Luh K-T, Hsieh W-C. Bacteremia due to *Aeromonas hydrophila*: a report of 40 episodes. *J Formas Med Assoc* 1986; **85**: 123–32.

4   Duthie R, Ling TW, Cheng AFB, French GL. *Aeromonas* septicaemia in Hong Kong: species distribution and associated disease. *J Infect* 1995; **30**: 241–4.

5   Harris RL, Fainstein V, Elting L, Hopfer RL, Bodey GP. Bacteremia caused by *Aeromonas* species in hopitalized cancer patients. *Rev Infect Dis* 1985; **7**: 314–20.

6   Dryden M, Munro R. *Aeromonas* septicemia: relationship of species and clinical features. *Pathology* 1989; **21**: 111–14.

7   Picard B, Arlet G, Goullet Ph. Septicemies à *Aeromonas hydrophila*. Aspects épidémiologiques. Quinze observations. *La Presse Med* 1984; **13**: 1203–5.

8   Janda JM, Brenden R. Importance of *Aeromonas sobria* in *Aeromonas* bacteremia. *J Infect Dis* 1987; **155**: 589–91.

9   Parras F, Diaz MD, Reina J, Guerrero C, Bouza E. Meningitis due to *Aeromonas* species: case report and review. *Clin Infect Dis* 1993; **17**: 1058–60.

10   Heckerling PS, Stine TM, Pottage JC, Levin S, Harris AA. *Aeromonas hydrophila* myonecrosis and gas gangrene in a nonimmunocompromised host. *Arch Intern Med* 1983; **143**: 2005–7.

11   Ruiz de Gonzales P, Escolona C, Rodriguez JC, Sillero C, Royo G. *Aeromonas sobria* spontaneous bacterial peritonitis and bacteremia. *Am J Gastroenterol* 1994; **89**: 290–1.

12   Munoz P, Fernandez-Baca V, Palaez T, Sanchez R, Rodriguez-Creixems M, Bouza E. *Aeromonas* peritonitis. *Clin Infect Dis* 1994; **18**: 32–7.

13   Slevin NJ, Oppenheim BA, Deakin DP. *Aeromonas hydrophila* septicaemia and

muscle abscesses associated with immunosuppression. *Postgrad Med J* 1988; **64**: 701–2.

14 Jesudason MV, Koshi G. *Aeromonas* species in human septicaemia and diarrhoea. *Indian J Med Res (A)* 1990; **91**: 174–6.

15 Krovacek K, Conte M, Galderisi P, Morelli G, Postiglione A, Dumontet S. Fatal septicaemia caused by *Aeromonas hydrophila* in a patient with cirrhosis. *Comp Immun Microbiol Infect Dis* 1993; **16**: 267–72.

16 von Graevenitz A, Mensch AH. The genus *Aeromonas* in human bacteriology. Report of 30 cases and review of the literature. *N Engl J Med* 1968; **268**: 245–9.

17 Sirinavan S, Likitnukul S, Lolekha S. *Aeromonas* septicemia in infants and children. *Pediatr Infect Dis* 1984; **3**: 122–5.

18 Jacob L, Carron DB, Haji TC, Roberts DW. An unusual case of pyogenic meningitis due to *Aeromonas sobria*. *Br J Hosp Med* 1988; **396**: 449.

19 Moses AE, Leibergal M, Rahav G, Perouansky M, Or R, Shapiro M. *Aeromonas hydrophila* myonecrosis accompanying mucormycosis five years after bone marrow transplantation. *Eur J Clin Microbiol Infect Dis* 1995; **14**: 237–40.

20 Golik A, Leonov Y, Schlaeffer R, Gluskin I, Lewinsohn G. *Aeromonas* species bacteremia in nonimmunocompromised hosts. Two case reports and a review of the literature. *Isr J Med Sci* 1990; **26**: 87–90.

21 Kovarik JL, Sides LJ, Becky JR. Fatal septicemia with *Aeromonas hydrophila*. *Rocky Mt Med J* 1973; **70**: 36–8.

22 Scott EG, Russell CM, Noell KT, Sroul AE. *Aeromonas hydrophila* sepsis in a previously healthy man. *JAMA* 1978; **239**: 1742.

23 Musher DM. Cutaneous and soft-tissue manifestations of sepsis due to gram-negative enteric bacilli. *Rev Infect Dis* 1980; **2**: 854–66.

24 Lineaweaver WC, Follansbee S, Hing DN. Cefotaxime-sensitive *Aeromonas hydrophila* infection in a revascularized foot. *Ann Plast Surg* 1988; **20**: 322–4.

25 Voss LM, Rhodes KH, Johnson KA. Musculoskeletal and soft tissue *Aeromonas* infection: an environmental disease. *Mayo Clin Proc* 1992; **67**: 422–7.

26 Baddour LM. Extraintestinal *Aeromonas* infections —looking for Mr. Sandbar. *Mayo Clin Proc* 1992; **67**: 496–8.

27 Siddiqui MN, Ahmed I, Farooqi BJ, Ahmed M. Myonecrosis due to *Aeromonas hydrophila* following insertion of an intravenous cannula: case report and review. *Clin Infect Dis* 1992; **14**: 619–20.

28 Raynor AC, Bingham HG, Caffee HH, Dell P. Alligator bites and related infections. *J Florida Med Assoc* 1983; **70**: 107–10.

29 Revord ME, Goldfarb J, Shurin SB. *Aeromonas hydrophila* wound infection in a patient with cyclic neutropenia following a piranha bite. *Pediatr Infect Dis J* 1988; **7**: 70–1.

30 Kelly KA, Koehler JM, Ashdown LR. Spectrum of extraintestinal disease due to *Aeromonas* species in tropical Queensland, Australia. *Clin Infect Dis* 1993; **16**: 574–9.

31 Janda JM, Abbott SL, Morris Jr JG. *Aeromonas, Plesiomonas* and *Edwardsiella*. In Blaser MJ *et al.* (eds) *Infections of the Gastrointestinal Tract*, New York: Raven Press, 1995; **5**: 905–17.

32 Werner SB, Rutherford GW. *Aeromonas* wound infection associated with outdoor activities – California. *Morb Mortal Wkly Rep* 1990; **39**: 334–41.

33 Isaacs RD, Pavious SD, Bunker DE, Lang SDR. Wound infection with aerogenic *Aeromonas* strains: A review of twenty-seven cases. *Eur J Clin Infect Dis* 1988; **7**: 355–60.

34 Joseph SW, Carnahan AM, Brayton PR, Fanning GR, Almazan R, Drabick C, Trudo EW, Colwell RR. *Aeromonas jandaei* and *Aeromonas veronii* dual

infection of a human wound following aquatic exposure. *J Clin Microbiol* 1991; **29**: 565–9.

35   Wells RG, Chertow GM, Marcantonio ER. Parapharyngeal soft-tissue infection with *Aeromonas hydrophila. Head Neck* 1991; **Nov/Dec**: 528–30.

36   Abrutyn E. Hospital-associated infection from leeches. *Ann Intern Med* 1988; **Sept**: 356–7.

37   Mercer NSG, Beere DM, Bornemisza AJ, Thomas P. Medical leeches as sources of wound infection. *BMJ* 1987; **294**: 937.

38   Lineaweaver WC, Hill MK, Buncke GM, Follansbee S, Bunck HJ, Wong RKM, Manders EK, Grotting JC, Anthony J, Mathes SJ. *Aeromonas hydrophila* infections following use of medicinal leeches in replantation and flap surgery. *Ann Plast Surg* 1992; **29**: 238–44.

39   Whitlock MR. The medicinal leech and its use in plastic surgery: a possible cause for infection. *Br J Plast Surg* 1983; **36**: 240–4.

40   Snower DP, Reuf CC, Kuritza AP, Edberg SC. *Aeromonas hydrophila* infection associated with the use of medicinal leeches. *J Clin Microbiol* 1989; **27**: 142–2.

41   Fedorko DP. The medicinal leech as a source of infection. *Clin Microbiol Newsl* 1993; **15**:164–5.

42   Dickson WA, Bootham P, Hare K. An unsual source of hospital wound infection. *BMJ* 1984; **289**: 1727–8.

43   Frieling JS, Rosenberg R, Edelstein M, Colby SD, Kopelowitz NN. Endogenous *Aeromonas hydrophila* endophthalmitis. *Ann Ophthalmol* 1989; **21**: 117–18.

44   Smith JA. Ocular *Aeromonas hydrophila. Am J Opthalmol* 1980; **89**: 449–51.

45   Sire JM, Ropert P, Donnio PY. *Aeromonas hydrophila* blepharoconjunctivitis. *Eur J Clin Microbiol Infect Dis* 1990; **9**: 904–5.

46   Carta F, Pinna A, Zanetti S, Carta A, Sotgiu M, Fadda G. Corneal ulcer caused by *Aeromonas* species. *Am J Ophthalmol* 1994; **118**: 530–1.

47   Cohen KL, Holyk PR, McCarthy LR, Peiffer RL. *Aeromonas hydrophila* and *Plesiomonas shigelloides* endophthalmitis. *Am J Ophthalmol* 1983; **96**: 403–4.

48   Janda JM, Duffey PS. Mesophilic aeromonads in human disease: current taxonomony, laboratory identification, and infectious disease spectrum. *Rev Infect Dis* 1988; **10**: 980–97.

49   Goncalves JR, Brum G, Fernandes A, Biscaia I, Correia MJS, Bastardo J. *Aeromonas hydrophila* fulminant pneumonia in a fit young man. *Thorax* 1992; **47**: 482–3.

50   Baddour LM, Baselki VS. Pneumonia due to *Aeromonas hydrophila*-complex: epidemiologic, clinical, and microbiologic features. *South Med J* 1988; **81**: 461–3.

51   Steciw A, Colodny SM. Empyema due to *Aeromonas hydrophila. Clin Infect Dis* 1994; **18**: 474–5.

52   Reines HD, Cook FV. Pneumonia and bacteremia due to *Aeromonas hydrophila. Chest* 1981; **80**: 264–7.

53   Hur T, Cheng K-C, Hsieh J-M. *Aeromonas hydrophila* lung abscess in a previously healthy man. *Scand J Infect Dis* 1995; **27**: 295.

54   Preutthipan A, Chantarojanasiri T, Suwanjutha S, Tanpowpong K. *Aeromonas hydrophila* epiglottitis: a case report. *J Med Assoc Thai* 1993; **76**: 225–7.

55   Karam GH, Ackley AM, Dismukes WE. Posttraumatic *Aeromonas hydrophila* osteomyelitis. *Arch Intern Med* 1983; **143**: 2073–4.

56   Blatz DJ. Open fracture of the tibia and fibula complicated by infection with *Aeromonas hydrophila. J Bone Joint Surg* 1979; **61(A)**: 790–1.

57   Chmel H, Armstrong D. Acute arthritis caused by *Aeromonas hydrophila*, clinical and therapeutic aspects. *Arthritis Rheum* 1976; **19**: 169–72.

58   Elcuaz R, del Pino J, Fernandez A, Bordes A, Lafarga B. Peritonitis caused by

Aeromonas caviae in a patient undergoing peritoneal dialysis. *Clin Microbiol Newsl* 1995; **17**: 5–6.

59 Al-wali W, Baillod R, Hamilton-Miller JMT, Brumfitt W. Houseplant peritonitis. *Lancet* 1988; **11**: 957.

60 Parsons WJ. Acute cholecystitis due to *Aeromonas hydrophila*. *NY State J Med* 1985; **Sept**: 564–5.

61 Haake DA, Hewitt WL, George WL. Suppurative thrombophlebitis due to *Aeromonas*. *Diagn Microbiol Infect Dis* 1988; **10**: 57–60.

62 Ong KR, Sordillo E, Frankel E. Unusual case of *Aeromonas hydrophila* endocarditis. *J Clin Microbiol* 1991; 29: 1056–7.

63 Cook MA, Nedunchezian D, Phillips M. Colonic carcinoma manifesting as *Aeromonas* colitis. *J. Clin Gastroenterol* 1994; **18**: 242–3.

64 Robson WLM, Leung AKC, Trevenen CL. Haemolytic-uremic syndrome associated with *Aeromonas hydrophila* enterocolitis. *Pediatr Nephrol* 1992; **6**: 221.

65 Janda JM, Clark RB. *Aeromonas* sepsis in a healthy college student. *Clin Microbiol Newslett* 1995; **17**: 61–2.

66 Francis FF, Richman S, Hussain S, Schwartz J. *Aeromonas hydrophila* infection. *NY State J Med* 1982; **82**: 1461–4.

67 Grant A, Hoddinott C. *Aeromonas hydrophila* infection of a scalp laceration (with synergistic gas-gangrene). *Arch Emerg Med* 1993; **10**: 232–4.

68 Newton JA, Kennedy CA. Wound infection due to *Aeromonas sobria*. *Clin Infect Dis* 1993; **17**: 1082–3.

69 Suthipintawongs C, Wanvaree S. Gas gangrene: an unusual manifestation of *Aeromonas* infection. *J Med Assoc Thai* 1982; **65**: 678–80.

70 Warrier RP, Ducos R, Azeemuddin S, Ruff A. *Aeromonas* in an infant with aplastic anemia. *Pediatr Infect Dis* 1984; **3**: 491.

71 Vukmir RB. *Aeromonas hydrophila*: myofascial necrosis and sepsis. *Intensive Care Med* 1992; **18**: 172–174.

72 Abbott SL, Serve H, Janda JM. Case of *Aeromonas veronii* (DNA group 10) bacteremia. *J Clin Microbiol* 1994; **32**: 3091–2.

7 *Aeromonas* Species in Disease of Animals

PETER J. GOSLING

Department of Health, London, England

7.1 INTRODUCTION

The ubiquitous nature of *Aeromonas* species in aquatic environments provides ample opportunity for animals, particularly fish and amphibians, to come into contact with, and to ingest organisms. Occasionally, such contact may lead to infection which, depending on the animal species and the virulence of the strains encountered, may have life-threatening consequences. Motile *Aeromonas* species have long been recognized as primary pathogens to a variety of cold-blooded animals (poikilotherms). They are also increasingly being recognized as aetiological agents of disease syndromes such as septicaemia, pneumonia and wound infections, in warm-blooded animals (homoiotherms). A wide spectrum of animal species harbour the organisms gastrointestinally and in some they have been incriminated as a cause of gastroenteritis[1-4].

Only limited studies into the pathogenesis of *Aeromonas* spp. to animals have been conducted to date. In the vast majority of these the term '*A. hydrophila*' is used to refer to organisms contained within the genus *Aeromonas*, and not necessarily to those contained within that specific biotype. This has made it difficult to compare the pathogenicity of strains amongst the various *Aeromonas* species.

7.2 PATHOGENESIS OF NATURALLY OCCURRING INFECTION

The most frequently observed and best described naturally occurring infections caused by aeromonads occur in poikilotherms, and an insight to the pathogenesis of *Aeromonas* infection may be gained by considering the various syndromes.

7.2.1 COLD-BLOODED ANIMALS (POIKILOTHERMS)

Members of the genus *Aeromonas* are pathogenic to many poikilotherms, including amphibians, fish and reptiles. They are the aetiological agents of

The Genus Aeromonas. Edited by B. Austin, M. Altwegg, P.J. Gosling and S. Joseph
© 1996 John Wiley & Sons Ltd

several diverse illnesses which may present as haemorrhage, osteomyelitis, pneumonia, septicaemia, stomatitis, and ulcerative disease.

7.2.1.1 Reptiles

Infections with *Aeromonas* species causing ulcerative or necrotic stomatitis occur commonly in several species of snake and occasionally in lizards. Cases of *Aeromonas* septicaemia in a variety of other reptiles have additionally been reported. A number of disease syndromes of snakes in which *Aeromonas* species are aetiological agents have been described, including an acute septicaemia manifested by sluggishness, weakness and terminal convulsions, a frequently fatal pneumonia, and ulcerative stomatitis.

Systemic infection in snakes has been shown to be transmitted by the snake mite, *Ophionyssus natricis*, with the mite acting as a mechanical carrier through injection of the bacteria during feeding[5,6]. On systemic invasion by the organism, septicaemia develops, and death follows rapidly in the absence of an observable preceding illness. Septicaemic carcasses have haemorrhages in most organs, the oral mucosa, and on the mucosal surfaces of the intestinal tract, with a serosanguineous exudate in the abdominal cavity. Organs and tissues are hyperaemic, the liver is mottled, and the lungs are filled with a bloody fluid that extends to the trachea[7].

Pneumonia in reptiles is common and due to infection by a variety of microorganisms including *A. hydrophila*. *Aeromonas* bacteraemia resulting from snake bite, or aspiration of infectious caseous material of necrotic stomatitis, are possible sources of infection leading to pneumonia. Clinical signs may include nasal discharge, gaping of the mouth, wheezing respiratory sounds, anorexia, and death within 5 days. Gross lesions at post mortem include pulmonary congestion, accompanied by a fibrinopurulent or caseous exudate.

It has been shown that mucosal lesions, often caused during poison extraction from captive snakes, together with the stress originated by their living conditions may be important factors involved in the aetiology of the disease[8].

The clinical signs and gross lesions of ulcerative stomatitis of reptiles are distinctive and ushered in by the appearance of frothy, fibrinous exudate about the lips and in the mouth, followed by the refusal to eat. Small, yellowish-white, caseous masses develop within the buccal cavity within a week of the initial signs. As ulceration develops, the necrotic tissue that forms is friable, and the affected areas bleed readily when traumatized. The ulcers fill with a more abundant caseous exudate that is periodically ejected to leave fresh, bleeding wounds. The acute stage is of variable but short duration and over a period of weeks enters a chronic phase. The affected snake becomes steadily weaker and tooth loss and osteomyelitis may result as the disease progresses to involve the sockets of the teeth and bones of the jaw. Exudate on the palate may obstruct lacrimal flow from the duct of the Harderian gland that lubricates the potential space between the cornea and the spectacle, resulting in the accumulation of secretion, distension of the corneo-spectacular space, and subsequent swelling to as much as one-half the volume of the head[5,7].

7.2.1.2 Aquatic species

Motile *Aeromonas* isolates have been reported from a variety of fish species including ayu, bass, carp, catfish, perch and salmon. Strains of *A. hydrophila* have been reported to cause haemorrhagic septicaemia in fish and to play either a secondary or primary role in the fish disease Ulcerative Disease Syndrome (UDS), which is an epizootic fish disease characterized by the presence of severe open dermal ulcers on the head, mid-body and dorsal regions of the fish[9,10]. A full review of the pathogenesis of aeromonads to fish is given in Chapter 8 and will not therefore be considered further in this section.

Aeromonads may also affect other aquatic organisms and are associated with outbreaks of haemorrhagic septicaemia in freshwater eel farms in temperate areas[11-13]. Recent experimental studies have demonstrated that *A. hydrophila* and *A. jandaei* may act as primary pathogens for European eels (LD_{50} $10^{5.4}$ – $10^{7.6}$ CFU per fish), whereas strains of *A. sobria, A. caviae,* and *A. allosaccharophila* are apparently non-pathogenic at doses exceeding $10^{8.4}$ CFU per fish[14,15]. Pathogenic strains of *A. hydrophila* and *A. jandaei* produced elastases and haemolysins active against fish erythrocytes. Additionally, extracellular proteins (ECP) from these strains caused degenerative changes in fish tissue culture cell lines and were strongly toxic for eels (LD_{50} 1.0–3.2 mcg (g fish)$^{-1}$), reproducing the symptoms associated with natural disease. These biological activities were heat labile. ECP from non-pathogenic species were inactive on fish tissue culture cell lines as well as requiring a high dose to be lethal for eels. These findings indicate differences between pathogenic and non-pathogenic *Aeromonas* species with respect to the expression of virulence factors, and suggest that elastases, haemolysins and exotoxins play a leading role in the pathogenicity of motile *Aeromonas* for eels.

7.2.1.3 Amphibians

Motile *Aeromonas* spp. are pathogenic to turtles, causing high mortality, and responsible for the disease syndrome 'red-leg' in frogs[16,17]. Considerable information has accumulated regarding *Aeromonas* infection in frogs and is considered below.

At the onset of red-leg disease, the frogs are sluggish, muscle tone decreases, and within two days, foci of haemorrhage appear on the ventral surface of the body accompanied by extensive oedema of the abdomen and thighs, and minute ulcers on the toes. The haemorrhages and congestion turn the infected areas various shades of red and give the disease its common name. Other signs include bleeding from the nictitating membrane and haemorrhages within the eyeball. An acute systemic form of the disease may result in death without obvious signs or lesions, but *Aeromonas* spp. may be isolated from heart, blood and visceral organs on bacteriological examination.

At post-mortem oedematous fluid is typically present beneath the skin of the abdomen and thighs and multiple petechiae within the abdomen and thigh muscles. Small haemorrhages also occur on the surface of the tongue. The peritoneal cavity contains variable amounts of serosanguineous fluid, which teems with the infecting *Aeromonas* strain. The heart is pale and flaccid, and the lungs are often congested. The gastrointestinal tract is congested and greatly distended, and the lumen is filled with a viscid bloody exudate. Splenic enlargement is common, and the splenic surface is roughened by focal protrusions of the capsule[7].

Histopathological changes are those of extensive vascular congestion, haemorrhage and vascular dilation. Numerous bacteria are present within the blood vessels. Myodegeneration and myositis are characterized by an oedematous separation of muscle bundles and fibres and by the loss of cross-striation in the fibres. Focal areas of necrosis are present in the liver and kidneys[7].

Haematological investigation shows the erythrocytes to be deformed, swollen and vacuolated. Anaemia, thrombocytopenia, leucocytopenia due to granulocytopenia, and increases in clotting time and sedimentation rate are other clinical pathological alterations in frogs infected with *A. hydrophila*[7].

Antibodies to *Aeromonas* have been demonstrated in the serum of experimentally infected frogs[18]. Further experimental studies on the pathogenesis of *Aeromonas* infection in frogs, however, indicate that a very great number of bacteria (1.5×10^9) are required to cause death. These results suggest that factors in addition to the intact bacteria are required for full virulence[19]. Rigney and co-workers, studying the biological effects of endotoxin and haemolysin, found that neither the endotoxin nor the haemolysin alone produced the typical red-leg disease symptoms or death in frogs, even at a very high dosage of 8000 mcg, but that a combination of endotoxin and haemolysin did. Further evidence for synergy was obtained by demonstrating that haemolytic activity of crude haemolysin preparations increased in cells that were pre-incubated with endotoxin. Additionally it was shown that histamine-stressed frogs died from haemolysin but not endotoxin. It was postulated that one of the primary effects of haemolysin *in vivo* is the activation of histamine-heparin-releasing factors in serum or the release of histamine by a direct action upon mast cells, and that the higher level of histamine might act synergistically with haemolysin to produce red-leg symptoms. It should be noted, however, that the amounts of haemolysin required for synergy with histamine were significantly greater than those required with endotoxin. The evidence therefore suggests that red-leg disease in frogs represents a complex interaction between endotoxin and haemolysin (i.e. the endotoxin acts synergistically with other proteins or toxins) and that stress-producing factors other than the endotoxin might trigger the disease process.

An unusual epizootic of anuran aeromoniasis has been reported recently involving a breeding colony of the wholly aquatic African clawed frog, *Xenopus laevis*[20]. The frogs developed cutaneous ulcerations, visceral inflammation and fatal septicaemia, following a feeding programme involving freshly caught salmon which had been ground and added to the beef heart–vitamin–mineral feed mix. Large numbers of *Escherichia coli* and small numbers of *Enterococcus* spp., *Streptococcus* spp. and *A. salmonicida* were isolated from the water of the frog tanks and trays. Moderate numbers of *A. salmonicida* subsp. *masoucida* and small numbers of *A. hydrophila* were isolated from heart, blood and liver tissues. *A. salmonicida* was isolated from coelomic fluid. The recovery of *A. salmonicida*, typically a pathogen of teleost fish, from the visceral tissues and blood of these frogs is of interest and may broaden the species to which these organisms may be considered pathogenic.

7.2.2 WARM-BLOODED ANIMALS (HOMOIOTHERMS)

Aeromonads have been shown to be pathogenic for an increasing number of diverse species of homoiotherms including avians, aquatic and terrestrial mammals, such as cat, cattle, dog, dolphin, horse, fox, guinea-pig, mouse, monkey and rabbit. The illnesses have manifested in the form of septicaemia, pneumonia, peritonitis, and various localized infections frequently leading to death.

7.2.2.1 Mammals

Naturally occurring *Aeromonas* septicaemia has been described in dogs and a captive caracal lynx (*Felis caracal*), and there have been reports of pneumonia and dermatitis in dolphins, black rot in hens' eggs, abortion in cattle and water buffalo, and peritonitis in primates[21–30].

Additionally, septic arthritis in a calf and a foal has been reported, in which *Aeromonas* spp. were readily recoverable from synovial fluid. In the case involving the calf, it is postulated that the organism gained entry via a traumatized umbilicus, when the animal waded through a water channel shortly after birth. No localized cellulitis or abscess formation was evident, however, at this site or at any of its superficial abrasions and no evidence of internal lesions caused by the organism was found[31,32].

7.2.2.2 Avians

Aeromonas spp. appear widely disseminated in captive and free-living avian species and their environment. Naturally occurring *Aeromonas* infections, including epidemics, occur in avians, and reports include an epidemic death of feral birds, with symptoms similar to botulism; a septicaemic condition of commercial turkeys aged 3–16 weeks with flock morbidity ranging from 10 to 30% and mortality of 1.0 to 5.0%; conjunctivitis in a pet parrot (*Amazona versicolor*); and haemorrhagic septicaemia in a ground hornbull (*Bucorvus abyssinicus*)[33–36].

Additionally, isolation of *A. hydrophila* in pure culture has been reported from the viscera of companion birds with enteritis, including canaries, a toucan, and a cockatiel, and has been incriminated as the aetiological agent[37]. Faecal carriage is common, however, but rates vary amongst birds of differing species and environmental habitat. A recent report observed that birds from aquatic habitats harboured aeromonads more frequently (14.8%) than those from terrestrial habitats (1.8%). Furthermore, an influence of the kind of food on the carriage rate can be assumed: granivorous and herbivorous birds harboured aeromonads at a rate of 1.5% or 6.8%, omnivorous birds at 2.7% or 11.4%, and carnivorous and insectivorous birds at 4.0% or 21.0%, in terrestrial or aquatic habitats, respectively. *A. hydrophila* is isolated more frequently from wild birds during summer (12.9%) than in winter (8.9%)[38,39].

Variation in susceptibility of avians to *Aeromonas* infections has been demonstrated experimentally. For example, 2–4-day-old chickens and poults are highly susceptible to exposure to *A. hydrophila* (approx 10^9) introduced by the yolk-sac, intracerebral or intramuscular route; ducklings, however are generally refractory to exposure[40].

Additionally, a synergistic relationship between *Salmonella infantis* and *A. hydrophila* has been demonstrated experimentally in newly hatched poults: neither organism produced mortality when inoculated individually, but death occurred in 30% of the subjects that received both pathogens[41].

Predisposing factors such as previous debilitation of birds by environmental stress, intercurrent infection, or injury, seem to be necessary to provide the outbreak of disease[42].

7.3 PATHOGENESIS OF EXPERIMENTAL INFECTION

7.3.1 LABORATORY ANIMALS

The experimental inoculation of the usual laboratory animals such as mice and rabbits, has resulted in irregular strain-dependent outcomes.

7.3.1.1 Mice

Daily and co-authors[43] demonstrated, using 24 strains, that *Aeromonas* spp inoculated intraperitoneally into mice yielded LD_{50} values ranging from $10^{3.7}$ to $10^{8.5}$, while other workers[44] reported that not all strains of anaerogenic aeromonads they studied were pathogenic for mice.

Mice surviving infection following intraperitoneal injection of 10^7 cells of a strain of *A. hydrophila* were shown by Chakaborty and colleagues to display a rapid antibody response to β-haemolysin, which indicates a role for the toxin in the infection process[45].

Studies have been conducted into the pathophysiology of experimental *A. hydrophila* infection in mice[46]. In a study of 511 mice, intramuscular inoculation of a bacterial suspension (0.1 ml; 3×10^7 CFU) of a virulent fish strain led during the first 3 days to septicaemia, tissue damage, endotoxic shock and produced 75% mortality within 6 days and 50% mortality within 3 days. Histological examination typical of systemic *Aeromonas* infection revealed: severe muscle necrosis at the injection site; oedema, haemorrhage and neutrophil infiltration of the lung; and focal parenchymal necrosis in the liver. Significant increases in aspartate aminotransferase, alanine aminotransferase, intestinal bilirubin and blood urea nitrogen were noted in blood and intestinal samples; decreased plasma glucose and haematological changes were also recorded. Ketones, increased protein, glucose, bilirubin and blood were detected in the urine. Endotoxaemia was demonstrated as early as 2 h after inoculation and persisted for more than 36 h. The results clearly indicate that *Aeromonas* endotoxin also contributes to the pathogenesis of experimentally infected mice.

The ability of *A. sobria* and *A. hydrophila* isolates to cause subcutaneous lesions in mice was investigated by Brook and colleagues using an inoculum of 10^{11} CFU/l injected subcutaneously[47]. Surviving animals developed a subcutaneous abscess and/or localized skin sloughing and loss of hair (alopecia). An abscess was induced by all *A. hydrophila* tested and by 20% of the *A. sobria* strains, demonstrating the pathogenic potential of such strains.

7.3.1.2 Rabbits

Marked strain-dependent outcomes following experimental inoculation of aeromonads into rabbits have also been observed. Shilkin and associates injected a strain of *A. hydrophila*, isolated from human ankle ulcer, intravenously into a single rabbit, and studied the outcome[48]. The organism killed the rabbit within 18 h, with widespread haemorrhages in the striated muscles and, to a lesser extent, in the cardiac muscle. The *Aeromonas* strain was freely recovered from most tissues. Histological examination revealed acute and severe muscle lesions similar to those seen in their human subject. Conversely, Slotnick reported that using eight human isolate strains of *A.*

hydrophila, despite of injection with cultures containing 10^9 cells per ml, by several routes, including intravenous, intraperitoneal, and intramuscular, little, if any pathogenicity for rabbits was observed[49]. He suggested that the age of the rabbit may also affect the outcome to exposure, adult rabbits being less susceptible to *Aeromonas* infection.

7.3.2   INTESTINAL LIGATION MODELS

Over the last three decades various procedures involving animals have been designed to detect enterotoxins. Several have been used both to determine the enterotoxigenicity of *Aeromonas* strains and to study the effect on the animal following introduction of aeromonads into the intestinal ligation model.

7.3.2.1   Rabbit ileal loop

The ligated rabbit ileal loop (RIL) test, developed for detection of cholera toxin, was the first test used and was subsequently used for detection of most enterotoxin-producing organisms[50]. Maximal response may occur after different holding times with different enterotoxins; e.g. 5 h for *E. coli* ST but 10–18 h for *E. coli* LT and cholera toxin CT. *Aeromonas* species were first incriminated as enteropathogens following inoculation into the RIL test of whole cells by Sanyal and co-workers, who used a holding time of 16–18 h and later by others inoculating cell-free culture filtrates[51-53]. In addition to humans, strains originating from cows, goat, buffalo and chicken have been shown to accumulate fluid in RIL when whole cells were used[52]. Several other investigators have used the RIL tests to detect enterotoxin production by *Aeromonas* strains[54-57].

Additional *in vivo* assays are the infant rabbit test (IRT), the infant mouse test, and the vascular permeability (skin) test. The IRT involves the introduction of test organisms by gastric intubation or by direct injection into the stomach and measurement of fluid accumulation after a 7 h holding period. This test has been found to be more sensitive than the RIL test for detecting enterotoxigenicity in enterotoxigenic *E. coli* from infants with diarrhoea[58,59]. However, aeromonads have not been tested in this model (to the author's knowledge), but it is in any case of limited use as it is also unsuitable for mass screening of strains.

7.3.2.2   Removable intestinal tie adult rabbit diarrhoea model

A modification of the RIL and IRT test in which the small intestine of an adult rabbit is temporarily ligated, the RITARD test was used by Pazzaglia and colleagues as a screening method for detecting toxigenic *Aeromonas* isolates and as a tool for studying possible mechanisms of pathogenesis[60].

Some isolates produced a bacteraemia in less than 12 h followed by death in less than 24 h, often accompanied by moderate amounts of diarrhoea, fluid accumulation in the small bowel and lesions of gut mucosal epithelium, from which they concluded that some *Aeromonas* strains are diarrhoeagenic. The RITARD model correlated with but offered no advantages over simpler mouse lethality and infant mouse assays for routine screening of *Aeromonas* isolates for enteropathogenesis[61,62]. The application of the RITARD model for intestinal challenge with *Aeromonas* strains has shown that some isolates are capable of invading the mucosa of rabbits, causing diarrhoea and bacteraemia. Death, diarrhoea and bacteraemia were all strongly strain-dependent. Gastrointestinal lesions varied from moderate focal enteritis to severe multifocal necrosis and haemorrhage of the ileal mucosa, often accompanied by hepatic and splenic lesions. Intestinal colonization assays performed after infection indicated that the ileum was the most heavily colonized portion of the gut and the probable site of invasion[60].

7.3.2.3 Other intestinal ligation models

The rat intestinal perfusion assay is another *in vivo* loop method. This assay, which involves measurement of fluid changes by monitoring the concentration of a non-absorbable marker, has been used to detect enterotoxin from several species of organism but is known to show positive reponses while other systems remain negative[63]. Millership and co-authors[64] successfully used this technique to demonstrate the effect in support of the existence of an enterotoxic *A. sobria* β-haemolysin.

7.3.3 SUCKLING MOUSE MODEL

The suckling mouse assay has proved to be a reliable system for detecting enterotoxigenic strains of *Aeromonas* when standard conditions for growth and toxin testing are used. Enterotoxins are produced by bacteria grown in tryptone soya broth supplemented with yeast extract and aerated by shaking in an environmental incubator or waterbath. When culture supernatants together with dye are administered intragastrically to mice less than 6 days old, the presence of enterotoxin indicated by fluid accumulation is assessed on the basis of a scoring system that incorporates the ratio intestinal weight/remaining bodyweight[56,65].

7.3.4 EPICUTANEOUS MODELS

Epicutaneous models offer alternative systems to the *in vivo* ligated loop tests such as the RIL test, the RITARD test and the rat intestine perfusion assay, which are cumbersome, unpleasant, expensive and are not suitable for screening large numbers of strains.

7.3.4.1 Vascular permeability

The rabbit vascular permeability assay is a non-gastrointestinal test for detection of enterotoxigenicity shown to correlate well with the RIL test when used to detect CT, and subsequently shown to be valid for certain cytotoxic enterotoxins such as that of *Clostridium perfringens* and *Bacillus cereus*[66–71]. The test involves the intradermal injection of a small amount of test material (0.1 ml) followed after an appropriate time interval by an intravenous injection of approximately 1.5 ml/kg bodyweight of 2% Evans Blue dye. Increased permeability of the cutaneous vessels to blood plasma around the test injection site is indicated by the formation of a zone of blueing presumed to represent action of the toxins on the endothelial cells of the skin capillaries analogous to that which occurs in the intestinal epithelial cells. Despite being appropriate for CT the skin test has not been found to be applicable to LT, which is immunologically related and which also acts through the adenylate cyclase/cAMP system[72–75].

Skin tests have been used in studies on enterotoxins of *A. hydrophila* although they suffer from the problem of differentiating the action of haemolysin from enterotoxic response. The usefulness of this test for detection of *Aeromonas* enterotoxigenicity is therefore doubtful.

The ability to invade tissues assayed by the Sereny method (inoculation into adult guinea-pigs' eyes to see if keratoconjunctivitis develops over 8 days) have failed to be positive strains of *A. hydrophila*[57].

7.3.4.2 Tissue culture

A number of *in vitro* tests for detection of enterotoxigenic activity have been developed in recent years. The first was a tissue culture assay in which characteristic cytopathic effect was induced in Chinese hamster ovary (CHO) or Y-1 mouse adrenal tumour cell lines. In the case of CT, these are sensitive to as little as 10 picograms[76]. Characteristic elongation of CHO cells is also observed with LT. *A. hydrophila* is among the several species of organisms reported to cause similar cytopathic effects in the CHO assay but apart from CT and LT the correlation between this cytopathogenic effect (CPE) and typical enterotoxic responses in animal models has not been validated[77]. CPE in tissue culture is probably inappropriate for detection of enterotoxin in culture filtrates of organisms which produce other cytolytic toxins during growth, *Aeromonas* being a good example of this.

7.4 PATHOGENIC MECHANISMS

Aeromonas species have been isolated from a wide variety of animal species but their pathogenic role in many cases remains controversial. It is not possible in this chapter to consider comprehensively the pathogenic mechanisms of all disease syndromes reported to occur. Additionally, as relatively little is known about the pathogenic mechanisms of aeromonads in most diseases in which they have been incriminated, some of the better studied are taken as reflecting the pathogenic mechanisms involved within a disease syndrome across many animal species.

The pathogenic mechanisms of *Aeromonas* species in fish and humans are considered in depth in Chapters 8 and 6 respectively, and consideration of evidence specifically relating only to other animal species will be carried out in this section.

To assign an aetiological role to an organism there is a need to distinguish between carriage, or normal flora, and pathogen. In the case of *Aeromonas*, due to its ubiquitous nature in aquatic systems, this has been difficult; however, in some cases convincing evidence has been produced. Consideration is needed of what evidence would be acceptable in assigning pathogenicity, and this requires an evaluation of the organism's ability to cause disease and the mechanisms by which it does so.

7.4.1 SEPTICAEMIA

The pathogenicity of *Aeromonas* species in septicaemic diseases of animals is not in doubt. The source of the infecting aeromonads is generally considered to be of aquatic origin as this represents the largest pool of the organisms in the environment, and contact to the victim by either ingestion of the organism or by epithelial contact. Gut carriage may therefore play a role in resulting septicaemic episodes as it is assumed this is the major entry route to the bloodstream. However, pathogenicity cannot be assumed merely by isolating *Aeromonas* spp. from faeces or the epithelium of animal species, particularly if this is in intimate contact with water that may contain a large population of aeromonads. Clearly, strains capable of either damaging the gut mucosa and entering the bloodstream, or of entering already damaged mucosa and subsequently resisting the humoral defence systems, must possess certain virulence factors that augment their pathogenesis.

In septicaemic episodes it is apparent that the strains responsible are predominantly contained within species *A. hydrophila* and *A. sobria*, and these species characteristically produce a wide variety of extracellular enzymes and toxins, which are putative virulence factors (see Table 7.1).

The role of extracellular enzymes and toxins, in particular the various proteases, endotoxin and the potent β-haemolysin in the disease process, has been elucidated in septicaemia of frogs and mice. However the sensitivity of erthrocytes from different animal species to lysis by β-haemolysin varies, rabbit erythrocytes being considerably more sensitive than those from humans, horse and sheep, and this may indicate varying degrees of resistance to disease (Table 7.2)[95,100].

Table 7.1.   Putative virulence factors of *Aeromonas* strains in animal disease

Virulence factors	Reference	Experimental evidence
Adhesins	43, 78, 79	Adherence to rabbit intestine
Endotoxin	19, 43, 80	Toxic to mice, contributes to red-leg disease in frogs
Enterotoxins	51, 53, 56, 81, 82, 83	Fluid accumulation in rabbit, rat, and mouse intestinal loop experiments; cytopathogenic effects to CHO and Y-1 mouse adrenal tumour cell lines
Enzymes including: amylases, chitinase, elastase, lecithinase, nucleases, phospholipases, proteases	84–94	Putative cause of necrosis in snakes and contributes to muscle necrosis and degeneration of connective tissue
Haemolysins β-haemolysin (aerolysin)	81, 95	Dermonecrotic in rabbit skin; lethal to mice; complete haemolysis of erythrocytes from various species (rat erythrocytes being particularly susceptible)
α-haemolysin	96, 97	Partial haemolysis of erythrocytes from various species (including bovine and human)
Cytotoxin	43, 56, 81, 98, 89, 99	Membrane damage in rabbit intestinal loop experiments

Variation in relative sensitivity has been recorded, however; Wretlind and colleagues observed only a 16-fold difference in sensitivity between cells of the most sensitive and resistant species, and considered rat to be as resistant as sheep[101]. Other workers have also reported differing sensitivity patterns but it is probable that these differences reflect the state of purity of the haemolysin preparations[102,103]. Erythrocyte membranes of different species bind the toxin, and the efficiency of binding is a function of sensitivity to lysis[95].

It is probable that the ability to bind to cell membranes or mucus layers will play a critical role in pathogenesis of *Aeromonas* spp. Levett and Daniel showed that four of the 23 (17%) clinical *A. hydrophila* strains they studied possessed adhesins to the microvillous surfaces of rabbit brush borders[104]. Crichton and Walker reported that some *Aeromonas* strains were able to haemagglutinate erythrocytes of domesticated fowls (hens), guinea-pigs, horses, humans, pigs, rabbit, rat, sheep and oxen[105]. Stewart and colleagues[106] demonstrated, however, that while cell-bound agglutins could be demonstrated, soluble haemagglutinins were not detectable to sheep and cow erythrocytes.

The fate of *A. hydrophila* after bacteraemia in four healthy pigs

Table 7.2. Relative sensitivity to β-haemolysin (aerolysin) of erythrocytes of various animal species. Reproduced by permission of the American Society for Microbiology

Species	Sensitivity of erythrocytes compared with that of rats (%)
Rat	100.0
Mouse	78.0
Dog	18.0
Guinea-pig	18.0
Cat	16.0
Rabbit	3.1
Ox	2.4
Swine	1.1
Human	0.9
Horse	0.5
Goat	0.4
Sheep	0.3

From Bernheimer and Avigad[95]

intravenously inoculated with 10^9 and 10^{15} cells of the bacterium was investigated recently[107]. In two pigs slaughtered 30 min after inoculation the bacteria were found in blood (8×10^2/ml and 4.2×10^2/ml), meat and fatty tissues (1–10/g), but not in liver, spleen and superficial inguinal lymph nodes. *A. hydrophila* was not found in blood, meat, fatty tissue, liver, spleen, kidneys and superficial inguinal lymph nodes of the other two pigs slaughtered 5.5 h postinoculation, with the exception of one superficial inguinal lymph node of one pig. This study demonstrates that animal hosts may vary in their ability to resist systemic infection with *Aeromonas* spp.

Recent studies have indicated that a number of bacterial species produce a surface array protein that is arranged in an orderly fashion on the outermost surface of the bacterium as a paracrystalline 'S' layer[108,109]. While the S layer of *A. salmonicida* has been shown to be essential to the virulence of *A. salmonicida* for fish (salmon), little information currently exists regarding the function and potential pathogenic role that surface array proteins play in other S layer-containing bacteria[110,111]. It is of interest that an *A. sobria* strain (serogroup 0:11), with a 50% lethal dose (intraperitoneal injection) in mice, containing an S layer was recently isolated from a frog, and further investigations are required to ascertain its full pathogenic significance[112].

7.4.2 EVIDENCE FOR ENTEROPATHOGENICITY IN ANIMALS

Aeromonas species have been incriminated as naturally occurring enteropathogens in homoiotherms following isolation of aeromonads in diarrheotic faeces from birds and piglets[1,37], and there is some experimental evidence to support enteropathogenicity in the RITARD model. However, aeromonads are harboured intestinally apparently without signs of disease by many animal species, and the evidence to date for enteropathogenicity remains largely circumstantial. Despite this, strains of *Aeromonas* spp. appear to have several attributes in favour of its putative enteropathogenicity, including the ability to adhere to animal intestinal cells, to invade tissue culture cells, to produce extracellular mucus-degrading proteases and to produce several cytotonic and cytotoxic enterotoxins (see Chapter 9). Additionally, the evidence in support of its enteropathogenicity in humans is increasing, but attempts to induce experimental diarrhoea in rabbits, guinea-pigs, hamsters and rats have failed even after elimination of the resident intestinal flora by prior antibiotic therapy, neutralization of gastric pH and slowing of intestinal motility[54]. Similar attempts were also repeated in monkeys inoculated with up to 10^9 organisms, without success[55]. To date, the only attempt to induce experimental diarrhoeal disease in humans by oral challenge was also unsuccessful[57].

Faecal isolation of an organism from an animal with diarrhoea is clearly, by itself, insufficient evidence to consider that organism an enteropathogen. In order to assess the role of *Aeromonas* species in diarrhoeal disease of animals, in view of the apparent inability to induce the disease experimentally, it may therefore be necessary to consider the criteria that must be fulfilled in order to designate it enteropathogenic.

7.5   DISCUSSION

Aeromonas spp. are pathogenic to a large number of animal species, yet few systematic studies into the pathogenesis of naturally occurring infections have been conducted. It is clear that most animal infections prior debilitation, stress or injury, contributes to the onset of disease. The role of endotoxins, haemolysin and other extracellular proteins in pathogenesis has been well established and it is probable that the importance of adhesins and surface array proteins will also be established as more data accumulate. Information on the characterization of animal strains of *Aeromonas* is sparse, with the exception of strains carried by the medicinal leech (*Hirudinidae*). These are used medically to relieve venous insufficiency of grafted skin flaps, breast reconstructions, etc. in humans, and aeromonads carried as sole gut flora of the leeches have been responsible for infections[113-116].

The author knows of no collection of *Aeromonas* strains orginating from animal sources, other than fish, and few of such strains have been represented in taxonomic databases. It is interesting that, amongst the 68 strains used by Popoff and Véron in their taxonomic study, only five animal strains (four frog and one guinea-pig) were included[117]. It is probable that animal isolates will provide a source of new *Aeromonas* biotypes and serotypes, and characterization and study of virulence factors of such strains may provide information concerning the pathogenic mechanisms in human infections, including septicaemia and diarrhoeal disease.

It is possible that those animals carrying certain strains of *Aeromonas* may also act as a reservoir for transfer of humans via the food chain[118]. This may have important implications as human *Aeromonas* septicaemia is generally considered to have a gastrointestinal source, and the organisms are putative enteropathogens to humans.

The high potential of aeromonads to be animal enteropathogens (i.e. ease of contact, adherence to animal intestinal cells, production of mucus-degrading enzymes and production of enterotoxins) but the apparent inability to reproduce the disease is something of an enigma.

The classic concepts of pathogenicity and virulence are based on the hypothesis of Koch; 'Koch's postulates'[119]. Clearly from this approach the inability, to date, to reproduce *Aeromonas*-associated diarrhoeal disease in animals or humans would preclude its acceptance as an enteropathogen. Koch's postulates, however, were directed at a scientific definition of the relationship between microorganisms and disease in humans over a century ago, and have been considered to have significant limitations, particularly in regard to enteric infections, and to relate more closely to modern concepts of infectivity rather than pathogenicity[120]. Koch's postulates fail to address limitations of laboratory methods: portal of entry for assessing pathogenicity, e.g. mice do not respond to encapsulated aerolized *Steptococcus pneumoniae*, but intraperitoneal infection of these organisms is invariably lethal; the number of organisms required to initiate infection, e.g. *Shigella* species require very few bacteria whereas *Salmonella* species need many thousands; polymicrobial infections; and possible lack of animal models. Additionally, these postulates place the onus of disease solely on the microorganism, ignoring the role of the host, environment and risk factors not directly related to a particular health problem.

It is therefore ill-advised to regard *Aeromonas* species as non-enteropathogens on the grounds that they do not fully comply with Koch's postulates. A more recent appraisal of the criteria needed to be fulfilled when establishing an organism's enteropathogenicity, if adapted for animals, would suggest that: initially the organism should be isolated in faeces of the animal species with diarrhoea compared to an appropriate matched control group without diarrhoea; experimental disease should ideally be induced by the pathogen in the animal species concerned; and finally, the virulence traits of the putative enteropathogens should be investigated and the pathophysiology and immunological response to the specific traits should be evaluated in order to obtain evidence in support of an organism being a true enteropathogen[121]. An assessment of the role of *Aeromonas* species as enteropathogens of animals using such criteria should be conducted.

7.6 REFERENCES

1   Dobrescu L. Enterotoxigenic *Aeromonas hydrophila* from a case of piglet diarrhoea. *Zentralbl Vet Med* 1978; **25**: 713–18.
2   Gray SJ. *Aeromonas hydrophila* in livestock: incidence, biochemical characteristics and antibiotic susceptibility. *J Hyg Camb* 1984; **92**: 365–75.
3   Gray SJ, Stickler DJ. Some observations on the faecal carriage of mesophilic *Aeromonas* species in cows and pigs. *Epidemiol Infect* 1989; **103**: 523–37.
4   Stern NJ, Drazek ES, Joseph SW. Low incidence of *Aeromonas* sp. in livestock faeces. *J Food Protect* 1987; **50**: 66–9.
5   Marcus LC. *Veterinary Biology and Medicine of Captive Amphibians and Reptiles.* Philadelphia: Lea and Febiger, 1981: 83–95.
6   Camin JH. Mite transmission of a haemorrhagic septicaemia in snakes. *J Parasitol* 1948; **34**: 345–54.
7   Carlton WW, Hunt RD. Bacterial diseases. In Benirschke K, Garner FM, Jones TC (eds) *Pathology of Laboratory Animals.* New York: Springer-Verlag, 1978; 1373–7.
8   Hipolito M, Mavridis SC, Baldassi L. *Aeromonas hydrophila* and *Pseudomonas aeruginosa* isolated from stomatitis in *Bothrops alternatus* (*Serpente, Viperidea*). *Rev Microbiol Sao Paulo* 1987; **18**: 224–8.
9   Austin B. Diseases of fish. *Proc 1st International Workshop on* Aeromonas *and* Plesiomonas, Manchester, England, 1986: 16.
10 McGarey DJ, Foley DP, Reyer B Jr, Frye LC, Lim DV. The role of motile aeromonads in the fish disease, ulcerative disease syndrome (UDS). *Proc 3rd International Symposium on* Aeromonas *and* Plesiomonas, Helsingør, Denmark, 1990.
11 Rickards WL. *A Diagnostic Manual of Eel Diseases Occurring Under Culture Conditions in Japan.* North Carolina: Sea Grant Pub. UNC-SG-78-06, 1978.
12 Kou G-H. Some bacterial diseases of eel in Taiwan. In *Proc Republic of China–United States Cooperative Science Seminar on Fish Diseases. NSC Symp Series No 3*, Republic of China: National Science Council, 1981; 11–20.
13 Esteve C, Garay E. Heterotrophic bacterial flora associated with European eel *Anguilla anguilla* reared in fresh water. *Nippon Suisan Gakkaishi* 1991; **57**: 1369–75.
14 Esteve C, Biosca EG, Amaro C. Virulence of *Aeromonas hydrophila* and some other bacteria isolated from European eels *Anguilla anguilla* reared in fresh water. *Dis Aquatic Org* 1993: **16**: 15–20.
15 Esteve C, Amaro C, Garay E, Santos Y, Toranzo AE. Pathogenicity of live bacteria and extracellular products of motile *Aeromonas* isolated from eels. *J Appl Bacteriol* 1995; **78**: 555–62.
16 Brackee G, Gunther R, Gillett CS. Diagnostic exercise: high mortality (*Aeromonas hydrophila*) in red-eared slider turtles (*Pseudemys scripta elegans*). *Lab Animal Sci* 1992; **42**: 607–9.
17 Benirschke K, Garner FM, Jones TC. Bacterial diseases. In *Pathology of Laboratory Animals*, Vol 2. New York: Springer-Verlag, 1978.
18 Kulp WL, Borden DG. Further studies on *Proteus hydrophilus*, the aetiological agent in 'red-leg' disease of frogs. *J Bacteriol* 1942; **44**: 673–85.
19 Rigney M, Zilinsky J, Rouf M. Pathogenicity of *Aeromonas hydrophila* in red leg disease in frogs. *Curr Microbiol* 1978; **1**: 175–81.
20 Frye FL. An unusual epizootic of anuran aeromoniasis. *J Am Vet Med Assoc* 1985; **187**: 1223–4.
21 Pierce RL, Daley CA, Gates CE, Wohlgemuth K. *Aeromonas hydrophila* septicaemia in a dog. *J Am Vet Med Assoc* 1973; **162**: 469.

22 Andre-Fontaine G, Monfort P, Buggin-Daubie M, Filloneau C, Ganiere JP. Fatal disease mimicking letospirosis in a dog, caused by *Aeromonas hydrophila*. *Comp Immun Microbiol Infect Dis* 1995; **18**: 69–72.

23 Ocholi RA, Spencer THI. Isolation of *Aeromonas hydrophila* from a captive caracal lynx (*Felis caracal*). *J Wildlife Dis* 1989; **25**: 122–3.

24 Cusick PK, Bullock BC. Ulcerative stomatitis and pneumonia associated with *Aeromonas hydrophila* infection in the bottle-nosed dolphin. *J Am Vet Med Assoc* 1973; **163**: 578–9.

25 Miles AA, Halnan ET. A new species of micro-organism (*Proteus melanovogenes*) causing black rot in eggs. *J Hyg (Camb)* 1937; **37**: 79–97.

26 Wohlgemuth K, Pierce RL, Kirkbride CA. Bovine abortion associated with *Aeromonas hydrophila*. *J Am Vet Med Assoc* 1972; **160**: 1001–2.

27 Love RJ, Love DN. *Aeromonas hydrophila* isolated from polyarthritis in a calf. *Aust Vet J* 1984; **61**: 65.

28 Shane SM, Gifford DH. Prevalence and pathogenicity of *Aeromonas hydrophila*. *Avian Dis* 1985; **29**: 681–9.

29 Das AM, Paranjape VL. Aeromonas sobria in bubaline (water buffalo) abortion: growth requirement, biochemical characters, antibiotic susceptibility, experimental pathogenicity and serology. *Ind J Exp Biol* 1990; **28**: 341–5.

30 Califoux LV, Hajema EM, Lee-Parritz D. *Aeromonas hydrophila* peritonitis in a cotton-top tamarin (*Saguinus oedipus*), and retrospective study of infections in seven primates. *Lab Animal Sci* 1993; **43**: 355–8.

31 Glunder G. Zum Vorkommen von *Aeromonas hydrophila* bei Vogeln. *J Vet Med* 1988; **35**: 331–7.

32 Traub-Dargatz JL, Schlipf JW Jr, Atwell E, Bennett DG, Jones RL, Ehrhart EJ, Scultheiss PC. *Aeromonas hydrophila* septic arthritis in a neonatal foal. *Equine Pract* 1994; **16**: 15–17.

33 Korbel R, Kosters J. Seuchenhaftes Sterben von Wildvogeln nach *Aeromonas hydrophila* infektion. *Tierarztl Prax* 1989; **17**: 297–8.

34 Gerlach H, Bitzer K. Infection with *Aeromonas hydrophila* in young turkeys. *DTW Dtsch Tieraerztl Wochenschr* 1971; **78**: 593–608.

35 Garcia ME, Domenech A, Dominguez L, Ramiro F, Fernandez-Garayzabal JF. *Aeromonas hydrophila* conjunctivitis in a pet parrot (*Amazona versicolor*). *Avian Dis* 1992; **36**: 1110–11.

36 Ocholi RA, Kalejaiye JO. *Aeromonas hydrophila* as cause of haemorrhagic septicaemia in a ground hornbull (*Bucorvus abyssinicus*). *Avian Dis* 1990; **34**: 495–6.

37 Panigrahy BJJ, Mathewson CF, Hall CF, Grumbles LC. Unusual disease conditions in pet and aviary birds. *J Am Vet Med Assoc* 1981; **178**: 394–5.

38 Glunder G, Siegman O. Occurrence of *Aeromonas hydrophila* in wild birds. *Avian Pathol* 1989; **18**: 685–95.

39 Needham JR, Kirkwood JK, Cooper JE. A survey of the aerobic bacteria in the droppings of captive birds of prey. *Res Vet Sci* 1979; **27**: 125–6.

40 Shane SM, Gifford DH. Prevalence and pathogenicity of *Aeromonas hydrophila*. *Avian Dis* 1985; **29**: 681–9.

41 Saif YM, Busch WF. *Aeromonas* and *Salmonella* infections in turkey poults. *Rep Ohio Agr Res and Dev Centre* 1974; 119–120.

42 Shane SM, Harrington KS, Montrose MS, Roebuck RG. The occurrence of *Aeromonas hydrophila* in avian diagnostic submissions. *Avian Dis* 1984; **28**: 804–7.

43 Daily OP, Joseph SW, Coolbaugh JC, Walker RK, Merrell BR, Rollins DM, Seidler RJ, Colwell RR, Lissner CR. Association of *Aeromonas sobria* with human infection. *J Clin Microbiol* 1981; **13**: 769–77.

44 Schubert RHW. Die pathogenitat der *Aeromonaden* fur mensch und tier. *Arch Hyg Backteriol* 1967; **150**: 709.

45 Chakraborty T, Montenegro MA, Sanyal SC, Helmuth R, Bulling E, Timmis KN. Cloning of enterotoxin gene from *Aeromonas hydrophila* provides conclusive evidence of production of a cytotonic enterotoxin. *Infect Immun* 1984; **46**: 435–44.

46 Brenden RA, Huizinga HW. Pathophysiology of experimental *Aeromonas hydrophila* infection in mice. *J Med Microbiol* 1986; **21**: 311–17.

47 Brook I, Rogers J, Rollins DM, Coolbaugh JC, Walker RI. Pathogenicity of *Aeromonas*. *J Infect* 1985; **10**: 32–7.

48 Shilkin KB, Annear DI, Rowett LR, Lawrence BH. Infection due to *Aeromonas hydrophila*. *Med J Aust* 1968; **1**: 351.

49 Slotnick IJ. *Aeromonas* species isolates. *Ann N Y Acad Sci* 1970; 503–10.

50 De SN, Chatterjee DN. An experimental study of the intestinal mucous membrane. *J Pathol Bacteriol* 1953; **66**: 559.

51 Sanyal SC, Singh SJ, Sen PC. Enteropathogenicity of *Aeromonas hydrophila* and *Plesiomonas shigelloides*. *J Med Microbiol* 1975; **8**: 195–8.

52 Wadström T, Wretlind B. Enterotoxin, haemolysin and cytotoxic protein in *Aeromonas hydrophila* from human infections. *Acta Pathol Microbiol* 1976; **84**: 112–14.

53 Annapurna E, Sanyal SC. The characterization and significance of *Plesiomonas shigelloides* and *Aeromonas hydrophila* isolated from an epidemic of diarrhoea. *Indian J Med Res* 1977; **62**: 1051–60.

54 Ljungh A, Kronevi T. *Aeromonas hydrophila* toxins–intestinal fluid accumulation and mucosal injury in animal models. *Toxicon* 1982; **20**: 397–407.

55 Pitarangsi C, Escheverria P, Whitmire R, Tirapat C, Formal S, Dammin GJ, Tingtalapong M. Enteropathogenicity of *Aeromonas hydrophila* and *Plesiomonas shigelloides*: Prevalence among individuals with and without diarrhoea in Thailand. *Infect Immun* 1982; **35**: 666–73.

56 Turnbull PC, Lee JV, Miliotis MD, Van-De Walle S, Koornhof HJ, Jeffery L, Bryant TN. Enterotoxin production in relation to taxonomic grouping and source of isolation of *Aeromonas* species. *J Clin Microbiol* 1984; **2**: 175–80.

57 Morgan DR, Johnson PC, DuPont HL, Satterwhite TK, Wood LV. Lack of correlation between known virulence properties of *Aeromonas hydrophila* and enteropathogenicity for humans. *Infect Immun* 1985; **50**: 62–5.

58 Gorbach SL, Khurana CM. Toxigenic *Escherichia coli*: A cause of infantile diarrhoea in Chicago. *N Engl J Med* 1972; **287**: 791–5.

59 Levine MM, Berquist EJ, Nalin DR, Waterman DH, Hornick RB, Young CR, Sotman S, Rowe B. *Escherichia coli* strains that cause diarrhoea but do not produce heat-labile or heat-stable enterotoxins and are non-invasive. *Lancet* 1978; **i**: 119.

60 Pazzaglia G, Escalante JR, Sack RB, Rocca C, Benavides V. Transient intestinal colonization by multiple phenotypes of *Aeromonas* species during the first week of life. *J Clin Microbiol* 1990; **28**: 1842–6.

61 Janda JM, Clark RB, Brenden R. Virulence of *Aeromonas* species as assessed through mouse lethality studies. *Curr Microbiol* 1985; **12**: 163–68.

62 Burke V, Gracey M, Robinson J, Peck D, Beaman J, Bundell C. The microbiology of childhood gastroenteritis: *Aeromonas* species and other infective agents. *J Infect Dis* 1983; **148**: 68–74.

63 Klipstein FA, Guerrant RL, Wells JG, Short HB, Engert RF. Comparison of assay of coliform enterotoxins by conventional techniques versus *in vivo* intestinal perfusion. *Infect Immun* 1979; **25**: 146–52.

64 Millership SE, Barer MR, Mulla RJ, Maneck S. Enterotoxic effects of

Aeromonas sobria haemolysin in rat jejunal perfusion system identified by specific neutralization with a monoclonal antibody. *J Gen Microbiol* 1992; **138**: 261–7.

65 Burke V, Robinson J, Berry RJ, Gracey M. Detection of entertoxins of *Aeromonas hydrophila* by a suckling-mouse test. *J Med Microbiol* 1981; **14**: 401–8.

66 Craig JP. A permeability factor (toxin) found in cholera stools and culture filtrates and its neutralization by convalescent cholera sera. *Nature* 1965; **207**: 614–16.

67 Hauschild AHW. Erythemal activity of the cellular enteropathogenic factor of *Clostridium perfringens* type A. *Can J Microbiol* 1970; **16**: 651.

68 Niilo L. Mechanism of action of the enteropathogenic factor of *Clostridium perfringens* type A. *Infect Immun* 1971; **3**: 100.

69 Stark RL, Duncan CL. Transient increase in capillary permeability induced by *Clostridium perfringens* type A enterotoxin. *Infect Immun* 1972; **5**: 147.

70 Glatz BA, Spira WM, Goepfert JM. Alteration of vascular permeability in rabbits by culture filtrates of *Bacillus cereus* and related species. *Infect Immun* 1974; **10**: 299.

71 Turnbull PCB, Kramer JM, Jorgensen K, Gilbert RJ, Melling J. Properties and production characteristics of vomiting, diarrhoeal and necrotizing toxins of *Bacillus cereus*. *Am J Clin Nutr* 1979; **32**: 219.

72 Smith HW, Halls S. The transmissible nature of the genetic factor in *Escherichia coli* that controls enterotoxin production. *J Gen Microbiol* 1968; **52**: 319.

73 Sack RB, Gorbach SL, Banwell JG, Jacobs B, Chatterjee BD, Mitra RC. Enterotoxigenic *Escherichia coli* isolated from patients with severe cholera-like disease. *J Infect Dis* 1971; **123**: 378.

74 Moon HW, Whipp SC. System for testing the enteropathogenicity of *Escherichia coli*. *Ann NY Acad Sci* 1971; **176**: 197.

75 Singh A, Gaind S, Sethi SK. Differential biological characteristics of *Vibrio cholerae* and *Escherichia coli* enterotoxins. *Ind J Med Res* 1976; **64**: 410.

76 Donta ST, King M. Induction of steriodogenesis in tissue culture by cholera enterotoxin. *Nature N Biol* 1973; **243**: 246–7.

77 Ljungh A, Popoff M, Wadström T. *Aeromonas hydrophila* in acute diarrhoeal disease: detection of enterotoxin and biotyping of strains. *J Clin Microbiol* 1977; **6**: 96–100.

78 Atkinson HM, Trust TJ. Haemagglutination properties and adherence ability of *Aeromonas hydrophila*. *Infect Immun* 1980; **27**: 938–46.

79 Honna Y, Nakasone N. Pili of *Aeromonas hydrophila*: Purification, characterisation and biological role. *Microbiol Immunol* 1990; **34**: 83–98.

80 Shaw DH, Hodder HJ. Lipopolysaccharides of the motile aeromonads; core oligosaccharide analysis as an aid to taxonomy classification. *Can J Microbiol* 1978; **24**: 864–8.

81 Wadström T, Ljungh A, Wretlind B. Entertoxin, haemolysin and cytotoxic protein in *Aeromonas hydrophila* from human infections. *Acta Pathol Microbiol* 1976; **84**: 112–14.

82 Dubey RS, Sanyal SC. Enterotoxigenicity of *Aeromonas hydrophila*: skin responses and *in vivo* neutralization. *Zentralbl Bakteriol Hyg* 1978; **242**: 487–99.

83 Gosling PJ, Turnbull PCB, Lightfoot NF, Pether JVS, Lewis RJ. Isolation of *Aeromonas sobria* cytotonic enterotoxin and beta-haemolysin by fast protein liquid chromatography. *J Med Microbiol* 1992; **38**: 227–34.

84 Gobius KS, Pemberton JM. Molecular cloning, characterisation and nucleotide sequence of an extracellular amylase gene from *Aeromonas hydrophila*. *J Bacteriol* 1988; **170**: 1325–32.

194 THE GENUS *AEROMONAS*

85 Mitsutomi MA, Ohtakara A, Fukamizo T, Goto S. Action patterns of *Aeromonas hydrophila* chitinase on partially N-acetylated chitosan. *Agric Biol Chem* 1990; **54**: 871–7.
86 Roffey PE, Pemberton JM. Cloning and expression of an *Aeromonas hydrophila* chitinase gene in *Escherichia coli. Curr Microbiol* 1990; **21**: 329–37.
87 Yabuki M, Mizushina K, Amatatsu T, Ando A, Fujii T, Shimada M, Yamashita M. Purification and characterisation of chitinase and chitobiase produced by *Aeromonas hydrophila* subsp. *anaerogenes* A52. *J Gen Microbiol* 1986; **32**: 25–38.
88 Hasan JAK, Carnahan AM, Macaluso P, Joseph SW. Elastolytic activity among newly recognised *Aeromonas* spp. using a modified bilayer plate assay. *Abstr Annu Mtg ASM* 1990; 42.
89 Cumberbatch N, Gurwith MJ, Langston C, Sack RB, Brunton JL. Cytotoxic enterotoxin produced by *Aeromonas hydrophila*: relationship of toxigenic isolates to diarrhoeal disease. *Infect Immun* 1979; **3**: 829–37.
90 Nord CE, Wadström T, Wretlind B. Antibiotic sensitivity of two *Aeromonas* and nine *Pseudomonas* species. *Med Microbiol Immunol* 1975; **161**: 89–97.
91 Bernheimer AW, Avigad LS, Avigad G. Interaction between aerolysin, erythrocytes and erythrocytes membranes. *Infect Immun* 1975; **11**: 1312–19.
92 Rivero O, Anguita C, Paniagua C, Naharro G. Molecular cloning and characterisation of extracellular protease gene from *Aeromonas hydrophila. J Bacteriol* 1990; **172**: 3905–8.
93 Leung KY, Stevenson RMW. Characterisation and distribution of extracellular proteases from *Aeromonas hydrophila. J Gen Microbiol* 1988; **134**: 151–60.
94 Allan BJ, Stevenson RMW. Extracellular virulence factors of *Aeromonas hydrophila* in fish infections. *Can J Microbiol* 1981; **27**: 1114–22.
95 Bernheimer AW, Avigad LS. Partial characterisation of aerolysin, a lytic exotoxin from *Aeromonas hydrophila. Infect Immun* 1974; **9**: 1016–21.
96 Ljungh A, Wretlind B, Möllby R. Separation and characterization of enterotoxin and two haemolysins from *Aeromonas hydrophila. Acta Pathol Microbiol Scand* 1981; **89**: 387–97.
97 Thelström M, Ljungh A. Membrane-damaging and cytotonic effects on human fibroblast of alpha- and beta-haemolysin from *Aeromonas hydrophila. Infect Immun* 1981; **34**: 949–56.
98 Donta ST, Haddow AD. Cytotoxic activity of *Aeromonas hydrophila. Infect Immun* 1978; **11**: 989–93.
99 Ljungh A, Wadstrom T. Aeromonas toxins. In Dorner F, Drew J. (eds) *Pharmacology of Bacterial Toxins.* Oxford, Pergamon Press, 1987: 289–305.
100 Brenden R, Janda JM. Detection, quantitation and stability of the beta-haemolysin of *Aeromonas* spp. *J Med Microbiol* 1987; **24**: 247–51.
101 Wretlind B, Möllby R, Wadström T. Separation of two haemolysins from *Aeromonas hydrophila* by isoelectric focusing. *Infect Immun* 1971; **4**: 503–5.
102 Scholz D, Scharmann W, Blobel H. Leucocidic substances from *Aeromonas hydrophila. Zentralbl Bakteriol Hyg I Abt Orig A* 1974; **228**: 312–16.
103 Caselitz FH, Gunther R. Hamolysinstudien mit Aeromonas-Stammen. *Zentralbl Bakteriol I Abt Orig* 1960; **180**: 30–8.
104 Levett PN, Daniel RR. Adhesions of vibrios and aeromonads to isolated rabbit brush borders. *J Gen Microbiol* 1981; **125**: 167–72.
105 Crichton P, Walker JW. Methods for the detection of haemagglutinins in *Aeromonas. J Med Microbiol* 1985; **19**: 273–7.
106 Stewart GA, Bundell CS, Burke V. Partial characterization of a soluble haemagglutinin from human diarrhoeal isolates of *Aeromonas. J Med Microbiol* 1986; **21**: 319–24.

107 Bunic S, Panin J. Presence of *Aeromonas hydrophila* in slaughtered animals. *Int J Food Microbiol* 1994; **23**: 221–5.

108 Koval SF. Paracrystalline protein surface arrays on bacteria. *Can J Microbiol* 1988; **34**: 407–14.

109 Sleytr UB, Messner P. Crystalline surface layers on bacteria. *Ann Rev Microbiol* 1983; **37**: 311–39.

110 Ishiguro EE, Ainsworth T, Trust TJ, Kay WW. Congo red agar, a differential medium for *Aeromonas salmonicida*, detects the presence of the cell surface protein array involved in virulence. *J Bacteriol* 1985; **16**: 1233–7.

111 Munn CB, Ishiguro EE, Trust TJ. Role of surface components in serum resistance of virulent *Aeromonas salmonicida*. *Infect Immun* 1982; **36**: 1069–75.

112 Kokka RP, Vedros NA, Janda JM. Electrophoretic analysis of the surface components of autoagglutinating surface array protein-positive and surface array protein-negative *Aeromonas hydrophila* and *Aeromonas sobria*. *J Clin Microbiol* 1990; **28**: 2240–7.

113 Bickel KD, Lineweaver WC, Follansbee S, Feibel R, Jackson R, Buncke HJ. Intestinal flora of the medicinal leech *Hirudinaria manillensis*. *J Reconstr Microsurg* 1994; **10**: 83–5.

114 Schadow KH, Lineweaver WC. Clinical and microbiological aspects of leech-associated aeromonads. *Proc 4th International Symposium on* Aeromonas *and* Plesiomonas, 1993, Atlanta, USA.

115 Shotts E, Sawyer R, Starliper C, Lance J. The inter-relationship of *Aeromonas* and the medicinal leech (*Hirudo medicinalis*). *Proc 4th International Symposium on* Aeromonas *and* Plesiomonas, 1993, Atlanta, USA.

116 Wilken GB, Appleton CC. Bacteriological investigation of the occurrence and antibiotic sensitivities of the gut-flora of the southern African medicinal leech *Asiatiocobdella buntonensis* (*Hirudinidae*). *J Hosp Infect* 1993; **23**: 223–8.

117 Popoff M, Véron M. A taxonomic study of the *Aeromonas hydrophila–Aeromonas punctata* group. *J Gen Microbiol* 1976; **94**: 11–22.

118 Palumbo SA, Bencivengo MM, Corral FD, Williams AC, Buchanan RL. Characterization of the *Aeromonas hydrophila* groups isolated from retail foods of animal origin. *J Clin Microbiol* 1989; **27**: 854–7.

119 Koch R. Die Aetiologie der Tuberkulose. *Mitt Kaisel Gesundheitsamt* 1884; **2**: 1–88.

120 Isenberg HD. Pathogenicity and virulence: another view. *Clin Microbiol Rev* 1988; **1**: 40–53.

121 Ashkenazi S, Pickering LK. New causes of infectious diarrhoea. *Eur J Clin Microbiol Infect Dis* 1991; **10**: 1–3.

8 Fish Pathogens

BRIAN AUSTIN AND C. ADAMS
Heriot-Watt University, Edinburgh, Scotland

8.1 INTRODUCTION

To date, representatives of *Aeromonas allosaccharophila*, *A. hydrophila*, *A. salmonicida* and *A. sobria* have been associated with fish pathogenicity. In addition, *A. veronii* has been implicated as a potential fish pathogen but only in laboratory-based experiments when intramuscular injection of high numbers of cells, i.e. $\sim 10^7$/ml, resulted in muscle necrosis in Atlantic salmon. *A. allo-saccharophila* was described as a result of an examination of 16S rRNA sequences of two isolates of motile aeromonads, which were recovered from diseased elvers in Spain during 1988[1]. Pathogenicity of the isolates for fish has not yet been confirmed.

Organisms, identified as *A. sobria*, have been isolated from wild spawning gizzard shad in Maryland, USA during 1987[2]. However, moribund fish did not display any external or internal signs of disease although pure cultures were obtained from the kidney, liver and spleen of moribund animals, and identification achieved by phenotypic and genotypic tests[2]. Cultures were pathogenic to rainbow trout, and dead animals revealed the presence of haemorrhagic septicaemia. Thermolabile extracellular products (ECP) were cytotoxic and lethal to rainbow trout. Further work by Paniagua and colleagues[3] highlighted the role of caseinase, haemolysins and cytotoxins in the disease process.

8.2 *AEROMONAS HYDROPHILA*

A. hydrophila has been recovered from a wide range of freshwater fish species worldwide, and occasionally from marine fish, e.g. ulcer disease of cod[4]. However, conflicting views have been expressed over the precise role of *A. hydrophila* as a fish pathogen[5,6]. Some workers contend that the organism is only a secondary invader of previously weakened hosts, while others believe that *A. hydrophila* is a primary pathogen of freshwater fish.

A. hydrophila has been associated with several disease conditions, including tail and fin rot and haemorrhagic septicaemias. Haemorrhagic septicaemia (= motile aeromonas septicaemia) is characterized by the presence of small surface lesions, often leading to sloughing off of the scales,

The Genus Aeromonas. Edited by B. Austin, M. Altwegg, P.J. Gosling and S. Joseph
© 1996 John Wiley & Sons Ltd

haemorrhaging in the gills and anus, ulcers, abscesses, exophthalmia (= bulging eyes) and abdominal swelling. Internally, there may be the presence of ascitic fluid in the peritoneal cavity, anaemia and swelling of the kidney and liver[7]. Another condition, termed redsore disease, occurs in bass and is characterized by the presence of surface petechial (= pin prick) haemorrhages and scale erosion[8].

Although the characteristics of *A. hydrophila* have been discussed elsewhere, it is relevant to consider serology, which is considered to be an important aspect of fish pathogens. Santos and colleagues[9] serotyped 62 isolates from rainbow trout, and concluded that 89% belonged to 17 serogroups, of which O3, O6, O11 and O19 dominated. Although the specificity of serogroups has been questioned because of the antigenic cross-reactivity between *A. hydrophila, A. salmonicida* and *A. sobria*[10], Shaw and Hodder[11] reported that the core region of the lipopolysaccharide (LPS) of *A. hydrophila*, of relevance to O-serogroups, is distinct.

8.2.1 PATHOGENICITY OF *A. HYDROPHILA*

Fish-pathogenic isolates produce various toxins of largely undetermined function[12]. Interestingly, the presence of lateral flagella has been considered to be related to the presence of toxigenic properties[13].

Surface structures of *A. hydrophila* have been associated with autoaggregation, hydrophobicity and haemagglutination[14]. The surface array protein, i.e. the S-layer, is thought to influence the interaction between the bacterial cell and its environment. A major function of the S-layer concerns physical protection from lytic action by serum proteins and bacteriophages[15]. Ascencio and co-authors[16] investigated extracellular matrix protein binding, and concluded that binding of ^{125}I-labelled collagen, fibronectin and laminin is common to isolates from diseased fish. The binding was specific, with cultural conditions influencing expression of the bacterial cell surface binding structures. Experiments showed that calcium (in the growth medium) enhanced expression of the bacterial extracellular matrix protein surface receptors. It was concluded that success in infecting/colonizing fish depended on the ability of the pathogen to bind to specific cell surface receptors of the mucus layer, epithelial cells and subepithelial basement membranes.

Evidence suggests that *A. hydrophila* has the ability to attach to selected fish cells, namely erythrocytes, and tissue proteins via the action of 'adhesins'[16,17], which are extremely selective, recognizing D-mannose and L-fucose side chains on surface polymers.

Fish pathogenic isolates produce ECP, which contain considerable enzymic activity[18,19]. The relevance of the ECP was highlighted by Allan and Stevenson[20] and Stevenson and Allan[21], who induced pathology in fish following administration via injection. However, there are contrasting views about the role of ECP[3,22]. Stevenson and colleagues reported the presence of haemolytic (heat-labile) and proteolytic activity, the former of which was concluded to be of greater importance in pathogenesis[20,21]. Conversely, Kanai and Takagi[23] recovered a heat-stable α-type haemolysin, which was inactivated by EDTA, trypsin and papain. This crude preparation caused swelling and reddening of the body surface following injection into carp. The conclusion about the importance of haemolysins in pathogenicity was based on work with protease-deficient mutants, the ECP from which was more toxic to recipient fish than that from wild-type cultures. The haemolysins are iron-regulated, and access to iron in the haemolytic destruction of the host cells may be necessary[24]. Evidence suggests that the acquisition of iron from iron-transferrin in serum is dependent on the siderophore amonabactin. However, Thune and colleagues obtained a fish-toxic fraction which was proteolytic but not haemolytic[25,26]. In a comparison of ECP from virulent and weakly virulent isolates, Lallier and co-workers[27] noted that both were haemolytic, enterotoxigenic and dermonecrotic, but the weakly virulent isolate produced 20-fold more haemolysin than the virulent organism. However, only cell-free supernatants from virulent isolates produced toxic effects in fish. The conclusion was reached that factors, other than haemolysins and proteases, were relevant in fish pathogenicity. After studying numerous isolates, Hsu and colleagues[28,29] and Paniagua and co-authors[3] correlated virulence with extracellular proteolytic enzymes, particularly caseinase and elastase. Santos and colleagues[18] noted a relationship between virulence and elastase and haemolysin (of human erythrocytes) production and fermentation of arabinose and sucrose. Similarly, Hsu and colleagues[29] associated virulence with gas production from fructose, glucose, mannitol, mannose, salicin and trehalose, and the possession of resistance to colistin. Subsequently, acetylcholinesterase (a 15.5 kDa polypeptide) was found in the ECP, and deduced to be a major lethal factor, possibly interfering with the function of the central nervous system[30].

Extracellular metallo- and serine-proteases of *A. hydrophila* have been characterized, and found to be heat- (to 56°C)[31] and cold-stable (to −20°C)[32]. Most activity was inhibited by EDTA. Overall, there were many differences in the proteases (four or five were present) described by Nieto and Ellis[32] to reports from other workers. This may be explained by the work of Leung and Stevenson[31], who examined the proteases from 47 isolates, of which 27 produced both metallo- and serine-proteases, 19 produced only metallo-proteases, and ATCC 7966 produced only a serine-protease. The differences in these 47 isolates may well explain the apparently conflicting reports which result from the examination of only single isolates. It has been suggested that the proteases may be involved in protecting the pathogen against serum bacteriocidal effects, by providing nutrients for growth following the destruction of host tissues, and by enhancing invasiveness[31].

8.2.2 CONTROL

Plasmid-mediated resistance by means of 20–30 MDa plasmids is widespread in fish farms, e.g. eel ponds[33], and may reduce the potential benefit of some antimicrobial compounds[34]. Resistance in *A. hydrophila* has been recorded to a wide range of antimicrobial compounds, including ampicillin, chloramphenicol, erythromycin, nitrofurantoin, novobiocin, streptomycin, sulphonamides and tetracycline[35,36].

Some attention has been devoted to developing vaccines, although commercial products are still not available. Simple preparations of inactivated whole cells, which may be administered by immersion, injection or via the oral route, appear to work quite well[36–38], Schachte[36] and Loghothetis and Austin[39] recorded that the most convincing immune response, measured in terms of antibody titre, was achieved after using injection techniques. Using formalized whole cells applied by intraperitoneal injection, Ruangpan and colleagues[38] recorded complete protection in Nile tilapia within 2 weeks. Lamers and de Haas[37] deduced that heat-inactivated vaccines (60°C/1 h) gave superior results to formalized products (0.3% formalin). However, it was apparent that concentration of the vaccine, in terms of the numbers of cells, was very important in eliciting an immune response. Using carp, Lamers and de Haas[37] concluded that 10^7–10^9 cells generated a distinct agglutinating response, whereas 10^5 cells did not. Secondary doses of vaccine were shown to be beneficial. However, single doses of a formalin-inactivated vaccine (containing 10^7–10^9 cells), administered via intramuscular injection, were capable of eliciting an immune response that was maintained for 360 days. This demonstrated that fish have immunological memory[40]. Lamers and colleagues[41] vaccinated carp by bathing, and determined that a single immersion vaccination did not result in significant serum antibody levels. However, secondary vaccination after 1, 3 or 8 months gave rise to a dramatic immune response. Subcellular components, particularly LPS, offer promise as components of vaccines[39]. Indeed, evidence has been presented that LPS induces cell-mediated protection (= regulates T-cell-like macrophage system) in carp[42].

8.3 *AEROMONAS SALMONICIDA*

A. salmonicida, the aetiological agent of furunculosis, is one of the most important fish pathogens because of its economically devastating impact on cultivated fish, particularly salmonids. Atypical isolates have been increasingly associated with disease, notably ulceration, in marine fish. Furunculosis, named because of the subacute or chronic form of the disease, is recognized by the presence of boil-like lesions, termed furuncles, in the musculature.

The subacute or chronic form of furunculosis, which is common in older fish, is characterized by lethargy, slight exophthalmia, bloody fins, bloody discharge from the nares and anus, and multiple haemorrhages in the muscle. Internally, haemorrhaging in the liver, swelling of the spleen, and kidney necrosis occur[43]. This form of the disease usually results in low mortality rates, and the fish may well survive, although survivors have scar tissue, particularly in the vicinity of furuncles[44].

The acute form of furunculosis, which is of sudden onset, is characterized by a general septicaemia accompanied by darkening in colour (= melanosis), inappetance, lethargy, and the presence of small haemorrhages at the base of the fins. This form of the disease is of short duration, insofar as the fish usually die in 2–3 days. The pathogen is widespread in the blood, the tissues, and in lesions. Internally, haemorrhaging occurs in the abdominal walls, viscera and heart. The spleen may appear enlarged[44].

McCarthy and Roberts[43] discussed a third clinical form of furunculosis, termed peracute furunculosis, which is restricted to fingerling fish. The infected animals darken in colour, and may quickly die with only slight external symptoms, such as mild exophthalmia. Haemorrhages may occur at the base of the pectoral fin. Losses in farmed stock may be very high[45,46].

A fourth form of furunculosis, termed intestinal furunculosis, was discussed by Amlacher[47]. The signs included inflammation of the intestine, and anal inversion.

In addition to furunculosis, *A. salmonicida* has been implicated in several conditions of non-salmonids, such as carp erythrodermatitis (CE)[48]. Bootsma and colleagues[49] isolated a small, Gram-negative, rod-shaped organism from skin lesions in mirror carp. This organism was subsequently identified as an 'atypical' strain of *A. salmonicida*. CE was described as a subacute to chronic contagious skin disease, which varied in its morbidity and mortality[49]. Infection starts at the site of injury to the epidermis. Then, a haemorrhagic inflammatory process develops, principally between the epidermis and dermis. This red inflamed area gradually extends with the breakdown of tissue leading to the formation of a central ulcer. This may occur in any location on the body surface, although it is most frequently located on the flanks. Infected fish exhibit inappetance, and appear to be darker in colour. The healed ulcer is recognizable as a grey-black scar. Contraction of the collagen of the scar tissue may result in serious deformity, which reduces the value of the fish[48]. CE may also result in generalized septicaemia leading to death.

A. salmonicida has been reported to cause a cutaneous ulcerative disease in goldfish (= ulcer disease)[50] and ulceration in marine fish[51–54]. Disease signs included lethargy, loss of orientation, and abnormal swimming behaviour. The ulcers may be of various sizes and depths. Death may result.

A disease, termed 'head ulcer disease', has been described in Japanese eels for which the causal agent has been deduced to be atypical *A.*

salmonicida[54,55]. Interestingly, the pathogen is capable of causing mortalities solely as a result of localized growth in the musculature, with no evidence for the development of a generalized septicaemia[55,56].

8.3.1   ISOLATION OF *A. SALMONICIDA*

Apart from the techniques discussed in Chapter 2, an improvement in recovering *A. salmonicida* involved use of an enrichment procedure using tryptone soya broth (TSB)[57]. This entailed placing swabbed material, derived from the organs of diseased animals, into TSB with incubation at 26°C for 48 h. The resulting broth cultures were streaked for single colony isolation onto plates of tryptone soya agar (TSA). The results indicated that the recovery of *A. salmonicida* was twice that using the conventional direct plating of swabbed material onto TSA.

Media supported with blood or blood products, e.g. serum, have been employed for the isolation, especially of atypical forms of the pathogen such as found in CE and goldfish ulcer disease. We have experienced success with isolating atypical *A. salmonicida* using blood agar (blood agar base supplemented with 10% v/v horse, sheep or bovine blood), which is inoculated and incubated at 15–18°C for up to 7 days.

Iron limitation has a detrimental effect on the growth of *A. salmonicida*. Addition of the iron-chelating agent ethylenediamine dihydroxyphenylacetic acid (EDDA) to media caused a small decrease in growth[58], whereas 2,2′-dipyridyl (Dipy) and 8-hydroxyquinoline (8HQ) reduced cell growth by 50 and 90% respectively. However, there was a greater secretion of protein, in particular a 70 kDa protease.

The isolation of *A. salmonicida* from water has been unreliable. However, the concentration and detection of the pathogen from hatchery water have been achieved using M-MDS electropositive filters, and permitted the recovery of 35% of the seeded bacteria[59]. Similarly, Ford[60] used a filtration method to detect *A. salmonicida* at a concentration of 0.1–2.2 CFU/ml, before clinical signs of furunculosis became apparent.

8.3.2   CHARACTERISTICS OF *A. SALMONICIDA*

The traditional description of *A. salmonicida* is of non-motile, fermentative, Gram-negative rods, which produce a brown water-soluble pigment on tryptone-containing agar, which do not grow at 37°C, and which produce catalase and oxidase. However, non-pigmented and/or cytochrome oxidase-negative strains have been isolated with increasing regularity[51,54,61–66]. In addition, motility[67–69] and growth at 30–37°C[70] have been reported in some cultures.

An intriguing trait is the ability of *A. salmonicida* to dissociate into different colony types, coined as rough, smooth and G-phase (intermediate) colonies[71]. Electron microscopy has demonstrated that the 'rough' and 'smooth' forms reflect the presence or absence of the A-layer, respectively.

The presence of atypical strains from a wide range of fish hosts and geographical locations is increasing. McCarthy and Roberts[43] proposed that there should be three subspecies of *A. salmonicida*, namely subspecies *salmonicida*, subspecies *achromogenes* (incorporating subsp. *masoucida*) and subspecies *nova*. Belland and Trust[72], on the basis of DNA:DNA reassociation studies, also supported the opinion of McCarthy and Roberts[43], namely to create a new subspecies to accommodate atypical isolates from non-salmonids. However, the new subspecies was never formally proposed. Austin and colleagues[73] elevated a group of 18 non- or slowly pigmenting 'atypical' isolates into a new subspecies, as *A. salmonicida* subsp. *smithia*.

The biochemical properties of 105 *A. salmonicida* strains were examined in order to distinguish typical and atypical cultures[74]. Essentially, typical and atypical strains could be differentiated by biochemical tests, i.e. the production of acid from saccharose, salicin, α-methyl-D-glucoside, L-arabinose and arbutin, production of gas from glucose and maltose, hydrolysis of aesculin, haemolysis, and hydrolysis of Tween 80. Interestingly, the atypical strains could not be subdivided further by the biochemical tests.

Plasmids carried by typical (14 strains) and atypical (11 strains) isolates of *A. salmonicida* have provided additional genetic evidence for the classification of typical and atypical isolates into separate taxa[75]. Typical isolates possessed a very homologous plasmid content comprising a single large (70–145 kb) plasmid and three low-molecular-weight plasmids. In comparison, the atypical isolates possessed two to four different plasmid types[75]. Moreover, there was a correlation between plasmid composition and source of the atypical isolates.

Plasmid profiling was evaluated as an epidemiological marker within *A. salmonicida* subsp. *salmonicida* by Neilsen and co-workers[76], who examined 124 isolates, and determined the presence of a common large plasmid in the size range of 60 to 150 kb. In addition, all cultures possessed two low-molecular-weight plasmids of 5.2 and 5.4 kb, and two additional low-molecular-weight plasmids of 5.6 and 6.4 kb were frequently encountered. In all, 23 different plasmids were demonstrated, and 40 profiles were obtained. Although variations in plasmid content, relative to the country of isolation, were recorded, it was considered that plasmid profiling was limited as a method of epizootiological marking. This conclusion correlated with studies by Bast and associates[77], Toranzo and colleagues[78] and Sorum and colleagues[79]. Also, it was noted that the plasmid content of *A. salmonicida* subsp. *salmonicida* was very constant, and thus of little epizootiological value. Moreover, phenotypic and molecular analysis of *A. salmonicida* subsp. *salmonicida* isolated in southern and northern Finland did not correlate with plasmid profiles and randomly amplified polymorphic DNA (RAPD)[80]. Three strains isolated from feral Atlantic salmon and a whitefish in the Kemi river basin had identical RAPD profiles, but one of the strains had a different plasmid profile. Certainly, RAPD pattern analysis has proven to be very reliable, and can be used to differentiate between strains within a species. In particular, Hanninen and colleagues[80] showed that only two of 10 different primers were able to detect polymorphism between *A. salmonicida* subsp. *salmonicida* strains. Fifteen RAPD types were found among 28 Finnish strains. These types were detected when the results of the two primers were combined; RAPD typing was able to differentiate most 'foreign' strains from those of Finland. RAPD pattern analysis using a single randomly designed primer was utilized by Miyata and colleagues[81] to analyse the genetic differentiation of 13 strains of *A. salmonicida* subsp. *salmonicida* and seven strains of *A. hydrophila*. Here, different RAPD profiles were recorded for motile and non-motile aeromonads.

Ribotyping is capable of highlighting genetic diversity between strains of *A. salmonicida* but it is not distinctive enough for epizootiological typing. Nielsen and colleagues[82] characterized 124 strains of *A. salmonicida* subsp. *salmonicida* using three enzymes, and showed that Danish and North American strains were genetically very homogeneous whereas Norwegian and Scottish isolates showed some diversity. Hänninen and colleagues[80] produced five different ribotypes among strains. Dorsch and co-workers[83] used 16S rDNA targeted oligonucleotide primers to differentiate between *Aeromonas* spp. Using one sequence region, these workers discriminated between *A. hydrophila*, *A. jandaei*, *A. salmonicida*, *A. schubertii*, *A. sobria* and *A. veronii*. However, *A. salmonicida* subsp. *salmonicida* and *achromogenes* had identical sequences and, therefore, could not be distinguished. Multilocus enzyme electrophoresis (MLEE) is a proven method to reveal genetic relatedness of strains. A study by Boyd and colleagues[84] using a collection of 53 isolates of *A. salmonicida* and one *Haemophilus piscium* isolate, assayed with nine enzymes, showed that the population structure of *A. salmonicida* is clonal. Only two electrophoretic types (ET) were evident, which differed at only one enzyme locus, and the 42 typical *A. salmonicida* isolates were assigned to ET1, with atypical isolates to ET2. Therefore, it is argued that ribotyping is a more sensitive method than multilocus enzyme electrophoresis, whereas RAPD typing is significantly more sensitive than ribotyping, producing almost strain-specific profiles. Yet another method, restriction endonuclease fingerprinting (REF), was used to analyse atypical *A. salmonicida* isolates from goldfish and silver perch[85]. Thus, using one restriction enzyme (*Cfo*I), restriction endonuclease fragment patterns were successful in showing the epidemiological background and origin of the strains and their likely transfer across species barriers. Previous studies also showed that REF analysis permitted the identification of subgroups, allowing epidemiological studies[86].

Additional approaches to determining the intra- and interspecific relationships of *A. salmonicida* have been used. These include serology and bacteriophage typing. The antigenicity of *A. salmonicida* has been studied extensively because of its significance in vaccine development programmes. Ewing and co-workers[87] examined agglutinin absorption with reference to 'O' and 'H' antigens, and concluded that the 21 strains examined were related to each other, but also to a strain of *A. hydrophila* (O-antigen suspensions prepared using the 21 cultures reacted to ca. 25% of the titre of the O-antiserum prepared with *A. hydrophila*). Common antigens among two *A. salmonicida* strains, as determined by gel-diffusion, and cross-reactions between *A. salmonicida* antiserum and three out of four isolates of *A. hydrophila*, but not of *A. salmonicida* strains and *A. hydrophila* antiserum, was reported by Liu[88]. Karlsson[89], using the antigenic properties of a haemolysin from *A. salmonicida*, found no serological differences among six strains. Bullock[90] found evidence of cross-reactions between soluble antigens of *A. salmonicida* and *A. hydrophila*. However, Popoff[91] did not find any serological differences among large numbers of typical pigment-producing isolates. Indeed, Popoff[92] concluded that *A. salmonicida* is a serologically homogeneous species.

McCarthy and Rawle[93] examined thermolabile and thermostable somatic antigens of *A. salmonicida* and their relationship to other bacteria. From whole-cell agglutination and double cross-absorption of smooth strains, and passive haemagglutination and double cross-absorption of rough colony types, it was determined that cross-reaction titres for both antigen types were high, and cross-reactions between the *A. hydrophila* isolate and three out of six thermostable *A. salmonicida* antisera were extremely weak. By double-diffusion with cell-free extracts, prepared from the bacteria used for the somatic antigen study, strong cross-reactions among *A. salmonicida*, *A. hydrophila* and *Vibrio anguillarum* and, to a lesser extent, with *Pseudomonas fluorescens* occurred.

The first isolation of bacteriophage specific for *A. salmonicida* was by Todd[94], although the usefulness for typing purposes was not demonstrated until the work of Popoff[95,96]. Now a view holds that phage typing has great value in epizoological studies[75,92,97]. The bacteriophages have been divided into three morphological groups and 10 serological types[92]. Using a set of eight bacteriophages, Popoff[96] defined 14 phage types. Rodgers and co-workers[97] defined 27 groups of *A. salmonicida* on the basis of sensitivity patterns to 18 bacteriophage isolates. Significantly, the morphological characteristics of the host bacterium, i.e. rough, smooth or G-phase, influenced attachment of the bacteriophage. This was attributed to the varying quantities of LPS in the cell wall of the different morphological types.

Mention should be made of *Haemophilus piscium*, the causal agent of ulcer disease, a name coined by Snieszko and co-workers[98]. Examination of DNA, biochemical, serological and bacteriophage sensitivity data revealed that *H. piscium* was highly related to *A. salmonicida*[99]. Further, Trust and colleagues[100] concluded, on the basis of bacteriophage sensitivity, that *H. piscium* is an atypical form of *A. salmonicida*.

8.3.3 DIAGNOSIS/IDENTIFICATION OF *A. SALMONICIDA*

Diagnosis may be achieved by examination of the gross clinical signs of disease, culturing, serology or the use of DNA probes. Serology, whether on pure or mixed cultures or histological sections, may be used for confirmation of the presence of *A. salmonicida*. Slide agglutination is effective only for smooth (non-autoagglutinating) colonies[101]. This is a pity since the majority of isolates recovered from clinical cases of disease are rough and autoagglutinating[102]. For such cultures, McCarthy and Rawle[93] recommended the mini-passive agglutination test. This technique involves the use of sheep erythrocytes sensitized with *A. salmonicida* O-antigen (extracted with hot physiological saline), and antiserum to *A. salmonicida*. The latex agglutination test was introduced by McCarthy[103], and has found use in the detection of *A. salmonicida* in diseased fish tissue. The method has the advantage that diagnoses may result from tissues unsuitable for culturing, e.g. frozen or formalin-fixed material[103]. The India ink immunostaining reaction, developed initially by Geck[104], is a microscopic technique in which the precise mode of action is unknown, although Geck suggested that it could be regarded as an immunoadsorption method. Kimura and Yoshimizu[105,106] described a co-agglutination test, using anti-*A. salmonicida* antibody-sensitized staphylococci suspensions, for the rapid diagnosis of furunculosis. Monoclonal antibodies (MAb) have been developed[107,108] and used in enzyme-linked immunosorbent assays (ELISA) which have proven useful for the diagnosis of furunculosis on fish farms[107]. MAb have also been developed to iron-regulated outer membrane proteins (IROMP) and LPS of *A. salmonicida*[109]. Two MAb against the IROMP and four MAb against the LPS were obtained, and used in ELISA and the indirect fluorescent antibody test (iFAT). Only MAb directed against LPS reacted with whole bacteria. It was reasoned that these MAb should be valuable for detecting typical and atypical forms of *A. salmonicida*.

In a comparison of serological techniques, Sakai and colleagues[110] reported that iFAT and the peroxidase–antiperoxidase enzyme immunoassay (PAP) were more sensitive than the latex agglutination and co-agglutination techniques. Nevertheless, with latex agglutination and co-agglutination techniques, more positive samples were detected than by iFAT or PAP.

Bernoth[111] used an indirect dot blot immunoassay for the immunological identification of *A. salmonicida*, using antisera raised against outer membrane proteins (OMP). However, there was some cross-reactivity with other bacterial taxa. The problem was overcome by diluting the antisera.

Barry and associates[112] suggested that DNA probes have the potential to detect *A. salmonicida* in environmental and clinical samples. These workers found that specific probes could be developed, even if only two base pair differences existed in the target sequence. Hiney and colleagues[113] isolated a DNA fragment from a 6.4 kbp cryptic plasmid specific to *A. salmonicida*, which when incorporated into a polymerase chain reaction (PCR) technique enabled a sensitivity of detection of approximately two cells. The locus was associated with *A. salmonicida* subsp. *achromogenes*, *A. salmonicida* subsp. *masoucida* and some atypical strains of *A. salmonicida*.

Gustafson and colleagues[114] developed a PCR using a 421 kbp sequence from the 3' region of the surface array protein gene (*vapA*) of *A. salmonicida*. This PCR detected <10 colony forming units (CFU) of *A. salmonicida*/mg from fish tissues. Mooney and co-workers[115] used a PCR and a DNA probe to detect *A. salmonicida* in blood samples taken from wild Atlantic salmon in three Irish rivers. Results revealed that the pathogen was detected, albeit at a low level, in 87% of the 61 fish examined. These data suggested a widespread low-level infection of *A. salmonicida* in wild salmon stocks in Irish rivers.

8.3.4 EPIZOOTIOLOGY

The survival of *A. salmonicida* in water, has been thoroughly examined by numerous investigators[116-126]. It would appear that *A. salmonicida* is capable of surviving for a prolonged period in fresh, brackish and sea water, although contradictory results as to the exact time interval involved abound. Moreover, the temperature at which the experiments were run may well have also influenced the results.

The ability of *A. salmonicida* to persist in mud (sediment) or detritus in the fish farm environment has also been examined, with McCarthy[117] demonstrating survival in fish pond mud and detritus for up to 29 days. Michel and Dubois-Darnaudpeys[127] investigated the persistence of *A. salmonicida* in sediments, and reported survival and growth in sterilized river sediments for over 10 months. However, pathogenicity was lost after 8 or 9 months. These workers concluded that in natural conditions, this time would enable the pathogen to be released from sediment into the water, and that the behaviour of bottom-feeding fishes would allow direct contamination of fish, possibly becoming carriers. A reduction in pathogenicity, subsequent to prolonged incubation in river sediments, was also noted by Sakai[118,119]. He offered an explanation whereby avirulent cells (with a positive electrical charge), which originate from virulent cells (negatively charged) attached to sediment, spontaneously detach from the sediment particles (river sand in survival experiments) thus decreasing the number of virulent cells recovered. Michel and Dubois-Darnaudpeys[127] conceded that competition of *A. salmonicida* with large numbers of other bacteria in streams, some with an ability to synthesize bacteriocins, may act as a regulatory mechanism and limit the proliferation of the pathogen. Previous work by Dubois-Darnaudpeys[128] supported the concept that *A. salmonicida* is capable of survival and multiplication in natural sediments, hence providing a reservoir of infection, even though direct contact with diseased fish is likely to remain the primary route of transmission. Furthermore, Wiklund[51,52] suggested that atypical strains, when shed from ulcers of diseased flounders, survived in the bottom sediment of brackish water environments for prolonged periods.

Some investigators have examined animals, other than fish, as sources of infection. However, Cornick and co-workers[129] examined 2954 vertebrate and invertebrate specimens, collected from fish ponds during an epizootic of furunculosis, but could not recover *A. salmonicida*. However, *A. salmonicida* has been isolated in high numbers from marine plankton and salmon lice. Samples taken from a Norwegian fish farm that had experienced an outbreak of furunculosis revealed $\sim 10^4$ and 600 cells of *A. salmonicida* per salmon louse and per millilitre of homogenized plankton, respectively[130].

Sakai[119] proposed a mechanism for the long-term survival of *A. salmonicida* in the aquatic environment based on electrostatic charge differences on individual cells, with net negative charges reported on virulent, agglutinating cells, and net positive charges on avirulent non-agglutinating strains. He suggested that the negatively charged virulent form of *A. salmonicida* is able to persist, albeit under starvation conditions, retaining viability in river sediments. It was also proposed that the decline in negatively charged virulent cells in sediments over prolonged periods, also noted by other investigators[127], may be caused by the spontaneous occurrence of positively charged avirulent free-living cells of *A. salmonicida*. These cells originate from the virulent ones, attached to sediment particle surfaces, and subsequently detach from the sediment/sand particles. This free-living form could be considered to enter a dormant phase, according to Sakai[119], because the viability of these bacteria declines due to a lack of nutrients. It was further proposed that the free-living cells represent a transitional life-stage of the pathogen which would ultimately lose viability[119].

High cell-surface hydrophobicity may have ecological implications regarding the survival of the pathogen in the lipid-rich air–water interface[131]. In particular, by application of highly specific monoclonal antibodies and immunofluorescence techniques, the pathogen was found in high numbers (4.3×10^3 cell/ml) at the air–water interface and in the sediments beneath a fish farm stocked with Atlantic salmon that were suffering from furunculosis. It is argued that in the aquatic environment, hydrophobic bacteria that are suspended in the water column and on sediment, would be transported to the surface lipid layer by rising gas bubbles or lipid droplets[132].

Some workers have addressed the possible development of a dormant/non-culturable/changed phase of *A. salmonicida*. However, conflicting evidence has been published. Rose and co-workers[120] doubted the presence of a dormant/non-culturable phase of the pathogen in the marine environment. It has been contended that intact cells remaining in aquatic systems after culturing would indicate an absence of viability[122,123,125,133,134]. For example, Morgan and colleagues[122] assessed the survival of the pathogen in lake water, employing a wide range of techniques, including epifluorescence microscopy, respiration, cell culture, cell revival, flow cytometry, plasmid maintenance and membrane fatty acid analysis. Evidence was presented that *A. salmonicida* became unculturable in sterile lake water, but microscopic and flow cytometric methods revealed the continued presence of cells. However, attempts to revive these cells by the addition of nutrients were unsuccessful. Despite this, it was found that both genomic and plasmid DNA, and RNA, were maintained in the cells. It was clearly demonstrated that the cells remained intact, although their viability could not unfortunately be demonstrated. A comment was made that non-culturability of some bacteria from environmental samples may be as much a function of the ignorance of the parameters necessary for their recovery, as the occurrence of a truly non-culturable, specialized survival state.

Could the organism become altered in the aquatic environment such that specialized recovery methods would be necessary? Effendi and Austin[124,125] found that marine samples, considered to be devoid of the pathogen, contained cells which passed through 0.22 and 0.45 μm pore size porosity filters. Some of these cells grew on specialized media designed for the recovery of L-forms, and showed agreement with the characteristics of *A. salmonicida* L-forms as reported by McIntosh and Austin[135–137]. Subsequently, *A. salmonicida* colonies developed on basal marine agar (BMA) plates inoculated with material from turbid L-form broth medium. On this basis, Effendi and Austin[124] recorded populations of ca. 10^3 *A. salmonicida* cells/ml in the microcosms after corresponding enumeration of colonies on BMA had reached zero. Thus, they suggested that the existence of specialized forms, e.g. L-forms, of *A. salmonicida*, may be a factor in the difficulties previous researchers have experienced in attempts to recover the pathogen from environmental samples.

Undoubtedly fish are important in the transmission of the disease. Clinically diseased fish and asymptomatic carriers may spread infection. Secondly, carriers, which appear healthy, harbour the pathogen in their organs where it can be released if the fish eventually succumb to the disease. McCarthy[117] established that furuncle material contained ≤10^8 viable cells/ml of necrotic tissue. These cells remained viable in muscle for 32 days, and for 40 days in the tank water where the dead fish had been kept, thus providing a source of infection for healthy fish. McCarthy[117] emphasized that the prolonged survival of *A. salmonicida* in dead fish demonstrated the risk of using trash fish as food.

Another line of investigation examined L-forms (spherical, filterable cells) of *A. salmonicida*, which were induced experimentally, and were found to be unculturable by conventional plating methods[137]. However, attempts to infect fish using L-forms did not result in recovery of the pathogen from fish, even after the administration of immunosuppressants. In addition, it was reported that small numbers of 'natural' L-form colonies were observed, albeit infrequently, on specialized medium (containing horse serum and high quantities of sucrose), which had been inoculated with kidney and spleen samples taken from farmed Atlantic salmon suffering with furunculosis[137]. This suggested that L-forms or related morphological variants may have a role in the disease process. Such findings certainly merit further investigation to establish more conclusively the role of this form of *A. salmonicida* in disease processes. Obviously, if *A. salmonicida*, in a dormant, non-culturable or any other altered form, is genuinely unable to transmit infection, then control of the diseases is vastly simplified.

Because of the importance of carriers in the epizootiology of furunculosis, it is essential that effective methods are used to detect their presence. At present, a combination of increasing the water temperature to 18°C and the injection of corticosteroids is employed to activate the carrier state[138]. From experiments using immunofluorescence techniques, Klontz[139] concluded that the intestine is a primary site of infection, leading to the establishment of asymptomatic carriers. There is also some evidence to suggest that the kidney is involved[117].

Another problem of the carrier state is that antibiotic therapy to control furunculosis outbreaks does not necessarily completely remove the bacterium from the tissues of fish. However, McCarthy[117] reported that fish treated with tetracycline hydrochloride survived attempts to induce the disease. On the basis of these findings, McCarthy[117] warned that fish populations treated for furunculosis will undoubtedly remain carriers. Subsequently, other antibacterial compounds have been investigated for their ability to reduce or eliminate the carrier state of *A. salmonicida* in fish. Examples include erythromycin phosphate[140], flumequine, and an aryl-fluoroquinolone in combination with 0.01% Tween 80 to enhance the assimilation of the antimicrobial compound into fish[141]. The aryl-fluoroquinolone/surfactant combination was particularly effective in eliminating the asymptomatic carrier state of *A. salmonicida* within 48 h of treatment[141].

Several possible routes for the transmission of furunculosis have been investigated. It is commonly accepted that the disease is disseminated by lateral transmission of *A. salmonicida*, which includes contact with contaminated water and infected fish in addition to possible infection via the gastrointestinal tract. Also, vertical transmission has been considered in several investigations. A waterborne route where water contamination with *A. salmonicida* has occurred initially from moribund infected fish or from overtly healthy carriers shedding the pathogen is favoured as the common means of transmission. Once released into the aquatic environment, the organism is then able to persist for a prolonged period of time and the disease spread in this way. Early studies demonstrated that trout placed into water which had contained diseased fish contracted the infection[142]. The Furunculosis Committee of the UK concluded, on the basis of available data, that both food and diseased fish could constitute sources of infection. McCarthy[117] examined the likely transmission mechanisms, specifically with regard to the ability of *A. salmonicida* to penetrate fish tissues (invasiveness), a prerequisite for the occurrence of infection. Transmission by contaminated water was tested by McCarthy[117] in laboratory-based experiments by seeding water in a tank containing six brown trout with a suspension of *A. salmonicida* to a final concentraiton of 10^6 cells/ml. Five of the six fish died of furunculosis, and the sixth succumbed when given an injection of prednisolone acetate. In a subsequent large-scale experiment using 50 brown trout placed in a pond on a fish farm experiencing a summer epizootic of furunculosis, 41 fish had died within 28 days, and the remaining nine succumbed after injection with corticosteroid. McCarthy[117] concluded from these experiments that the disease is readily disseminated through water and also that brown trout surviving infection probably became carriers. However, Blake and Clark[142] and McCarthy[117] reported failure in attempts to infect rainbow trout by co-habitation with infected brown trout or the addition of *A. salmonicida*, respectively. It is known that rainbow trout are more resistant to the disease than are brown trout. In experiments which examined different routes of exposure to *A. salmonicida* subsp. *salmonicida* in Atlantic salmon in seawater, Rose and colleagues[143] noted that a minimum dose of 10^4 CFU/ml by bath was required to initiate infection.

Another unresolved aspect of the transmission of furunculosis is the uptake of *A. salmonicida* into a fish host. It is possible that the pathogen may gain entry to a new host through the gills, mouth, anus or a surface injury. McCarthy[117] demonstrated that rainbow trout that resisted the disease subsequently died from furunculosis after their flanks had been abraded with sandpaper. Lund[116] also found that infection was acquired by fish which had been scarified and experimentally challenged with the pathogen. However, these injuries were artificially induced. Unfortunately, the natural mode of uptake remains unresolved. Subsequently, it was observed that the pathogen could be detected in the blood and kidney within 5 min of immersion in a suspension containing 10^5 cells/ml[144]. Interestingly, it was also found that uptake of the pathogen was enhanced by the addition of particulates, e.g. latex, to the bacterial suspension. If latex was added, *A. salmonicida* was isolated from blood at 12 min, and from kidney, spleen and faeces at 4 h postchallenge. The organism was also cultured from the skin, gills, blood and faeces for up to 48 h. In the absence of latex, the pathogen could again be recovered at 12 min, but from a wider range of sites including kidney, spleen and the lower intestine. However by 24 h, the pathogen was no longer recovered from the fish. In addition, the method of challenge yielded different results. Thus, when the entire fish was immersed in the bacterial suspension, superior uptake occurred compared to exposing only the head or tail regions. The explanation of this phenomenon is unclear, but such results suggest that uptake may occur through several locations rather than a single site, e.g. mouth, nares, gill or anus. It is possible that the pathogen gains entry via all these sites or additionally through the lateral line and/or skin[144].

Perhaps the most significant observation resulting from these experiments was the rapidity with which *A. salmonicida* entered rainbow trout. Other investigators have not sampled so close to the initial time of challenge. McCarthy[117] reported uptake to occur from the oral route within 5 h, with the organism found in the kidney. However, the work indicated that the localization of *A. salmonicida* was principally within the reticuloendothelial systems of fish[145].

Another route of infection which has been proposed is via the gastrointestinal tract, due to intake of contaminated food. However, there is disagreement as to whether or not this occurs. Blake and Clark[142] reported success in experimentally infecting fish by feeding contaminated food. However, Krantz and colleagues[146] and McCarthy[117] failed to infect brown trout by feeding with food containing the pathogen. Despite the contradictory information available, infection via the gastrointestinal tract cannot be dismissed as a possibility, particularly in view of a report by Klontz and Wood[147] concerning clinical furunculosis in the sable fish, apparently caused by ingestion of carrier coho salmon. In addition, in laboratory-based experiments that compared various methods designed to induce the carrier state of *A. salmonicida* in juvenile spring chinook salmon, Markwardt and Klontz[141] observed that gastric intubation (of ca. 1×10^8 bacteria) resulted in a 65% carrier state. This result was a significantly higher percentage than those recorded for exposure to a broth culture as a bath, ingestion of broth culture-coated food, and exposure to intraperitoneally injected fish (40, 20 and 10% carrier rates, respectively). Markwardt and Klontz[141] commented that exposure to clinically diseased fish and bathing in broth cultures probably simulated the natural routes of infection.

Vertical or transovarian transmission of *A. salmonicida* as a possible route of infection has been widely studied, albeit with inconclusive results. The possibility that the pathogen could be transmitted at fertilization was examined by the Furunculosis Committee in the UK. The experimental evidence gathered for their report indicated that furunculosis was not transmitted in such a fashion. However, Lund[116] contended that since the conclusions were based on the results of a solitary set of experiments, further work was necessary to confirm this point. In a detailed series of experiments aimed at clarifying the situation regarding transmission of *A. salmonicida* at fertilization, Lund[116] examined ovaries, testes and ova of experimentally infected fish for the presence of the pathogen. *A. salmonicida* was recovered in pure culture from the ovaries and testes of infected fish. Results of isolation of the pathogen from ova were decisive as, of 500 ova sampled, only three obtained from the same fish yielded the organism in pure culture from the interior of the ovum. In further experiments using wild spawners, ova were contaminated with *A. salmonicida* on the external surface at the time fertilization was effected, and the eggs then planted out in a river bed. It was observed that these ova died quickly and were subjected to *Saprolegnia ferax* infection. *A. salmonicida* was not isolated from dead or living ova, and Lund[116] concluded that the experiment had been unsuccessful. Ova taken from parents experimentally infected with *A. salmonicida* also failed to yield the pathogen. Continuing the study of vertical transmission, McCarthy[117] examined the fertilized ova of mature brood stock brown trout taken from a known carrier population (5/8 proved to be carriers when challenged with prednisolone acetate). However, *A. salmonicida* was not recovered from the fertilized egg sample. When artificially infected brood stock were stripped as soon as signs of clinical furunculosis had developed, both fish organs and fertilized eggs were positive for *A. salmonicida*. However, the high numbers of viable cells initially present, i.e. 10^6 cells/ml of egg macerate, rapidly declined and could not be detected 5 days after incubation began. Based upon these experiments, McCarthy[117] concluded that vertical transmission of *A. salmonicida* was not a significant means of disseminating the disease, and, moreover, in the improbable event that overtly infected fish were used for stripping, the pathogen was unlikely to survive to the eyed-egg stage at which the eggs are marketed. Neither Lund[116] nor McCarthy[117] recovered *A. salmonicida* from fry derived from experimentally infected parents or of known carrier brood stock, respectively. However, both these authors pointed out the possibility that the negative results obtained may have been due to the inadequacy of techniques used for detection and isolation of the pathogen in the face of inhibition or overgrowth by commensal bacteria, or the presence of only small numbers of *A. salmonicida*.

A remaining aspect of the epizootiology of *A. salmonicida* diseases which requires consideration is the transmission of the infection in seawater. This is an important topic for the aquaculture industry, as salmonids are not infrequently placed in seawater for on-growing. In addition, early in the study of the pathogen, the possibility that migratory Salmonidae could spread the infection was considered. Lund[116] investigated the possibility that the disease could be carried by salmon or sea trout smolts (previously infected in freshwater) when they migrated to the sea. The Furunculosis Committee had not agreed with this theory because examination of large numbers of smolts taken from the River Coquet in 1928 and 1929 had given no evidence for the presence of *A. salmonicida*, although the reverse process, i.e. salmon or sea trout contracting the infection upon migration into rivers containing infected trout, had become generally accepted. In an examination of 234 smolts from the River Coquet, Lund[116] isolated and confirmed *A. salmonicida* from four smolts (two salmon and two sea trout), and believed the findings to be significant as such fish would possibly develop the disease on exposure to suitable conditions or remain resistant, possibly transmitting the infection upon contact with healthy fish in sea or brackish waters. Lund[116] could not offer a definitive reason for the results differing from those of Williamson and Anderson[148], who examined 1339 smolts taken from the Coquet without recovering any isolates of *A. salmonicida*. Certainly, mortalities attributed to

A. salmonicida in anadromous fish in seawater and in trout grown-on in seawater have been reported[149,150]. However, it has not been determined whether the disease outbreaks resulted from stress experienced by fish carrying a latent infection initially contracted in freshwater, or whether they represented a case of lateral transmission of the pathogen via seawater. For example, Smith[151] established that *A. salmonicida* survived in seawater for a prolonged period. It had also been demonstrated that *A. salmonicida* is capable of infecting sea and brown trout by contact with infected fish in sea and brackish waters[152]. She found that the infection was transmitted between salinities of 2.54 and 3.31% (w/v) at water temperatures ranging from 5.6 to 14.5°C. Smith and colleagues[153] reported on mortalities of Atlantic salmon from two marine fish farms in Ireland, presenting evidence for the lateral transmission of *A. salmonicida* in seawater to a group of fish not known to be carriers. They also provided data suggesting that subsequent to the stocking in spring 1978, and removal of carrier fish in summer 1979, at a marine fish farm, the pathogen became established and persisted in the fish farm environment for at least 6 months after the removal of the carrier fish. Thus, a carrier-free population placed on the site in the spring of 1980 was infected. Unfortunately, it was not determined whether the pathogen persisted in feral fish outside the cages or in the sediments under the cages. To lend support to a seawater transmission of furunculosis, Evelyn[149] documented isolation of *A. salmonicida* from a strictly marine host, the sable fish, although probably the route of infection was by ingestion of moribund or dead salmonid carrier fish[147]. Obviously, *A. salmonicida* has wider potential for causing disease problems than has been hitherto suspected. To study the epizootiology of *A. salmonicida* diseases in aquatic environments other than freshwater, again, as in so many other aspects of the pathogen, demands additional attention to unravel its complexities.

8.3.5   PATHOGENICITY MECHANISMS

Although the factors conferring pathogenicity on *A. salmonicida* strains have been speculative since early in the study of the pathogen, it is only relatively recently that the details concerning pathogenesis and virulence have begun to be elucidated. An important aspect of pathogenicity was discussed by Duff[71]. He reported a loss in pathogenicity among strains after 6 or more months of maintenance on artificial culture in the laboratory. The loss was accompanied by a change in the appearance of colonies on nutrient agar from glistening, convex and translucent to strongly convex, distinctly opaque and cream-coloured. Subsequently, Duff discovered that dissociation could be induced by culturing the pathogen in nutrient broth with the addition of either 0.25% lithium chloride or 0.1% phenol. Use of this procedure gave rise to several distinct colony forms. One of these resembled the original stock culture, a second corresponded to the 'new' type, and a third was intermediate between the other two forms. When the colony types were inoculated intra-peritoneally into goldfish, the blue-green, translucent dissociant caused the deaths of the fish and was accompanied by lesions typical of the disease. In contrast, the original type of colony did not adversely affect fish, which survived for the 30-day duration of the experiment without any signs of illness. Thus, Duff concluded that the cream-opaque form which produced friable colonies was non-pathogenic, and more stable on prolonged storage. Duff designated this colony type as 'rough'. The 'smooth' form (i.e. the blue-green-translucent dissociant, which produced butyrous colonies on agar media) was pathogenic, but less stable in prolonged storage. In the subsequent study, Duff[154] also reported the presence of an extra antigen in the rough strains. Although Duff[71] was the first worker to report the ability of *A. salmonicida* to dissociate into several distinct colony types with differences in pathogenicity, it is curious that he ascribed pathogenicity to the smooth colony type. This is in contrast to the view currently held that the rough colony type is, in fact, virulent. Interestingly, the Furunculosis Committee had also reported a variation in colony morphology among isolates but contended that this phenomenon was not accompanied by a difference in virulence. It is regrettable that this initial confusion over dissociation occurred, preventing an earlier realization of its significance. In fact, the relevance of dissociation of *A. salmonicida* colonies and the relationship to virulence were not made apparent until the work of Udey[155], almost 40 years later. Early studies provided tentative evidence for a variety of possible pathogenic mechanisms, but there is no doubt that progress in the understanding of *A. salmonicida* pathogenesis and virulence has been accelerated by rapid advances in the knowledge of cell biology and the development of sophisticated biochemical techniques. It is the application of such techniques which continues to yield considerable new information about the manner in which *A. salmonicida* may affect its disease processes in fish.

A variety of pathogenicity mechanisms and virulence factors have been proposed for diseases caused by *A. salmonicida*, namely possession of an extracellular (A) layer and the production of ECP, although there is confusion and even contradiction about the relative merits of the various components in pathogenicity[156]. Yet, ironically, fish may mount an antibody response during infection[157]. Munro[158] grouped the virulence/pathogenicity factors into cell-associated and extracellular components. The best-studied cell-associated factor is the A-layer, which is now thought to be the product of a single chromosomal gene[159]. First reported by Udey and Fryer[160], and resulting from detailed electron microscopic studies, the A-layer was determined to be correlated with virulence. Thus, it was observed that virulent strains possessed the A-layer, whereas avirulent isolates did not. In addition, the presence of the A-layer corresponded with strong autoagglutinating properties of the organism, and to the adhesion to fish tissue culture cells. The presence of the A-layer may confer protection against phagocytosis and thus destruction by macrophages[161,162]. Essentially, these workers noted that avirulent cells, i.e. those without an A-layer, were phagocytosed and destroyed when virulent cells with A-layer were more resistant.

For its formation, Belland and Trust[159] reasoned that the A-layer subunits pass through the periplasm and across the outer membrane for assembly on the cell surface. A requirement for the presence of O-polysaccharide chains on the LPS was reported as necessary for the assembly of A-layer[163]. These virulent, autoagglutinating forms produce characteristic deep blue colonies on Coomassie brilliant blue agar[164,165]. Sakai[118,119] postulated that a possible mechanism for autoagglutination and adhesion could be attributed to the presence of net negative electrical charge in the interiors or on the surfaces of cells. In particular, pathogenic cultures were highly adhesive[166]. It was observed that fresh isolates, obtained from epizootics, were of the aggregating type. Udey and Fryer[160] contended that more work was needed to establish whether or not the A-layer alone could confer virulence. The discovery of the A-layer generated much interest, resulting in further study of its chemical composition and its specific role in fish pathology. Kay and co-workers[167] succeeded in purifying the A-layer, and concluded that it was composed of a surface-localized protein with a molecular weight of 49 kDa. Phipps and colleagues[168] determined that it was hydrophobic in nature, present on the entire cell surface, did not possess any enzymic activity, but instead constituted a macromolecular refractive protein barrier which was essential for virulence. Meanwhile, an independent parallel investigation of Evenberg and colleagues[169] highlighted the relationship between auto-agglutination and the presence of the A-layer. This group examined the cell envelope protein patterns of a variety of isolates obtained from a wide range of geographical locations and different fish species (i.e. carp, minnow, goldfish and salmonids). These fish were suffering from either furunculosis, CE or ulcer disease. A major protein (molecular weight = 54 kDa) was found in all autoaggregating strains, but little or no trace occurred in isolates which were not autoagglutinating. When examinations for the presence of the protein were carried out after a change of growth medium, i.e. replacement of horse serum by synthetic sea salt, it was observed that an almost complete loss of the additional cell envelope and the autoagglutinating ability of the isolate had occurred. Using gel immunoradioassays, it was also determined that the extra cell envelope proteins of all the isolates, irrespective of fish host, type of infection or geographical source, were immunologically related. Evenberg and Lugtenberg[170] pursued this topic, and described the protein as water-insoluble with an amino acid composition similar to those of the additional surface layers of other bacteria, e.g. the adhesive K88 fimbriae of enteropathogenic strains of *E. coli*. It is particularly relevant that the findings of Evenberg and co-workers[169], concerning the autoagglutinating ability of 'atypical' strains from cases of CE and ulcer disease, and the presence of the

A-layer, were in excellent agreement with the work of Trust and associates[17] and Hamilton and colleagues[171]. These earlier studies deduced the presence of an outer layer protein, which was estimated to have a molecular weight of 50 kDa. Evidence was provided by Ishiguro and co-workers[172] that loss of the A-layer and loss of autoagglutinating properties resulted in decreased virulence. After examining the effects of temperature on the growth of A. salmonicida, it was shown that in cells cultured at 30°C virulence was restricted to <10% of the population. The avirulent attenuated cells, which resulted from use of the higher growth temperature, did not autoagglutinate and did not possess the A-layer. It is interesting to note that higher maximum growth temperatures were recorded for the attenuated strains, in comparison to their virulent counterparts. Perhaps this is explained by their selection at high temperatures. Because of this observation, Ishiguro and co-workers[172] hypothesized that the A-layer is important in determining physical properties of the cell envelope, and that these properties undergo a change when the A-layer is lost, permitting growth at higher than normal temperatures. If the A-layer is a prerequisite for virulence, it may be assumed that its presence confers advantages on the bacterial cell in its role as a pathogen. Indeed, several prime functions for the A-layer have been proposed. Thus, evidence exists that the extracellular layer protects A. salmonicida cells from the action of protease[173] and bacteriophage, by shielding its phage receptors[172]. In addition, the layer may protect the cell from serum complement, insofar as Munn and Trust[174] demonstrated that virulent strains (with A-layer) were resistant to complement bacteriocidal activity in the presence (and indeed absence) of specific antibody in rainbow trout serum. Other investigations have revealed that hydrophobicity is conferred upon the bacterial surface by the A-layer[175,176]. These workers reported that the hydrophobic A-layer provided A. salmonicida cells with an affinity for fatty acid esters of polyethylene glycol and an enhanced ability to associate with rainbow trout and mouse phagocytic monocytes (macrophages), in the absence of opsonizing antibody. Although Trust and colleagues[175] conceded that the advantages to the pathogen of the increased association with macrophages remained to be determined, they suggested the possibility that A. salmonicida is a facultative intracellular pathogen able to survive within phagocytes. Indeed, Munn and Trust[174] demonstrated that A-layer+ bacteria (i.e. bacteria with A-layer) were able to multiply within the principal phagocytic organs, e.g. the spleen, following experimental infection.

The A-layer has also been implicated in a role concerning adhesion to fish tissues. By means of in vitro experiments, Parker and Munn[177] examined the ability of avirulent (A-layer−) cells to adhere to cells of baby hamster kidney and rainbow trout gonad in tissue culture. Attachment of A-layer+ A. salmonicida to both types of cells was greater than for the A-layer− derivative. As a result, Parker and Munn[177] proposed that since attachment to epithelial cells may be the primary step in the pathogenic process, their observations could account for the association of virulence with the presence of an extra outer membrane layer.

Physiological studies revealed that the presence of A-layer inhibited growth of *A. salmonicida* at 30°C, enhanced cell filamentation at 37°C and enhanced uptake of the hydrophobic antibiotics streptonigrin and chloramphenicol[178]. These traits were not observed when A-layer was missing or its arrangement altered such as in Ca^{2+}-limited or Ca^{2+}- and Mg^{2+}-limited cells, in A-layer⁻ cells with an artificially reconstituted A-layer, or in mutants unable to correctly assemble this extracellular layer. A-layer⁺ cells were far more sensitive to streptonigrin and chloramphenicol than A-layer⁻ cells. It would appear that the A-layer influences these physiological changes in the bacterial cell, due to a specific interaction of the (A-layer) subunits with the outer membrane. Certainly, the utilization of haem compounds is reliant on the presence of the A-layer[179]. Thus, A-layer⁺ and A-layer⁻ strains were placed under conditions of iron-restriction, and assessed for their ability to utilize haem sources. Binding was associated with the cell-surface, as shown by trypsin-digestion of whole cells which abolished haem-binding. Competitive binding studies indicated that all haem compounds were bound by a common receptor, which was not iron-regulated, and was associated specifically with the presence of the 49 kDa A-layer protein. It was concluded that the mechanism of haem utilization was not siderophore-mediated because both typical A-layer⁺ (siderophore-positive) and atypical A-layer⁺ (siderophore-negative) strains utilized haem compounds. Therefore, it would appear that *A. salmonicida* possessed both siderophore-dependent and siderophore-independent mechanisms to overcome the iron-restricted conditions which are encountered *in vivo*.

S-layer-mediated association of *A. salmonicida* with murine macrophages indicated that A-layer⁺ cells interacted with the macrophages in phosphate-buffered saline[180] whereas A-layer⁻ mutants did not. Latex beads coated with a partially assembled A-layer were taken up more efficiently than uncoated or A-protein-coated beads, which indicated that an organized A-layer was essential for murine macrophage uptake. In the presence of tissue culture medium, A-layer⁻ bacteria were internalized by macrophages. The A-layer⁺ cells, but not A-layer⁻ cells, were markedly cytotoxic to macrophages. The A-layer⁻ cells showed less cytotoxicity. Similarly, when using rainbow trout macrophages, A-layer⁺ cells readily adhered, with the addition of haem increasing adherence and cytotoxicity[181].

The gene (*vapA*) encoding for the A-layer protein has been cloned and sequenced[159,182]. The *vapA* gene was expressed at a low level in *E. coli* from an *A. salmonicida* insert containing *vapA* and 62 bp of the immediate upstream flanking sequence. Poor expression was thought to be due to the absence of additional control sequences. Additional upstream sequences flanking *vapA* were examined by primer extension to map the predominant promoter of *vapA* in *A. salmonicida*[183]. Northern blot analysis showed that *vapA* transcription in *A. salmonicida* was directed predominantly by a distal promoter P1. Also, Northern blot analysis showed that the mRNA transcribed from *vapA* is unit length, whereas S-layer-deficient strains and mutants produced undetectable levels of the transcript.

Garduno and colleagues[184] used various methods to reconstitute the A-layer in A-layer-defective mutants, and showed that purified A-protein monomers bound directly to LPS and may self-assemble into oligomers of an organized conformation depending on the availability of Ca^{2+}. Also, oligomers may bind to LPS and assemble into loose or interlocked arrays which also depend on the availability of Ca^{2+}. Reconstituted A-layers were functionally more competent than A-protein (purified A-layer protein) directly reattached to LPS.

The repeated passage of *A. salmonicida* through rainbow trout affected surface characteristics and virulence. Thus, Fernandez and colleagues[185] passaged 11 strains, and demonstrated that not all A-layer+ and LPS+ cultures were virulent. Indeed, only four of the strains were virulent for rainbow trout (LD_{50} dose = <10^6 cells/fish with an average weight of 8 g). A relationship was not demonstrated between virulence and the existence of hydrophobic surfaces, insofar as all of the isolates exhibited the ability to auto-aggregate and to agglutinate yeast cells.

Another conserved property of the *A. salmonicida* surface is the production of LPS with antigenically cross-reactive O-polysaccharides of homogeneous chain length. Evidence suggests that LPS with homogeneous length O-polysaccharide chains is important for the physical association of A-layer with the bacterial cell surface. A gene (*abcA*), immediately downstream of *vapA*, was identified as encoding for an ATP-binding cassette transport protein required for biogenesis of smooth LPS[186]. As well as containing an *N*-terminal ATP-binding domain, *abcA* encodes a *C*-terminal leucine zipper domain, which is needed for an increase in *vapA* expression, but not for O-polysaccharide side-chain synthesis, indicating that *abcA* encodes for a bifunctional protein[187]. Immunoelectron microscopy revealed that Tn5 mutagenesis and interruption by the endogenous insertion sequence element IS*AS1* prevented the accumulation of O-polysaccharides at the inner membrane–cytoplasm interface. Noonan and Trust[188] cloned a gene (*asoA*), which is approximately 7 kb downstream of the A-layer structural gene *vapA*. Marker exchange mutagenesis produced an *asoA* mutation (strain A449L-MB) in a low virulence strain and a disrupted altered surface morphology was observed. Protruding surface material was shown to contain LPS and A-layer subunit protein. Intraperitoneal injection of the mutant A449L-MB exhibited increased virulence. Moreover, the homology of the gene appeared to correlate to polytopic membrane proteins involved in translocation across the cytoplasmic membrane. Characterization of the extracellular secretion pathway has been made feasible by examining an *A. salmonicida* Tn5 mutant, A449-TM1, which is unable to secrete the A-protein through the outer membrane[189]. However, the mutant still maintained the ability to secrete haemolysin and protease but was avirulent for fish. The mutated gene (*apsE*) showed some homology to that of an ATP-binding secretion protein from *A. hydrophila*, although hybridization did not occur, proving that the gene is phylogenetically distinct from other species while being conserved among typical and atypical *A. salmonicida* strains.

The precise role of ECP has not been resolved due to the complexity of the substance(s), which include several proteases, phospholipase, haemolysins, a leucocidin and LPS[190–194]. Ellis and colleagues[195] reported that, in fish experiments, ECP reproduced the lesions normally associated with the chronic form of furunculosis, e.g. muscle necrosis and oedematous swelling at the site of injection, leading to mortalities[190]. This suggested that the ECP is responsible for some of the pathology of furunculosis. In fish cells, ECP exhibited cytotoxic, leucocytolytic and haemolytic effects. A conclusion was reached that most of the virulence factors were produced extracellularly, with most strains of *A. salmonicida* producing similar compounds[190,195].

Other studies indicated that injection of ECP led to the development of a pathological condition resembling furunculosis[192,196]. Cipriano and colleagues[192] correlated virulence with ECP fractions (four fractions were recovered). Mortalities, accompanied by haemorrhaging at the anus and fins, and inflammation at the site of injection, occurred in Atlantic salmon and brook trout which received Fractions I and II. Also, Fraction III caused haemorrhaging at the base of the fins, injection site, and in the mouth, but the majority of the fish survived. In contrast, rainbow trout were relatively resistant to the effects of all four fractions, without any mortalities resulting. Administration of Fraction II resulted in the development of furuncle-like lesions at the inoculation site. Fractions I, III and IV did not cause any obvious pathology in rainbow trout. The results of Cipriano and colleagues[192] supported the findings of Sakai[196], who considered that a protease was the most pathogenic substance in the ECP. However, Fyfe and co-workers[197] observed that protease preparations were less effective than equivalent amounts of ECP (with similar amount of proteolytic activity) at causing lesions in Atlantic salmon. Three major components were identified with molecular weights of 70 kDa (a serine protease), 56 kDa (a haemolysin) and 100 kDa (unidentified protein)[198].

Sheeran and colleagues[199] described two extracellular proteolytic activities that differed in their susceptibility to inhibitors and substrate specificity. One of the enzymes, designated P1, hydrolysed casein, elastin and gelatin, and showed a low non-specific activity against collagen. The second enzyme (P2) hydrolysed collagen and gelatin, but not casein or elastin, a pattern which suggested that it is a specific collagenase. Rockey and co-workers[193] described two proteases, coined P1 and P2, and a haemolysin (T-lysin) in the ECP. P1 and T-lysin were shown to work separately in the complete lysis of (rainbow trout) erythrocytes. T-lysin interacted with the outer membrane of the erythrocytes, whereas P1 destroyed the nuclear membrane. A novel metallo-protease has been recovered from a strain of *A. salmonicida* subsp. *achromogenes*[200]. A metalloprotease is a bacterial collagenolytic protease, which functions in the degradation of connective tissue thereby supporting the growth of *A. salmonicida*[201].

Intramuscular injection of proteases into brown trout resulted in the development of gross symptoms similar to those occurring in natural outbreaks of furunculosis, i.e. muscle liquefaction along the flanks adjacent to the lesion and swelling at the site of injection[199]. Sakai[202] considered a role for proteases in reproduction of the pathogen by making available small peptides and amino acids from proteolysis. Salmonid serum is capable of neutralizing lethal doses of proteases[195], possibly through the action of an α-migrating antiprotease[203].

A LPS-free phospholipase has been recovered from the ECP, and demonstrated to cause disease signs in Atlantic salmon, including lethargy, melanosis and erythema (= reddening) on the undersurfaces of fish, particularly around the anus, at the bases of the pectoral and pelvic fins, and head[194].

The major lethal toxin in the ECP is reportedly a lethal cytolysin[204], glycerophospholipid:cholesterol acyltransferase (GCAT) complexed with LPS, which under denaturing conditions in sodium dodecyl sulphate–polyacrylamide gel electrophoresis (SDS–PAGE) forms a 25 kDa single protein band. GCAT–LPS acts as a phospholipase when no acyl acceptor is present and also acts as a lysophospholipase. Whole salmon serum was shown to enhance the phospholipase activity of the purified toxin and the serum components were identified as low-density lipoprotein-like (LDL-like), high-density lipoprotein (HDL) and serum albumin. Intravascular injection of GCAT–LPS led to consumptive coagulopathy in rainbow trout, whereas administration of the 70 kDa serine protease to Atlantic salmon also led to circulatory failure. Salte and colleagues[205] found that the plasma antithrombin (AT) and α_2-macroglobulin inhibited the protease, and were consumed *in vivo* in response to the administration of the enzymes. *In vitro* and *in vivo* haemolysis of salmon erythrocytes by GCAT–LPS was demonstrated by Rosjo and colleagues[206], who suggested that the accumulation of lysophospholipids caused by the enzymic activity of GCAT–LPS led to lysis of salmonid red blood cells. Also, this suggests that the lysophospholipase activity is inhibited or suppressed.

The gene encoding the 70 kDa serine protease (caseinase) was cloned by Whitby and co-workers[207]. (This is homologous to the AHH1 haemolysin of *A. hydrophila*.) A 70 kDa serine protease gene probe was constructed, which consisted of 70% of the serine protease gene, and was used in hybridization studies with 26 typical and 18 atypical strains of *A. salmonicida*[208]. All of the typical strains secreted a phenol methyl sulphonyl fluoride (PMSF)-sensitive proteolytic activity, and most strains reacted positively. In comparison, the atypical strains produced various combinations, and it has been proposed that the possession of the 70 kDa serine protease gene places *A. salmonicida* strains in the subspecies *salmonicida*. Lee and Ellis[204] investigated the histopathological effect of the purified GCAT–LPS complex with or without the 70 kDa serine protease (caseinase). When injected singly, via the intramuscular route, GCAT–LPS produced a limited coagulative necrosis, whereas the purified 70 kDa protease induced a limited liquefaction and haemorrhaging. In combination, they induced the extensive liquefactive haemorrhagic 'furuncle' characteristic of the whole ECP.

Production of haemolysins may also contribute to the disease[190]. Titball and Munn[191] determined that the supernatant from unshaken broth cultures contained haemolytic ('H') activity against erythrocytes from a wide range of vertebrates, with maximal activity against horse red blood cells. However, if cultures were shaken, the resulting supernatant yielded an activity ('T' activity) against trout erythrocytes only. The H lysin was unstable in culture supernatants, sensitive to heat after exposure to 56°C for 5 min, and became membrane bound when solutions were filtered. In contrast, T lysin, which was stable in supernatant and inactivated by normal rainbow trout serum, could be separated into two factors, i.e. a caseinase and a membrane-bound (T_1) activity[209,210]. Complete lysis of trout erythrocytes occurred only in the presence of T_1 activity and caseinase. Nomura and Saito[211] observed that the production of haemolysin was stimulated by the addition of enzymic hydrolysates of protein, but suppressed by carbohydrates, such as glucose or sucrose. Bivalent metal ions, e.g. Ca^{2+}, Co^{2+} and Mn^{2+}, and phosphate ion (($HPO_4)^{2-}$) were found to be necessary for production of the haemolysin. Nomura and Saito[211] concluded that the haemolysin was produced during the stationary phase of growth, and was relatively heat-labile, being inactivated at 60°C. Incidentally, a 200 kDa haemolytic toxin, coined salmolysin, was described by Nomura and colleagues[212], but the relationship to other haemolysins remains to be established.

The possible relationship between haemolysins and proteases has been addressed. Hackett and colleagues[213] studied a possible plasmid-encoded origin for these extracellular enzymes, but concluded that the loss of proteolytic and haemolytic activity in variants of wild-type *A. salmonicida*. Moreover, there was no change in the LD_{100} between the virulent wild-type strain and a protease–haemolysin deficient variant. This implied that the extracellular activities were not essential for virulence, at least with regard to the acute form of furunculosis in rainbow trout. Two clones derived from a virulent strain, one of which was negative for protease and haemolysin production whereas the second derivative was positive for these attributes, were avirulent. Overall, it was apparent that attenuation of a virulent strain was without loss of the A-layer, plasmids or extracellular proteolytic and haemolytic activities. Thus, Hackett and co-authors[213] concluded that virulence was attributable to unknown factors. Titball and Munn[210] studied the effects of differences in virulence on the production of potential toxins, and concluded that release of proteases and haemolysins by virulent strains and their avirulent attenuated derivatives (differing only in the presence or absence of the A-layer) was not linked directly to virulence.

Of current interest is the ability to scavenge successfully for iron in iron-limited conditions. Initially, Chart and Trust[214] demonstrated that typical strains of *A. salmonicida* were capable of sequestering iron. Kay and colleagues[215] determined that the A-layer was implicated as a component of an iron-uptake mechanism by functioning as the initial stage of iron-uptake, being a binding site for porphyrins, i.e. haemin and protoporphyrin. Hirst and co-workers[216] described a siderophore-dependent iron-chelating system in typical strains and a siderophore-independent system in atypical *A. salmonicida*. The siderophores, which were soluble low-molecular-weight iron-chelators, behaved as 2,3-diphenol-catechol.

Amylase activity in culture filtrates was examined by Campbell and co-workers[217], who determined that three bands of activity, which corresponded to M_r values of 46, 43 and 38 kDa, were present on polyacrylamide gels. However, these M_r values did not agree with the value of 23 kDa obtained from the elution volume of a Sephacryl column. Amylase activity corresponded closely with glycogen-hydrolysing activity. Both activities were inhibited similarly by $HgCl_2$. This suggests that both activities relate to the same enzyme. Fish muscle contains glycogen reserve, and it has been proposed that amylase has glycogen-hydrolysing activity with glycogen being its natural substrate in furunculosis. Certainly, glycogen stimulated the highest level of α-amylase formation, being 1.6 times higher than in the presence of maltose[218]. In contrast, enzyme formation was not detected in a culture grown in the presence of glucose. It has been proposed that the 'maltose' system, for which the gene for the maltoporin has been cloned and sequenced, is present in Gram-negative bacteria, and functions for the utilization of linear and branched polyglucosides of the maltose series.

Ellis and colleagues[195] formulated a tentative explanation for the lesions caused by *A. salmonicida*, insofar as they reported that nearly all of the damage normally associated with the disease may be achieved by intraperitoneal or intramuscular injection of ECP in artifically high doses. Ellis and Grisley[219] concluded that normal trout serum inhibited ECP protease, but neutralization was effected by different antiproteases and less efficiently than trypsin. Ellis and colleagues[195] assumed that, in natural infections, lesions would be produced after the ECP had exhausted any inhibiting factors, either locally or systemically. They thought that the various symptoms of furunculosis were explained by the colonization of different host tissues by the pathogen. Overall, it was concluded that the pathological effects were probably caused by the ECP, with a leucocytolytic component acting against leucocytes. This would eventually result in leucopenia, preventing the destruction of the bacterial colonies and thus allowing the pathogen to be transmitted to other organs via the circulatory system. It was further considered that lesions and mortalities were actually due to the collagenolytic activity of the ECP, with haemorrhaging resulting in the vicinity of bacterial colonization. Generalized circulatory failure could ensue if the ECP subsequently entered the circulatory system.

The susceptibility of *A. salmonicida* to superoxide anion may be an indicator of the relative virulence of different strains, such as in terms of the presence of A-layer and the production of proteases[220]. Here, photoreduced riboflavin and methionine were used to generate superoxide anion in a cell-free system. Thus, using 11 strains of *A. salmonicida*, a correlation between susceptibility to killing by superoxide anion and 'virulence factors' was demonstrated. In the presence of an inhibitor of superoxide dismutase (sodium nitroprusside), the bactericidal effect of superoxide was increased, demonstrating that superoxide dismutase in combination with catalase or glutathione peroxidase to detoxify hydrogen peroxide could form a significant virulence factor.

A. salmonicida was rapidly killed *in vivo* inside a specialized intraperitoneal chamber implanted in rainbow trout. However, survival of *A. salmonicida* in the peritoneal chamber was higher than survival of implanted cells. The free cells were capable of penetrating peritoneal or tissue macrophages, and were found to replicate inside head kidney macrophages[221]. After a period of regrowth, the survivors within the diffusion chamber had acquired resistance to bacteriolysis and phagocytosis, and this was associated with a newly acquired capsular layer[222]. Also, resistance to oxidative killing was evident, and was not associated with the presence of capsular material. These resistant strains were examined for the presence of unique antigens, some of which were duly isolated[223]. With the exception of LPS, these novel antigens were destroyed after treatment with proteinase K. However, the antigens were not induced *in vitro* in response to either iron limitation or anaerobiosis. Moreover, antiserum raised against *in vitro*-grown cells recognized only the LPS expressed *in vitro*.

In a study of 130 strains, eight outer membrane proteins (OMPs) of 49, 40, 38, 37, 33, 31, 30 and 29 kDa were demonstrated[82]. Strains deficient in the 38 kDa protein but containing a 37 kDa OMP showed multiple low-level antibiotic resistance towards cephalothin, penicillin, chloramphenicol, tetracyline and quinolones. The bacteria were unable to degrade bovine and trout serum proteins, and displayed delayed haemolytic activity and an attack on casein. Pathology experiments revealed that the strains possessing the 37 kDa OMP produced virtually no pathological effects, whereas the strain with a normal protein profile generated typical furuncles. These data correlated with the study of Barnes and colleagues[224], who reported that resistance to a wide range of antimicrobials including 4-quinolones, β-lactams and tetracycline was associated with increased expression of a 37 kDa protein and a corresponding decrease in a 43 kDa protein (rather than a 38 kDa protein). The selection for oxytetracycline resistance simultaneously selected for resistance to the 4-quinolones including oxolinic acid. Clinical isolates resistant to oxolinic acid and oxytetracycline were screened for the two outer membrane alterations associated with cross-resistance. It was demonstrated that an oxolinic acid-susceptible isolate did not exhibit OMP changes, whereas a low-level resistant strain showed alterations in the OMP. However, a further nine isolates, which were resistant to oxolinic acid, did not display changes in the OMP. This study showed that changes in the OMP, which were associated with *in vitro* resistance to antibiotic, could occur in wild-type strains. Certainly, the development of antibiotic resistance would be an excellent way of retaining pathogenicity. Oxytetracycline, oxolinic acid and potentiated sulphonamides have been used successfully in treatment of furunculosis, but their value has been decreased by the development of resistant strains. The value of amoxycillin, which has recently become available for the treatment of furunculosis[225], has already been reduced by the rapid emergence of resistant strains, notably of *A. salmonicida* subsp. *achromogenes*. Hayes and colleagues[226] detected three β-lactamase activities, which corresponded to penicillinase, cephalosporinase and carbapenemase.

8.3.6   DISEASE CONTROL

Methods of disease control include adequate husbandry practices such as the maintenance of good water quality, disinfection of fish farm equipment and utensils especially when disease outbreaks occur, and routine disinfection policies for eggs upon arrival at receiving sites[43]. Stocking of fish farms with disease-resistant strains of fish, and the development of effective vaccines would also greatly improve the health of stock. If all else fails and fish become diseased, there may be recourse to antimicrobial compounds.

A variety of antimicrobial compounds have been used with varying degrees of success in the treatment of furunculosis. Initially, sulphonamides, notably sulphamerazine, were successful in controlling furunculosis when administered as a food additive[227]. Snieszko[227] showed the usefulness of oxytetracycline, which is still used extensively for the control of furunculosis. Additional effective chemotherapeutants include flumequine[228], oxolinic acid[229] and potentiated sulphonamides[230]. Potentiated sulphonamide and oxytetracycline are generally effective against the pathogen, regardless of the disease manifestation and assuming that treatment begins at an early stage in the disease cycle[231].

With the extensive use of chemotherapeutants in aquaculture, resistant strains of *A. salmonicida* have emerged[232]. It is disquieting that plasmids coding for antibiotic resistance (R factors) have been isolated from *A. salmonicida*[233,234]. However, a second generation of 4-quinolones/fluoroquinolones, notably enrofloxacin and sarofloxacin are effective at inhibiting the pathogen, and offer promise for the future[235-238].

Vaccine development started with the work of Duff[239], who produced an orally administered, chloroform-inactivated whole-cell preparation. However, it appears that *A. salmonicida* is an inefficient antigen, in terms of its overall capability of stimulating a protective immune response[240]. There is some controversy over the effectiveness of formulations based on ECP. Some studies indicate that they may well be immunosuppressive[241], whereas others describe their benefit in terms of immunogenicity[242]. Total ECP have been found to be more effective than purified GCAT–LPS preparations[208]. Kawahara and co-workers[243] showed that inactivated salmolysin was fairly protective. The application of ECP to polystyrene beads improved protection from 40% to 67%[244]. Good results have been found with passive immunization and the use of attenuated live vaccines[245,246] The latter was particularly effective in Atlantic salmon, in which use resulted in 12.5% mortality in the vaccinated group compared with 87.5% mortality among control fish, after challenge with a virulent culture of *A. salmonicida*. Avirulent cells, with altered A-layer, have also been suggested as candidates for live vaccines[247,248]. However, it is curious that Norqvist and colleagues[249] used live attenuated cells of *V. salmonicida*, and reported effectiveness at controlling furunculosis.

Promising results have followed the study of OMP[250]. Immunoblotting techniques revealed that four iron-regulated outer membrane proteins (IROMPs) were immunogenic in Atlantic salmon. The OMPs of *A. salmonicida* were grown in the presence of 2,2′-dipyridyl to induce IROMPs of 82, 77, 72 and 70 kDa molecular weight. The resulting vaccines gave considerably enhanced protection. Lutwyche and co-workers[251] isolated an outer membrane porin (28 kDa) from *A. salmonicida* which was heat-modifiable but not peptidoglycan-associated. The *N*-terminal amino acid sequence of the porin was determined and showed some homology to an *A. hydrophila* OMP.Vaccine trials with the purified porin produced RPS values of 53%. Comparison with a commercial whole-cell vaccine showed that the porin was more efficient than the conventional vaccine when administered by immersion (RPS = 19.4%) but less efficient than when injected intraperitoneally (RPS = 72.6%). When given intraperitoneally, the purified porin protected fish to the same level as a 15-minute immersion in the live attenuated parental strain of *A. salmonicida*.

Vaccine development based on the genetic manipulation of *A. salmonicida* was first reported by Bennett and colleagues[252], when a subunit vaccine was constructed by cloning a 587 kbp restriction fragment from the 70 kDa serine protease gene, containing the 'active serine' site sequence of the enzyme. Expression in *Escherichia coli* resulted in a high yield of a 142 kDa hybrid protein, which resulted from the fusion between an essentially complete β-galactosidase subunit and approximately one-third of the serine protease[252].

Using the same fragment, a further 42 kDa hybrid was constructed and fused to a truncated β-galactosidase subunit. This hybrid permitted the identification of smaller epitopes that may have been masked by a large β-galactosidase domain. Both fusion proteins were shown to possess recognizable epitopes by dot blotting against rabbit anti-*A. salmonicida* 70 kDa serine protease antibody. Subsequently, Coleman and co-authors[253] isolated, cloned and sequenced the gene of a maltoporin, which is the second most abundant OMP. A hybrid protein was constructed from a large fragment of its *C*-terminus possessing immunoaccessible loops inserted into dihydrofolate reductase. Atlantic salmon, which were injected intraperitoneally with these hybrid proteins, were protected against subsequent challenge with a virulent culture of *A. salmonicida*. Genetically attenuated live vaccines have also been developed using aromatic-dependent mutants[254]. These mutants were constructed by inserting kanamycin- and tetracycline-resistant determinants into the coding sequence of the *aro* gene. The *aro* A::Kar region of pAAK2 was cloned on a 4.2 kbp *Eco*R1 fragment into the *Eco*R1 site of the broad-host-range plasmid suicide vector pSUP202, forming pSUP202K2, which was transformed into *E. coli*. The *aro* A::Ka^r mutation was introduced by allele replacement into the chromosome of virulent *A. salmonicida* by conjugation, using a suicide plasmid delivery system. In comparison with the wild-type strains, *aro*-mutants were avirulent and conferred substantial protection to brown trout, which is normally very sensitive to *A. salmonicida*. Fish administered a single dose of vaccine demonstrated a 253-fold increase in LD_{50} (lethal dose 50%; the number of cells required to kill half of the fish population) values. Moreover, a booster dose led to a further 16-fold increase in LD_{50} values. Following administration at 10^7 cells/ml, the vaccine could be detected in tissues for up to 12 days. However, it should be emphasized that mortalities did not occur in the recipient fish. Further work has demonstrated that antigen-specific cell-mediated immunity and humoral responses occurred in rainbow trout following administration of an aromatic-dependent mutant, termed BRIVAX[255]. However, there is an advantage with the previously mentioned live attenuated streptomycin pseudorevertant of Thornton and colleagues[247] insofar as the lack of tissue persistence lowers the perceived risk of reversion to a virulent form and reduces the chance of establishing an asymptomatic carrier state. However, it is conceded that the rapid elimination of the pseudorevertant may well limit the development of a protective immune response. In contrast, the aromatic-dependent mutant is non-reverting due to its diverse metabolic deficiency and is effectively avirulent. Yet the retention of virulence determinants, specifically the A-layer, and the persistence of the mutant in tissues suggest an ideal vaccine candidate. Furthermore, the potential value of attenuated live vaccines has been reinforced by the recent work of Noonan and colleagues[256].

Michel[257] noted that there was no difference in the effectiveness of vaccines prepared with either rough or smooth cultures, when administered orally or via intraperitoneal injection to rainbow trout. However, neither type of vaccine was protective. McCarthy and co-workers[258] reported that only rough strains conferred protection. Similarly, Olivier and co-workers[259] determined that avirulent cells were less effective immunogens than their virulent counterparts. Both suggested that the A-layer protein was probably the protective antigen. Hastings and Ellis[260] recorded that rainbow trout responded to A-protein and LPS O-antigen and some of the components of the ECP (including proteases)[246]. Furthermore, Lund and colleagues[261] suggested that the dominant antigens included a caseinolytic protease, the A-protein and high- and low-molecular-weight LPS.

Olivier and associates[262] recorded protection in coho salmon after intraperitoneal injection of formalized cells and Freund's complete adjuvant. Presumably, the adjuvant stimulated non-specific immunity, probably involving macrophage activity. Indeed, from this and the subsequent study of Norqvist and colleagues[249], it is necessary to question the need for incorporation of *A. salmonicida* cells or their cellular components in furunculosis vaccines. Apart from the obvious benefits of Freund's complete adjuvant[262], the use of β-1,3-glucan, chitinosan[263], lentinan and formalin-killed cells of *Renibacterium salmoninarum* have enhanced the effectiveness of whole-cell vaccines[264]. Evidence indicated that β-1,3-glucan enhanced the production of antibodies in Atlantic salmon to key antigens, including LPS[265]. However, the recipient fish were not highly protected against *A. salmonicida*, insofar as the relative percentage survival did not exceed 54%.

Rodgers[266] reported the benefits of using inactivated whole cells, toxoided ECP and LPS. Moreover, the vaccinated salmonids grew better than the unvaccinated controls. Work has indicated that the duration of the immersion vaccination process does not affect the uptake of the vaccine, providing that the antigens are not in low concentrations[267].

The value of breeding disease-resistant fish was indicated originally by Embody and Hayford[268], who increased the level of resistance in brook trout to furunculosis by selective breeding. Cipriano[269] examined the role of serum components in the protection of salmonids, and concluded that serum from rainbow trout, which are naturally resistant to furunculosis, could protect passively immunized brook trout. In contrast, the injection of serum from susceptible Atlantic salmon was unsuccessful in conferring resistance upon brook trout. The protective effect of rainbow trout serum was believed to be attributed to the neutralization of toxic components produced by the pathogen. Stromsheim and colleagues[270] used the A-layer as the antigen in genetic studies which attempted to relate levels of antibody to the survival of fish. Furthermore, Lund and co-authors[271] evaluated various humoral parameters as possible indirect markers for breeding improved resistance to furunculosis. Thus, the level of antibodies against the A-layer protein and LPS and the total amount of immunoglobulin (IgM) were measured. The total amount of IgM explained 17% of the survival rates, but when correlated with the level of anti-A-protein antibodies, the two parameters explained 37% of the survival rates. Two genetic lines of carp produced differing degrees of resistance to CE[272]. Indeed, two transferrin genotypes were identified, but did not appear to influence resistance. Certainly, this aspect of disease control merits further research.

The possibility of biological control of furunculosis has been indicated by use of fluorescent pseudomonads which are inhibitory to *A. salmonicida*. Thus, Smith and Davey[273] showed that the fluorescent pseudomonad F19/3 was capable of inhibiting the growth of *A. salmonicida* in culture media, due to competition for free iron. On a similar trend Austin and colleagues[274] reported a probiotic strain of *Vibrio alginolyticus* capable of inhibiting the pathological effects of *A. salmonicida*.

8.4 REFERENCES

1 Martinez-Murcia AJ, Esteve C, Garay E, Collins MD. *Aeromonas allosaccharophila* sp. nov., a new mesophilic member of the genus *Aeromonas*. *FEMS Microbiol Lett* 1992; **91**: 199–206.

2 Toranzo AE, Baya AM, Romalde, JJ, Hetrick FM. Association of *Aeromonas sobria* with mortalities of adult gizzard shad, *Dorosoma cepedianum* Lesueur. *J. Fish Dis* 1989; **12**: 439–48.

3 Paniagua C, Rivero O, Anguita J, Naharro G. Pathogenicity factors and virulence for rainbow trout (*Salmo gairdneri*) of motile *Aeromonas* spp. isolated from a river. *J Clin Microbiol* 1990; **28**: 350–5.

4 Larsen JL, Jensen NJ. An *Aeromonas* species implicated in ulcer-disease of the cod (*Gadus morhua*). *Nord Veterinaermed* 1977; **29**: 199–211.

5 Eurell TE, Lewis DH, Grumbles LC. Comparison of selected dianostic tests for detection of motile *Aeromonas* septicaemia in fish. *Am J Vet Res* 1978; **39**: 1384–6.

6 Michel C. A bacterial disease of perch (*Perca fluviatilis* L.) in an Alpine lake: isolation and preliminary study of the causative organism. *J Wildlife Dis* 1981; **17**: 505–10.

7 Miyazaki T, Kaige N. A histopathological study on motile aeromonad disease of Crucian carp. *Fish Pathol* 1985; **21**: 181–5.

8 Hazen TC, Raker ML, Esch GW, Fliermans CB. Ultrastructure of red-sore lesions on largemouth bass (*Micropterus salmoides*); association of the ciliate *Epistylis* sp. and the bacterium *Aeromonas hydrophila*. *J Protozool* 1978; **25**: 351–5.

9 Santos, Y, Bandín I, Núñez S, Nieto TP, Toranzo AE. Serotyping of motile *Aeromonas* species in relation to virulence phenotype. *Bull Eur Assoc Fish Pathol* 1991; **11**: 153–5.

10 Leblanc D, Mittal KR, Olivier G, Lallier R. Serogrouping of motile *Aeromonas* species isolated from healthy and moribund fish. *Appl Environ Microbiol* 1981; **42**: 56–60.

11 Shaw DH, Hodder HJ. Lipopolysaccharides of the motile aeromonads; core oligosaccharide analysis as an aid to taxonomic classification. *Can J Microbiol* 1978; **24**: 864–8.

12 Cumberbatch N, Gurwith MJ, Langston C, Sack RB, Brunton JL. Cytotoxic

enterotoxin produced by *Aeromonas hydrophila*; relationship of toxigenic isolates to diarrhoeal disease. *Infect Immun* 1979; **23**: 829–37.

13   Nzeako BC. Variation in *Aeromonas hydrophila* surface structures. *Bull Eur Assoc Fish Pathol* 1991; **11**: 176–9.

14   Paula SJ, Duffey PS, Abbott SL, Kokka RP, Oshio LS, Janda JM, Shimada T, Sakazaki R. Surface properties of autoagglutinating mesophilic aeromonads. *Infect Immun* 1988; **56**: 2658–65.

15   Dooley JSG, McCubbin WD, Kay CM, Trust TJ. Isolation and biochemical characterization of the S-layer protein from a pathogenic *Aeromonas hydrophila* strain. *J Bacteriol* 1988; **170**: 2631–8.

16   Ascencio F, Ljungh Å, Wadstrom T. Comparative study of extracellular matrix protein binding to *Aeromonas hydrophila* isolated from diseased fish and human infection. *Microbios* 1991; **65**: 135–46.

17   Trust TJ, Howard PS, Chamberlain JB, Ishiguro EE, Buckley JT. Additional surface protein in autoaggregating strains of atypical *Aeromonas salmonicida*. *FEMS Microbiol Lett* 1980; **9**: 35–8.

18   Santos, Y, Toranzo AE, Dopazo CP, Nieto TP, Barja JL. Relationship among virulence for fish, enterotoxigenicity, and phenotypic characteristics of motile *Aeromonas*. *Aquaculture* 1987; **67:** 29–39.

19   Shotts EB, Hsu TC, Waltman WD Extracellular proteolytic activity of *Aeromonas hydrophila* complex. *Fish Pathol* 1984; **20**: 37–44.

20   Allan BJ, Stevenson RMW. Extracellular virulence factors of *Aeromonas hydrophila* in fish infections. *Can J Microbiol* 1981; **27**: 1114–22.

21   Stevenson RMW, Allan BJ. Extracellular virulence factors in *Aeromonas hydrophila* disease processes in salmonids. *Dev Biol Stand* 1981; **49**: 173–80.

22   Karunasagar I, Segar K, Karunasagar I, Ali PKMM, Jeyasekaran G. Virulence of *Aeromonas hydrophila* strains from fish ponds and infected fishes. In: Chang S-T. Chan K-Y, Norman YSW(eds) *Recent Advances in Biotechnology and Applied Biology*. The Chinese University of Hong Kong, Hong Kong: 1990.

23   Kanai K, Takagi Y. Alpha-hemolytic toxins of *Aeromonas hydrophila* produced in *in vitro*. *Fish Pathol*. 1986; **21**: 245–50.

24   Massad G, Arceneaux JEL, Byers BR. Acquisition of iron from host sources by mesophilic *Aeromonas* species. *J Gen Microbiol* 1991; **137**: 237–41.

25   Thune RL, Graham TE, Riddle LM, Amborski RL. Extracellular products and endotoxin from *Aeromonas hydrophila*: effect on age-0 channel catfish. *Trans Am Fish Soc* 1982; **11** 404–8.

26   Thune RL, Graham TE, Riddle LM, Amborski RL. Extracellular proteases from *Aeromonas hydrophila*: partial purification and effects on age-0 channel catfish. *Trans Am Fish Soc* 1982; **11**: 749–54.

27   Lallier R, Bernard F, Lalonde G. Difference in the extracellular products of two strains of *Aeromonas hydrophila* virulent and weakly virulent for fish. *Can J Microbiol* 1984; **30**: 900–4.

28   Hsu TC, Waltman WD, Shotts EB. Correlation of extracellular enzymatic activity and biochemical characteristics with regard to virulence of *Aeromonas hydrophila*. *Dev Biol Stand* 1981; **49**: 101–111.

29   Hsu TC, Shotts EB, Waltman WD. Quantitation of biochemical and enzymatic characteristics with pathogenicity of *Aeromonas hydrophila* complexes in fish. *Proc Rep China-Japan SympFish Dis* 1983; 205–9.

30   Nieto TP, Santos Y, Rodriguez LA, Ellis AE. An extracellular acetylcholinesterase produced by *Aeromonas hydrophila* is a major lethal toxin for fish. *Microb Pathog* 1991; **11**: 101–10.

31   Leung K-Y, Stevenson RMW. Characteristics and distribution of extracellular proteases from *Aeromonas hydrophila*. *J Gen Microbiol* 1988; **134**: 151–60.

32 Nieto TP, Ellis AE. Characterization of extracellular metallo- and serine-proteases of *Aeromonas hydrophila* strain B_{51}. *J Gen Microbiol* 1986; **132**: 1975–9.

33 Aoki T. Drug-resistant plasmids from fish pathogens. *Microbiol Sci* 1988; **5**: 219–23.

34 Toranzo AE, Barja JL, Colwell RR, Hetrick FM. Characterization of plasmids in bacterial fish pathogens. *Infect Immun* 1983; **39**: 184–92.

35 De Paola A, Flynn P, McPhearson RM, Levy SB. Phenotypic and genotypic characterization of tetracycline- and oxytetracycline-resistant *Aeromonas hydrophila* from cultured channel catfish (*Ictalurus punctatus*) and their environment. *Appl Environ Microbiol* 1988; **54**: 1861–3.

36 Schachte JH. Immunization of channel catfish, *Ictalurus punctatus*, against two bacterial diseases. *Mar Fish Rev* 1978; **40**: 18–19.

37 Lamers CHL, de Haas MJM. The development of immunological memory in carp (*Cyprinus carpio* L.) to a bacterial antigen. *Dev Comp Immunol* 1983; **7**: 713–14.

38 Ruangpan L, Kitao T, Yoshida T. Protective efficacy of *Aeromonas hydrophila* vaccines in nile tilapia. *Vet Immunol Immunopathol* 1986; **12**: 345–50.

39 Loghothetis PN, Austin B. Immune response of rainbow trout (*Oncorhynchus mykiss*, Walbaum) to *Aeromonas hydrophila*. *Fish Shellfish Immunol* 1994; **4**: 239–54.

40 Lamers CHJ, de Haas MJM, van Muiswinkel WB. Humoral response and memory formation in carp after injection of *Aeromonas hydrophila* bacterin. *Dev Comp Immunol* 1985; **9**: 65–75.

41 Lamers CHJ, de Haas MJM, van Muiswinkel WB. The reaction of the immune system of fish to vaccination: development of immunological memory in carp, *Cyprinus carpio* L., following direct immersion in *Aeromonas hydrophila* bacterin. *J Fish Dis* 1985; **8**: 253–62.

42 Baba T, Imamura J, Izawa K, Ikeda K. Cell-mediated protection in carp *Cyprinus carpio* L., against *Aeromonas hydrophila*. *J Fish Dis* 1988; **11**: 171–8.

43 McCarthy DH, Roberts RJ. Furunculosis of fish – the present state of our knowledge. In Droop, MA, Jannasch HW (eds) *Advances in Aquatic Microbiology*. London: Academic Press, 1980: 293–341.

44 McCarthy DH. Fish furunculosis. *J Inst Fish Manage* 1975; **6**: 13–18.

45 Davis HS. Care and diseases of trout. *Res Rep U S Fish Wildl Service* 1946; **12**: 98.

46 Austin B, Austin DA. *Bacterial Fish Pathogens, Disease of Farmed and Wild Fish*, 2nd edn. Simon and Schuster, 1993.

47 Amlacher E. *Taschenbuch der Fischkrankheiten*. Jena: Gustav Fisher Verlag, 1961.

48 Fijan NN. Infectious dropsy of carp – a disease complex. *Proc Sym Zool Soc London* 1972; **30**: 39–57.

49 Bootsma R, Fijan N, Blommaert J. Isolation and identification of the causative agent of carp erythrodermatitis. *Vet Arch* 1977; **47**: 291–302.

50 Shotts EB, Talkington FD, Elliott DG, McCarthy DH. Aetiology of an ulcerative disease in goldfish, *Carassius auratus* (L): characterization of the causative agent. *J Fish Dis* 1980; **3**: 181–6.

51 Wiklund T. Virulence of 'atypical' *Aeromonas salmonicida* isolated from ulcerated flounder *Platichthys flesus*. *Dis Aquat Org* 1995; **21**: 145–50.

52 Wiklund T. Survival of 'atypical' *Aeromonas salmonicida* in water and sediment microcosms of different salinities and temperatures. *Dis Aquat Org* 1995; **21**: 137–43.

53 Kitao T, Yoshida T, Aoki T, Fukudome M. Characterization of an atypical

Aeromonas salmonicida strain causing epizootic ulcer disease in cultured eel. *Fish Pathol* 1985; **20**: 107–14

54 Kitao T, Yoshida T, Aoki T, Fukodome M. Atypical *Aeromonas salmonicida*, the causative agent of an ulcer disease of eel occurred in Kagoshima Prefecture. *Fish Pathol* 1984; **19**: 113–17.

55 Ohtsuka H, Nakai T, Muroga K, Jo Y. Atypical *Aeromonas salmonicida* isolated from diseased eels. *Fish Pathol* 1984; **19**: 101–7.

56 Nakai T, Miyakawa M, Muroga K, Kamito K. The tissue distribution of atypical *Aeromonas salmonicida* in artificially infected Japanese eels, *Anguilla japonica*. *Fish Pathol* 1989; **24**: 23–8.

57 Daly JG, Stevenson RMW. Importance of culturing several organs to detect *Aeromonas salmonicida* in salmonid fish. *Trans Am Fish Soc* 1985; **114**: 909–10.

58 Neelam B, Robinson RA, Nicholas CR, Stevens L. The effect of iron limitation on the growth of *Aeromonas salmonicida*. *Microbios* 1993; **74**: 59–67.

59 Maheshkumar S, Goyal SM, Economon PP. Concentration and detection of *Aeromonas salmonicida* from hatchery water. *J Fish Dis* 1990; **13**: 513–18.

60 Ford LA. Detection of *Aeromonas salmonicida* from water using a filtration method. *Aquaculture* 1994; **122**: 1–7.

61 Wiklund T. Atypical *Aeromonas salmonicida* isolated from ulcers of pike, *Esox lucius* L. *J Fish Dis* 1990; **13**: 541–4.

62 Wiklund T, Bylund G. A cytochrome oxidase negative bacterium (presumptively an atypical *Aeromonas salmonicida*) isolated from ulcerated flounders (*Platichthys flesus* (L.)) in the northern Baltic Sea. *Bull Eur Assoc Fish Pathol* 1991; **11**: 74–6.

63 Teska JK, Cipriano RC, Schill WB. Molecular and genetic characterization of cytochrome oxidase-negative *Aeromonas salmonicida* isolated from coho salmon (*oncorhynchus kisutch*). *J Wildlife Dis* 1992; **28**: 515–20.

64 Pedersen K, Kofod H, Dalgaard I, Larsen JL. Isolation of oxidase-negative *Aeromonas salmonicida* from diseased turbot *Scophthalmus maximus*. *Dis Aquat Org* 1994; **18**: 149–54.

65 Wilson BW, Holliman A. Atypical *Aeromonas salmonicida* isolated from ulcerated chub *Lerciscus cephalis*. *Vet Rec* 1994; **135**: 185–6.

66 Wiklund T, Dalsgaard I. Atypical *Aeromonas salmonicida* associated with ulcerated flatfish species in the Baltic Sea and the North Sea. *J Aquat Anim Health* 1995; **7**: 218–24.

67 Bryant TN, Lee JB, West PA, Colwell RR. Numerical classification of species of *Vibrio* and related genera. *J Appl Bacteriol* 1986; **61**: 437–67.

68 Lee JB. Identification of *Aeromonas* in the routine laboratory. *Experientia* 1987; **43**: 355–7.

69 McIntosh D, Austin B. Atypical characteristics of the salmonid pathogen *Aeromonas salmonicida*. *J Gen Microbiol* 1991; **137**: 1341–3.

70 Austin B. Recovery of 'atypical' isolates of *Aeromonas salmonicida*, which grow at 37°C, from ulcerated non-salmonids in England. *J Fish Dis* 1993; **16**: 165–8.

71 Duff DCB. Dissociation in *Bacillus salmonicida* with special reference to the appearance of the G form of culture. *J Bacteriol* 1937; **34**: 49–67.

72 Belland RJ, Trust TJ. DNA:DNA reassociation analysis of *Aeromonas salmonicida*. *J Gen Microbiol* 1988; **134**: 307–15.

73 Austin DA, McIntosh D, Austin B. Taxonomy of fish associated *Aeromonas* spp., with the description of *Aeromonas salmonicida* subsp. *smithia* subsp. nov. *Syst Appl Microbiol* 1989; **11**: 277–90.

74 Hirvela-Koski V, Koski P, Niiranen J. Biochemical properties and drug resistance of *Aeromonas salmonicida* in Finland. *Dis Aquat Org* 1994; **20**: 191–6.

75 Belland BJ, Trust TJ. *Aeromonas salmonicida* plasmids: plasmid-directed

synthesis of proteins *in vitro* and in *Escherichia coli* minicells. *J Gen Microbiol* 1989; **135**: 513–24.

76 Neilsen B, Olsen, JE, Larsen JL. Plasmid profiling as an epidemiological marker within *Aeromonas salmonicida*. *Dis Aquat Org* 1993; **15**: 129–35.

77 Bast L, Daly JG, DeGrandis SA, Stevenson RMW. Evaluation of profiles of *Aeromonas salmonicida* as epidemiological markers of furunculosis infections in fish. *J Fish Dis* 1988; **11**: 133–45.

78 Toranzo AE, Santos Y, Nunez S, Barja JL. Biochemical and serological characteristics, drug resistance and plasmid profiles of Spanish isolates of *Aeromonas salmonicida*. *Fish Pathol* 1991; **26**: 55–60.

79 Sorum H, Kvello JJ, Hastein T. Occurrence and stability of plasmids in *Aeromonas salmonicida* ss *salmonicida* isolated from salmonids with furunculosis. *Dis Aquat Org* 1993; **16**: 199–206.

80 Hanninen ML, Ridell J, Hirvela-Koski V. Phenotypic and molecular characteristics of *Aeromonas salmonicida* subsp. *salmonicida* isolated in southern and northern Finland. *J Appl Bacteriol* 1995; **79**: 12–21.

81 Miyata M, Aoki T, Inglis V. Yoshida T, Endo M. RAPD analysis of *Aeromonas salmonicida* and *Aeromonas hydrophila*. *J Appl Bacteriol* 1995; **79**: 181–5.

82 Nielsen B, Dalsgaard E, Brown DJ, Larsen JL. *Aeromonas salmonicida* subsp. *salmonicida*: correlation of protein patterns, antibiotic resistance, exoprotease activity, haemolysis and pathological lesions produced *in vivo*. *J Fish Dis* 1994; **17**: 387–97.

83 Dorsch M, Ashbolt NJ, Cox PT, Goodman AE. Rapid identification of *Aeromonas* species using 16S rDNA targeted oligonucleotide primers: a molecular approach based on screening of environmental isolates. *J Appl Bacteriol* 1994; **77**: 722–6.

84 Boyd EF, Hiney BP, Peden JF, Smith PR. Assessment of genetic diversity among *Aeromonas salmonicida* isolates by multilocus enzyme electrophoresis. *J Fish Dis* 1994; **17**: 97–8.

85 Whittington RJ, Djordjevic SP, Carson J, Callinan RB. Restriction endonuclease analysis of atypical *Aeromonas salmonicida* isolates from goldfish *Carassius auratus*, silver perch *Bidyanus bidyanus*, and greenback flounder *Rhombosolea tapirina* in Australia. *Dis Aquat Org* 1995; **22**: 185–91.

86 McCormick WA, Stevenson RMW, MacInnes JI. Restriction endonuclease fingerprinting analysis of Canadian isolates of *Aeromonas salmonicida*. *Can J Microbiol* 1989; **36**: 24–32.

87 Ewing WH, Hugh R, Johnson JG. *Studies on the* Aeromonas *group*. U S Dept of Health, Education and Welfare, Public Health Services, Communicable Diseases Center, 1961: Atlanta, USA.

88 Liu PV. Observations on the specificities of extra-cellular antigens of the genera *Aeromonas* and *Serratia*. *J Gen Microbiol* 1961; **24**: 145–53.

89 Karlsson KA. An investigation of the *Aeromonas salmonicida* haemolysin. Ninth Nordic Veterinary Congress, Kobenhavn, 1962.

90 Bullock GL. Precipitins and agglutinin reactions of aeromonads isolated from fish and other sources. *Bull Office Int Epizoot* 1966; **65**: 805–24.

91 Popoff M. Étude sur les *Aeromonas salmonicida*. I. Caractéres biochemiques et antigéniques. *Rech Vétérin* 1969; **3**: 49–57.

92 Popoff M. Genus III. *Aeromonas* Kluyver and van Niel 1936, 398[AL]. In Krieg NR, Holt JG (eds), *Bergey's Manual of Systematic Bacteriology*, vol I. Baltimore: Williams and Wilkins, 1984: 545–8.

93 McCarthy DH, Rawle CT. Rapid serological diagnosis of fish furunculosis caused by smooth and rough strains of *Aeromonas salmonicida*. *J Gen Microbiol* 1975; **86**: 185–7.

94 Todd C. The presence of a bacteriophage for *B. salmonicida* in river waters. *Nature (London)* 1993; **131**: 360.

95 Popoff M. Étude sur les *Aeromonas salmonicida*. II. Caracterisation des bactériophages actifs sur les '*Aeromonas salmonicida*' et lysotypie. *Ann Rech Vétérin* 1971; **2**: 33–45.

96 Popoff M. Interest diagnostique d'un bactériophage specifique des *Aeromonas salmonicida*. *Ann Rech Vétérin* 1971; **2**: 137–9.

97 Rodgers CJ, Pringle JH, McCarthy DH, Austin B. Quantitative and qualitative studies of *Aeromonas salmonicida* bacteriophage. *J Gen Microbiol* 1981; **125**: 335–45.

98 Snieszko SF, Griffin P, Friddle SB. A new bacterium (*Haemophilus piscium* n. sp) from ulcer disease of trout. *J Bacteriol* 1950; **59**: 699–710.

99 Paterson WD, Douey D, Desautels D. Relationships between selected strains of typical and atypical *Aeromonas salmonicida, Aeromonas hydrophila*, and *Haemophilus piscium*. *Can J Microbiol* 1980; **26**: 588–98.

100 Trust TJ, Ishiguro EE, Atkinson HM. Relationship between *Haemophilus piscium* and *Aeromonas salmonicida* revealed by *Aeromonas hydrophila* bacteriophage. *FEMS Microbiol Lett* 1990; **9**: 199–201.

101 Rabb L, Cornick JW, MacDermott LA. A macroscopic slide agglutination test for the presumptive diagnosis of furunculosis in fish. *Prog Fish Cult* 1964; **26**: 118–20.

102 McCarthy DH. Laboratory techniques for the diagnosis of fish-furunculosis and whirling disease. *MAFF, Fish Res Tech Rep* 1976; **23**: 5.

103 McCarthy DH. Detection of *Aeromonas salmonicida* antigen in diseased fish tissue. *J Gen Microbiol* 1975; **88**: 384–6.

104 Geck P. India ink immuno-reaction for the rapid detection of enteric pathogens. *Acta Microbiol Acad Sci Hung* 1971; **18**: 191–6.

105 Kimura T, Yoshimizu M. Coagglutination test with antibody-sensitized staphylococci for rapid and simple serological diagnosis of fish furunculosis. *Fish Pathol* 1983; **17**: 259–62.

106 Kimura T, Yoshimizu M. Coagglutination test with antibody-sensitized staphylococci for rapid serological identification of rough strains of *Aeromonas salmonicida*. *Bull Jap Soc Sci Fish* 1984; **50**: 439–42.

107 Austin B, Bishop I, Gray C, Watt B, Dawes J. Monoclonal antibody based enzyme linked immunosorbent assays for the rapid diagnosis of clinical cases of enteric redmouth and furunculosis in fish farms. *J Fish Dis* 1986; **9**: 469–74.

108 Yoshimizu M, Diridbusarakom A, Ezura Y, Kimura T. Monoclonal antibodies against *Aeromonas salmonicida* for serological diagnosis of furunculosis. *Bull Jap Soc Sci Fish* 1993; **59**: 333–8.

109 Neelam B, Thompson KD, Price NC, Tatner MF, Adams A, Ellis AE, Stevens L. Development of monoclonal antibodies to iron-regulated outer membrane proteins and lipopolysaccharide of *Aeromonas salmonicida*. *Dis Aquat Org* 1995; **21**: 201–8.

110 Sakai M, Atsuta S, Kobayashi M. Comparative sensitivities of several diagnostic methods to detect fish furunculosis. *Kitasato Arch Exp Med* 1986; **59**: 43–8.

111 Bernoth E-M. Autoagglutination, growth on tryptone-soy-Coomassie-agar, outer membrane protein patterns and virulence of *Aeromonas salmonicida* strains. *J Fish Dis* 1990; **13**: 145–55.

112 Barry T, Powell R, Gannon F. A general method to generate DNA probes for microorganisms. *Biotechnology* 1990; **8**: 233–6.

113 Hiney M, Dawson MT, Heery DM, Smith PR, Gannon F, Powell R. DNA probe for *Aeromonas salmonicida*. *Appl Environ Microbiol* 1992; **58**: 1039–42.

114 Gustafson CE, Thomas CJ, Trust TJ. Detection of *Aeromonas salmonicida* from

fish by using polymerase chain reaction amplification of the virulence surface array protein gene. *Appl Environ Microbiol* 1992; **58**: 3816–25.

115 Mooney J, Powell E, Clabby C, Powell R. Detection of *Aeromonas salmonicida* in wild Atlantic salmon using a specific DNA probe test. *Dis Aquat Org* 1995; **21**: 131–5.

116 Lund M. A study of the biology of *Aeromonas salmonicida* (Lehmann and Neumann 1896) Griffin 1954. *MSc Thesis*, University of Newcastle upon Tyne, UK. 1967.

117 McCarthy DH. Some ecological aspects of the bacterial fish pathogen – *Aeromonas salmonicida*. *Aquat Microbiol Symp Soc Appl Bacteriol* 1980; **6**: 299–324.

118 Sakai DK. Kinetics of adhesion associated with net electrical charges in agglutinating *Aeromonas salmonicida* cells and their spheroplasts. *Bull Jap Soc Sci Fish* 1986; **52**: 31–6.

119 Sakai DK. Electrostatic mechanism of survival of virulent *Aeromonas salmonicida* strains in river water. *Appl Environ Microbiol* 1986; **51**: 1343–9.

120 Rose AS, Ellis AE, Munro ALS. Evidence against dormancy in the bacterial fish pathogen *Aeromonas salmonicida* subsp. *salmonicida*. *FEMS Microbiol Lett* 1990; **68**: 105–8.

121 Rose AS, Ellis AE, Munro ALS. The survival of *Aeromonas salmonicida* subsp. *salmonicida* in seawater. *J Fish Dis* 1990; **13**: 205–14.

122 Morgan JAW, Cranwell PA, Pickup RW Survival of *Aeromonas salmonicida* in Lake water. *Appl Environ Microbiol* 1991; **57**: 1777–82.

123 Morgan JAW, Rhodes G, Pickup RW. Survival of nonculturable *Aeromonas salmonicida* in lake water. *Appl Environ Microbiol* 1993; **59**: 874–80.

124 Effendi I, Austin B. Survival of the fish pathogen *Aeromonas salmonicida* in seawater. *FEMS Microbiol Lett* 1991; **84**: 103–6.

125 Effendi I, Austin B. Survival of the fish pathogen *Aeromonas salmonicida* in the marine environment. *J Fish Dis* 1994; **17**: 375–85.

126 O'Brien F, Mooney J, Ryan D, Powell E, Tiney M, Smith PR, Powell R. Detection of *Aeromonas salmonicida* causal agent of furunculosis in salmonid fish, from the tank effluent of hatchery-reared Atlantic salmon smolts. *Appl Environ Microbiol* 1994; **60**: 3874–7.

127 Michel C, Dubois-Darnaudpeys A. Persistence of the virulence of *Aeromonas salmonicida* kept in river sediments. *Ann Rech Vétérin* 1980; **11**: 375–86.

128 Dubois-Darnaudpeys A. Epidemiologie de la furonculose des salmonides. III. Ecologie de *Aeromonas salmonicida* proposition d'un modele epidemiologique. *Bull Français Piscicult* 1977; **50**: 21–32.

129 Cornick JW, Chudyk RV, McDermott LA. Habitat and viability studies on *Aeromonas salmonicida*, causative agent of furunculosis. *Prog Fish Cult* 1969; **31**: 90–3.

130 Nese L, Enger O. Isolation of *Aeromonas salmonicida* from salmon lice *Lepeophtheirus salmonis* and marine plankton. *Dis Aquat Org* 1993; **16**: 79–81.

131 Enger O, Thorsen BK. Possible ecological implications of the high cell surface hydrophobicity of the fish pathogen *Aeromonas salmonicida*. *Can J Microbiol* 1991; **38**: 1048–52.

132 Enger O, Gunnlaugsdottir B, Thorsen BK, Hjeltnes B. Infectious load of *Aeromonas salmonicida* subsp. *salmonicida* during the initial phase of a cohabitant infection experiment with Atlantic salmon, *Salmo salar* L. *J Fish Dis* 1993; **15**: 425–30.

133 Allen-Austin D, Austin B, Colwell RR. Survival of *Aeromonas salmonicida* in river water. *FEMS Microbiol Lett* 1984; **21**: 143–6.

134 Effendi I, Austin B. Dormant/unculturable cells of the fish pathogen *Aeromonas*

salmonicida. Microb Ecol 1995; **30**: 183–92.

135 McIntosh D, Austin B. Comparison of methods for the induction, propagation and recovery of L-phase variants of *Aeromonas* spp. *J Diarrh Dis Res* 1988; **6**: 131–6.

136 McIntosh D, Austin B. Recovery of cell wall deficient forms (L-forms) of the fish pathogens *Aeromonas salmonicida* and *Yersinia ruckeri. Syst Appl Microbiol* 1990; **13**: 378–81.

137 McIntosh D, Austin B. The role of cell wall deficient bacteria (L-forms; sphaeroplasts) in fish diseases. *J Appl Bacteriol* 1991; **70**: 1S–7S.

138 Bullock GL, Stuckey HM. *Aeromonas salmonicida*: detection of asymptomatically infected trout. *Prog Fish Cult* 1975; **37**: 237–9.

139 Klontz GW. Oral immunization of coho salmon against furunculosis. *Progr Sport Fish Res* 1968; **39**: 81–2.

140 Roberts SD. A method of reducing the carrier state of *Aeromonas salmonicida* in juvenile Pacific salmon. *MS Thesis*, University of Idaho, USA; 1980.

141 Markwardt NM, Klontz GW. A method to eliminate the asymptomatic carrier state of *Aeromonas salmonicida* in salmonids. *J Fish Dis* 1989; **12**: 317–22.

142 Blake I, Clark JC. Observations on experimental infection of trout by *B. salmonicida*, with particular reference to 'carriers' of furunculosis and to certain factors influencing susceptibility. *Fish Bd of Scotland, Salmon Fish* 1931; **7**: 1–13.

143 Rose AS, Ellis AE, Munro ALS. The infectivity of different routes of exposure and shedding rates of *Aeromonas salmonicida* subsp. *salmonicida* in Atlantic salmon, *Salmo salar* L., held in sea water. *J Fish Dis* 1989; **12**: 573–8.

144 Hodgkinson JJ, Bucke D, Austin B. Uptake of the fish pathogen, *Aeromonas salmonicida*, by rainbow trout (*Salmo gairdneri* L.). *FEMS Microbiol Lett* 1987; **40**: 207–10.

145 Tatner MF, Johnson CM, Horne MT. The tissue localization of *Aeromonas salmonicida* in rainbow trout, *Salmo gairdneri* Richardson, following three methods of administration. *J Fish Biol* 1984; **25**: 95–108.

146 Krantz GE, Reddecliff JM, Heist CE. Immune response of trout to *Aeromonas salmonicida*. Part 2. Evaluation of feeding techniques. *Prog Fish Cult* 1964; **26**: 65–9.

147 Klontz GW, Wood JW. Observations on the epidemiology of furunculosis disease in juvenile coho salmon (*Oncorhynchus kisutch*). *FI:EIFAC 72/SC II Symp* 1972; **27**: 1–8.

148 Mackie TJ, Arkwright JA, Pryce-Tannatt TE, Motram JC, Johnstone WR. *Interim, Second and Final Reports of the Furunculosis Committee*. Edinburgh: HMSO 1930, 1933, 1935.

149 Evelyn TPT. An aberrant strain of the bacterial fish pathogen *Aeromonas salmonicida* isolated from a marine host, the sable fish (*Anoplopoma fimbria*) and from two species of cultured Pacific salmon. *J Fish Res Bd Can* 1971; **28**: 1629–34.

150 Novotny AJ. Vibriosis and furunculosis in marine cultured salmon in Puget Sound. *Washington Mar Fish Rev* 1978; **40**: 52–5.

151 Smith IW. Furunculosis in kelts. *Freshwater Salmon Fish Res* 1962; **27**: 1–5.

152 Scott M. The pathogenicity of *Aeromonas salmonicida* (Griffin) in sea and brackish water. *J Gen Microbiol* 1968; **50**: 864–68.

153 Smith PR, Brazil GM, Drinan EM, O'Kelly J, Palmer R, Scallan A. Lateral transmission of furunculosis in sea water. *Bull Eur Assoc Fish Pathol* 1982; **3**: 41–2.

154 Duff DCB. Some serological relationships of the S, R and G phases of *Bacillus salmonicida. J Bacteriol* 1939; **38**: 91–103.

155 Udey LR. Pathogenic, antigenic and immunogenic properties of *Aeromonas*

salmonicida studied in juvenile coho salmon (*Oncorhynchus kisutch*). PhD Thesis, Oregon State University, Corvallis 1977.

156   Ellis AE, Burrows, AS, Stapleton KJ. Lack of relationship between virulence of *Aeromonas salmonicida* and the putative virulence factors: A-layer extracellular proteases and extracellular haemolysins. *J Fish Dis* 1988; **11**: 309–23.

157   Hamilton AJ, Fallon MJM, Alexander J, Canning EU. A modified enzyme linked immunosorbent assay (ELISA) for monitoring antibody production during experimental *Aeromonas salmonicida* infection in rainbow trout (*Salmo gairdneri*). *Dev Comp Immunol* 1986; **10**: 443–8.

158   Munro ALS. A furunculosis vaccine – illusion or achievable objective. In de Kinkelin PK (ed) *Symposium on Fish Vaccination*. Paris: Office International des Epizootics, 1984; 97–120.

159   Belland RJ, Trust T. Synthesis, export, and assembly of *Aeromonas salmonicida* A-layer analyzed by transposon mutagenesis. *J Bacteriol* 1985; **163**: 877–81.

160   Udey LR, Fryer JL. Immunization of fish with bacterins of *Aeromonas salmonicida*. *Mar Fish Rev* 1978; **40**: 12–17.

161   Olivier G, Eaton CA, Campbell N. Interaction between *Aeromonas salmonicida* and peritoneal macrophages of brook trout (*Salvelinus fontinalis*). *Vet Immunol Immunopathol* 1986; **12**: 223–34.

162   Graham S, Jeffries AH, Secombes CJ. A novel assay to detect macrophage bacterial activity in fish: factors influencing the killing of *Aeromonas salmonicida*. *J Fish Dis* 1988; **11**: 389–96.

163   Dooley JSG, Engelhardt H, Baumeister W, Kay WW, Trust TJ. Three-dimensional structure of an open form of the surface layer from the fish pathogen *Aeromonas salmonicida*. *J Bacteriol* 1989; **171**: 190–7.

164   Wilson A, Horne MT. Detection of A-protein in *Aeromonas salmonicida* and some effects of temperature on A-layer assembly. *Aquaculture* 1986; **56**: 23–7.

165   Bernoth E-M. Identification of cultured *Aeromonas salmonicida* by an indirect dot blot immunoassay. *J Fish Dis* 1991; **14**: 419–22.

166   Sakai DK. Adhesion of *Aeromonas salmonicida* strains associated with net electrostatic charges to host tissue cells. *Infect Immun* 1987; **55**: 704–10.

167   Kay WW, Buckley JT, Ishiguro EE, Phipps BM, Monette JPL, Trust TJ. Purification and disposition of a surface protein associated with virulence of *Aeromonas salmonicida*. *J Bacteriol* 1981; **147**: 1077–84.

168   Phipps BM, Trust TJ, Ishiguro EE, Kay WW. Purification and characterization of the cell surface virulent A protein from *Aeromonas salmonicida*. *Biochemistry* 1983; **22**: 2934–9.

169   Evenberg D, van Boxtel R, Lugtenberg B, Frank S, Blommaert J, Bootsma R. Cell surface of the fish pathogenic bacterium *Aeromonas salmonicida*. 1. Relationship between autoagglutination and the presence of a major cell envelope protein. *Biochem Biophys Acta* 1982; **684**: 241–8.

170   Evenberg D, Lugtenberg B. Cell surface of the fish pathogenic bacterium *Aeromonas salmonicida*. II. Purification and characterization of a major cell envelope protein related to autoagglutination. *Biochem Biophys Acta* 1982; **684**: 249–54.

171   Hamilton RC, Kalnins H, Ackland NR, Ashburner LD. An extra layer in the surface layers of an atypical *Aeromonas salmonicida* isolated from Australian goldfish. *J Gen Microbiol* 1981; **122**: 363–6.

172   Ishiguro EE, Kay WW, Ainsworth T, Chamberlain JB, Austen RA, Buckley JT, Trust TJ. Loss of virulence during culture of *Aeromonas salmonicida* at high temperature. *J Bacteriol* 1981; **148**: 333–40.

173   Kay WW, Trust, TJ. Form and functions of the regular surface array (S-layer) of *Aeromonas salmonicida*. *Experientia* 1991; **47**: 412–14.

174 Munn CB, Trust TJ. Role of additional protein layer in virulence of *Aeromonas salmonicida*. In Acuigrup (ed), *Fish Diseases, Fourth COPRAQ-Session*. Madrid: Editora ATP, 1984; 69–73.

175 Trust TJ, Kay WW, Ishiguro EE. Cell surface hydrophobicity and macrophage association of *Aeromonas salmonicida*. *Curr Microbiol* 1983; **9**: 315–18.

176 Van Alstine JM, Trust TJ, Brooks DE. Differential partition of virulent *Aeromonas salmonicida* and attenuated derivatives possessing specific cell surface alterations in polymer aqueous-phase systems. *Appl Environ Microbiol* 1986; **51**: 1309–13.

177 Parker ND, Munn CB. Cell surface properties of virulent and attenuated strains of *Aeromonas salmonicida*. In Ellis AE (ed) *Fish and Shellfish Pathology*. London: Academic Press, 1985: 97–105.

178 Garduno RA, Phipps BM, Kay WW. Physiological consequences of the S-layer of *Aeromonas salmonicida* in relation to growth, temperature, and outer membrane permeation. *Can J Microbiol* 1994; **40**: 622–9.

179 Hirst ID, Hastings TS, Ellis AE. Utilization of haem compounds by *Aeromonas salmonicida*. *J Fish Dis* 1994; **17**: 365–73.

180 Garduno RA, Lee EJY, Kay WW. S-layer-mediated association of *Aeromonas salmonicida* with murine macrophages. *Infect Immun* 1992; **60**: 4373–82.

181 Garduno RA, Kay WW. Interaction of the fish pathogen *Aeromonas salmonicida* with rainbow trout macrophages. *Infect Immun* 1992; **60**: 4612–20.

182 Chu S, Cavaignac S, Feutrier J, Phipps BM, Kostrzynska M, Kay WW, Trust TJ. Structure of the tetragonal surface virulence array protein and gene of *Aeromonas salmonicida*. *J Biol Chem* 1991; **266**: 15258–65.

183 Chu S, Gustafson CE, Feutrier F, Cavaignac S, Trust TJ. Transcriptional analysis of the *Aeromonas salmonicida* S-layer protein gene *vapA*. *J Bacteriol* 1993; **175**: 7968–75.

184 Garduno RA, Phipps BM, Kay WW. Physical and functional S-layer reconstitution in *Aeromonas salmonicida*. *J Bacteriol* 1995; **177**: 2684–94.

185 Fernandez AIG, Perez MJ, Rodriguez LA, Nieto TP. Surface phenotypic characteristics and virulence of Spanish isolates of *Aeromonas salmonicida* after passage through fish. *Appl Environ Microbiol* 1995; **61**: 2010–12.

186 Chu S, Noonan B, Cavaignac S, Trust TJ. Endogenous mutagenesis by an insertion sequence element identifies *Aeromonas salmonicida* AbcA as an ATP-binding cassette transport protein required for biogenesis of smooth lipopolysaccharide. *Proc Natl Acad Sci USA* 1995; **92**: 5754–8.

187 Noonan B, Trust TJ. The leucine zipper of *Aeromonas salmonicida* AbcA is required for the transcriptional activation of the P2 promoter of the surface-layer structural gene, *vapA*, in *Escherichia coli*. *Mol Microbiol* 1995; **17**: 379–86.

188 Noonan B, Trust TJ. Molecular characterization of an *Aeromonas salmonicida* mutant with altered surface morphology and increased systemic virulence. *Mol Microbiol* 1995; **15**: 65–75.

189 Noonan B, Trust TJ. Molecular analysis of an A-protein secretion mutant of *Aeromonas salmonicida* reveals a surface layer-specific protein secretion pathway. *J Mol Biol* 1995; **248**: 316–27.

190 Munro ALS, Hastings TS, Ellis AE, Liversidge J. Studies on an ichthyotoxic material produced extracellularly by the furunculosis bacterium *Aeromonas salmonicida*. In Ahne W (ed) *Fish Diseases – Third COPRAQ Session*. Berlin: Springer-Verlag, 1980: 98–106.

191 Titball RW, Munn B. Evidence for two haemolytic activities from *Aeromonas salmonicida*, *FEMS Microbiol Lett* 1981; **12**: 27–30.

192 Cipriano RC, Griffin BR, Lidgerding BC. *Aeromonas salmonicida*: relationship

between extracellular growth products and isolate virulence. *Can J Fish Aquat Sci* 1981; **38**: 1322–6.

193   Rockey DD, Fryer JL, Rohovec JS. Separation and in vivo analysis of two extracellular proteases and the T-hemolysin from *Aeromonas salmonicida*. *Dis Aquat Org* 1988; **5**: 197–204.

194   Huntly PJ, Coleman G, Munro ALS. The nature of the lethal effect on Atlantic salmon, *Salmo salar* L., of a lipopolysaccharide-free phospholipase activity isolated from the extracellular products of *Aeromonas salmonicida*. *J Fish Dis* 1992; **15**: 99–102.

195   Ellis AE, Hastings TS, Munro ALS. The role of *Aeromonas salmonicida* extracellular products in the pathology of furunculosis. *J Fish Dis* 1981; **4**: 41–51.

196   Sakai DK. Causative factors of *Aeromonas salmonicida* in salmonid furunculosis: extracellular protease. *Sci Rep Hokkaido Fish Hatchery* 1977; **32**: 61–89.

197   Fyfe L, Finley A, Coleman G, Munro ALS. A study of the pathological effect of isolated *Aeromonas salmonicida* extracellular protease on Atlantic salmon, *Salmo salar* L. *J Fish Dis* 1986; **9**: 403–9.

198   Fyfe L, Coleman G, Munro ALS. Identification of major common extracellular proteins secreted by *Aeromonas salmonicida* strains isolated from diseased fish. *Appl Environ Microbiol* 1987; **53**: 722–6.

199   Sheeran B, Drinan E, Smith PR. Preliminary studies on the role of extracellular proteolytic enzymes in the pathogenesis of furunculosis. In Acuigrup (ed) *Fish Diseases, Fourth COPRAQ-Session*. Madrid: Editora ATP, 1984; 89–100.

200   Gudmundsdottir BK, Hasting TS, Ellis AE. Isolation of a new toxic protease from a strain of *Aeromonas salmonicida* subspecies *achromogenes*. *Dis Aquat Org* 1990; **9**: 199–208.

201   Arnesen JA, Bjornsdottir R, Jorgensen TO, Eggset G. Immunological responses in Atlantic salmon, *Salmon salar* L., against purified serine protease and haemolysins from *Aeromonas salmonicida*. *J Fish Dis* 1993; **16**: 409–23.

202   Sakai DK. Significance of extracellular protease for growth of a heterotrophic bacterium, *Aeromonas salmonicida*. *Appl Environ Microbiol* 1985; **50**: 1031–7.

203   Grisley MS, Ellis AE, Hastings TS, Munro ALS. An alpha-migrating anti-protease in normal salmonid plasma and its relationship to the neutralization of *Aeromonas salmonicida* toxins. In: Acuigrup (ed) *Fish Diseases, Fourth COPRAQ-Session*. Madrid: Editora ATP, 1984: 77–82.

204   Lee KK, Ellis AE. Interactions between salmonid serum components and the extracellular haemolytic toxin of *Aeromonas salmonicida*. *Dis Aquat Org* 1991; **11**: 207–16.

205   Salte R, Norberg K, Arnesen JA, Odegaard OR, Eggset G. Serine protease and glycerophospholipid:cholesterol acyltransferase of *Aeromonas salmonicida* work in concert in thrombus formation: in vitro the process is counteracted by plasma antithrombin and α_2-macroglobulin. *J Fish Dis* 1992; **15**: 215–27.

206   Rosjo C, Salte R, Thomassen MS, Eggset G. Glycerophospholipid:cholesterol acyltransferase complexed with lipopolysaccharide (GCAT–LPS) of *Aeromonas salmonicida* produces lysophospholipids in salmonid red cell membranes: a probable haemolytic mechanism. *J Fish Dis* 1993; **16**: 87–99.

207   Whitby PW, Landon M, Coleman G. Characteristics of *Aeromonas salmonicida* 70kDa serine protease deduced from the nucleotide sequence of its gene. *Abstr Fifth Int Conf Eur Assoc Fish Pathol*, Budapest, Hungary, 1991.

208   Whitby PW, Delany SG, Coleman G, Munro ALS. The occurrence of a 70 kDa serine protease gene in typical and atypical strains of *Aeromonas salmonicida*. *J Fish Dis* 1992; **15**: 529–35.

209   Titball RW, Munn CB. Partial purification and properties of a haemolytic

activity (T-lysin) from *Aeromonas salmonicida*. *FEMS Microbiol Lett* 1983; **20**: 207–10.

210 Titball RW, Munn CB. The purification and some properties of H-lysin from *Aeromonas salmonicida*. *J Gen Microbiol* 1985; **131**: 1603–9.

211 Nomura S, Saito H. Production of the extracellular hemolytic toxin by an isolated strain of *Aeromonas salmonicida*. *Bull Jap Soc Sci Fish* 1982; **48**: 1589–97.

212 Nomura S, Fujino M, Yamakawa M, Kawahara E. Purification and characterization of salmolysin, an extracellular hemolytic toxin from *Aeromonas salmonicida*. *J Bacteriol* 1988; **170**: 3694–702.

213 Hackett JL, Lynch WH, Paterson WD, Coombs DH. Extracellular protease, extracellular haemolysis and virulence in *Aeromonas salmonicida*. *Can J Fish Aquat Sci* 1984; **41**: 1354–60.

214 Chart H, Trust TJ. Acquisition of iron by *Aeromonas salmonicida*. *J Bacteriol* 1983; **156**: 758–64.

215 Kay WW, Phipps BM, Ishiguro EE, Trust TJ. Porphyrin binding by the surface array virulence protein of *Aeromonas salmonicida*. *J Bacteriol* 1985; **164**: 1332–6.

216 Hirst ID, Hastings TS, Ellis AE. Siderophore production by *Aeromonas salmonicida*. *J Gen Microbiol* 1991; **137**: 1185–92.

217 Campbell CM, Duncan D, Price NC, Stevens L. The secretion of amylase, phospholipase and protease from *Aeromonas salmonicida*, and the correlation with membrane-associated ribosomes. *J Fish Dis* 1990; **13**: 463–74.

218 Coleman G, Collighan RJ, Dodsworth SJ. Glycogen, the preferred energy source for extracellular protein formation and growth of *Aeromonas salmonicida*. *J Fish Dis* 1995; **18**: 221–8.

219 Ellis AE, Grisley MS. Serum antiproteases of salmonids: studies on the inhibition of trypsin and the proteolytic activity of *Aeromonas salmonicida* extracellular products. In Ellis AE (ed) *Fish and Shellfish Pathology*. London: Academic Press, 1985: 85–96.

220 Karczewski JM, Sharp GJE, Secombs CJ. Susceptibility of strains of *Aeromonas salmonicida* to killing by cell-free generated superoxide anion. *J Fish Dis* 1991; **14**: 367–73.

221 Garduno RA, Thornton JC, Day WW. Fate of the fish pathogen *Aeromonas salmonicida* in the peritoneal cavity of rainbow trout. *Can J Microbiol* 1993; **39**: 1051–8.

222 Garduno RA, Thornton JC, Day WW. *Aeromonas salmonicida* grown *in vivo*. *Infect Immun* 1993; **61**: 3854–62.

223 Thornton JC, Garduno RA, Carlos SJ, Kay WW. Novel antigens expressed by *Aeromonas salmonicida* grown *in vivo*. *Infect Immun* 1993; **61**: 4582–9.

224 Barnes AC, Lewin CS, Hastings TS, Amyes SGB. Alterations in outer membrane proteins identified in a clinical isolate of *Aeromonas salmonicida* subsp. *salmonicida*. *J Fish Dis* 1992; **15**: 279–82.

225 Inglis B, Richards RH. The *in vitro* susceptibility of *Aeromonas salmonicida* and other fish-pathogenic bacteria to 29 antimicrobial agents. *J Fish Dis* 1991; **14**: 641–50.

226 Hayes MV, Thomson CJ, Amyes SGB. Three Beta-lactamases isolated from *Aeromonas salmonicida*, including a carbapenemase not detectable by conventional methods. *Eur J Clin Microbiol Infect Dis* 1994; **13**: 805–11.

227 Snieszko SF. *Fish Furunculosis. Fishery Leaflet 467*. Washington DC: U S Fish and Wildlife Service, 1958.

228 Michel C, Gerard J-P, Fourbet B, Collas R, Chevalier R. Emploi de la flumequine contre las furonculose des salmonides: essais therapeutiques et perspectives pratiques. *Bull Français Piscicult* 1980; **52**: 154–62.

229   Endo T, Ogishima K, Hayasaki H, Kaneko S, Ohshima S. Application of oxolinic acid as a chemotherapeutic agent for treating infectious diseases in fish. 1. Antibacterial activity, chemotheraputic effect and pharmacokinetic effect of oxolinic acid in fish. *Bull Jap Soc Sci Fish* 1973; **2**: 165–71.

230   McCarthy DH, Stevenson JP, Salsbury AW. Therapeutic efficacy of a potentiated sulphonamide in experimental furunculosis. *Aquaculture* 1974; **4**: 407–10.

231   Gayer EK, Bekesi L, Csaba G. Some aspects of the histopathology of carp erythrodermatitis (CE). In Ahne W (ed) *Fish Diseases, Third COPRAQ-Session*. Berlin: Springer-Verlag, 1980: 127–36.

232   Barnes AC, Lewin CS, Hastings TS, Amyes SGB. Cross resistance between oxytetracycline and oxolinic acid in *Aeromonas salmonicida* associated with alterations in outer membrane proteins. *FEMS Microbiol Lett* 1990; **72**: 337–40.

233   Aoki T, Egusa S, Kimura T, Watanabe T. Detection of R factors in naturally occurring *Aeromonas salmonicida* strains. *Appl Microbiol* 1971; **22**: 716–17.

234   Aoki T, Kitao T, Iemura N, Mitoma Y, Nomura T. The susceptibility of *Aeromonas salmonicida* strains isolated in cultured and wild salmonids to various chemotherapeutants. *Bull Jap Soc Sci Fish* 1983; **49**: 17–22.

235   Barnes AC, Lewin CS, Hastings TS, Amyes SGB. *In vitro* activities of 4-quinolones against the fish pathogen *Aeromonas salmonicida*. *Antimicrob Agents Chemother* 1990; **34**: 1819–20.

236   Barnes AC, Amyes SGB, Hastings TS, Lewin CS. Fluoroquinolones display rapid bactericidal activity and low mutation frequencies against *Aeromonas salmonicida*. *J Fish Dis* 1991; **14**: 661–7.

237   Hsu H-M, Bowser PR, Schachte JH, Scarlett JH, Babish JG. Winter field trials of enroflaxacin for the control of *Aeromonas salmonicida* infections in salmonids. *J World Aquacult Soc* 1995; **26**: 307–14.

238   Martinsen B, Myhr E, Reed E, Håstein T. *In vitro* antimicrobial activity of sarafloxacin against clinical isolates of bacteria pathogenic to fish. *J Aquat Animal Health* 1991; **3**: 235–41.

239   Duff DB. The oral immunization of trout against *Bacterium salmonicida*. *J Immunol* 1942; **44**: 87–94.

240   Tatner MF. The antibody response of intact and short term thymectomised rainbow trout (*Salmo gairdneri*) to *Aeromonas salmonicida*. *Dev Comp Immunol* 1989; **13**: 387.

241   Sövényi JF, Yamamoto H, Fujimoto S, Kusuda R. Lymphomyeloid cells, susceptibility to erythrodermatitis of carp and bacterial antigens. *Dev Comp Immunol* 1990; **14**: 185–200.

242   Kawahara E, Oshima S, Nomura S. Toxicity and immunogenicity of *Aeromonas salmonicida* extracellular products to salmonids. *J Fish Dis* 1990; **13**: 495–503.

243   Kawahara E, Ueda T, Nomura S. *In vitro* phagocytic activity of white-spotted Char blood cells after injection with *Aeromonas salmonicida* extracellular products. *Fish Pathol* 1991; **26**: 213–14.

244   Tatner MR. Modified extracellular product antigens of *Aeromonas salmonicida* as potential vaccines for the control of furunculosis in Atlantic salmon, *Salmo salar* L. *J Fish Dis* 1991; **14**: 395–400.

245   Cipriano RC, Starliper CE. Immersion and injection vaccination of salmonids against furunculosis with an avirulent strain of *Aeromonas salmonicida*. *Prog Fish Cult* 1982; **44**: 167–9.

246   Ellis AE, Burrows AS, Hastings TS, Stapleton KJ. Identification of *Aeromonas salmonicida* extracellular proteases as a protective antigen against furunculosis by passive immunization. *Aquaculture* 1988; **70**: 207–18.

247   Thornton JC, Garduño RA, Newman SG, Kay WW. Surface-disorganized,

attenuated mutants of *Aeromonas salmonicida* as furunculosis live vaccines. *Microb Pathog* 1991; **11**: 85–99.

248 Thornton JC, Garduno RA, Kay WW. The development of live vaccines for furunculosis lacking the A-layer and O-antigen of *Aeromonas salmonicida. J Fish Dis* 1994; **17**: 195–204.

249 Norqvist A, Hagström Å, Wolf-Watz H. Protection of rainbow trout against vibriosis and furunculosis by the use of attenuated strains of *Vibrio salmonicida. Appl Environ Microbiol* 1989; **55**: 1400–5

250 Hirst ID, Ellis AE. Iron-regulated outer membrane proteins of *Aeromonas salmonicida* are important protective antigens in Atlantic salmon against furunculosis. *Fish Shellfish Immunol* 1994; **4**: 29–45.

251 Lutwyche P, Exner MM, Hancock REW, Trust TJ. A conserved *Aeromonas salmonicida* porin provides protective immunity to rainbow trout. *Infect Immun* 1995; **63**: 3137–42.

252 Bennett AJ, Whitby PW, Coleman G. Retention of antigenicity by a fragment of *Aeromonas salmonicida* 70 kDa serine protease which includes the primary substrate binding site expressed as β-galactosidase hybrid proteins. *J Fish Dis* 1992; **15**: 473–84.

253 Coleman G, Dodsworth SJ, Whitley PW, Bennett AJ, Bricknell I. Studies on the protective effect of genetically engineered antigens against furunculosis in Atlantic salmon. *Abstract, Int Symp Aquat Animal Health*, Seattle, Washington, 1994.

254 Vaughan LM, Smith PR, Foster TJ. An aromatic-dependent mutant of the fish pathogen *Aeromonas salmonicida* is attenuated in fish and is effective as a live vaccine against the salmonid disease furunculosis. *Infect Immun* 1993; **61**: 2172–81.

255 Marsden MJ, Secombes CJ. Antigen specific immune response in rainbow trout, *Oncorhynchus mykiss*, immunised with an aromatic mutant of *Aeromonas salmonicida* (BRIVAX) as a live vaccine for furunculosis. *Abstract. Int Symp Aquat Animal Health*, Seattle, Washington, 1994.

256 Noonan B, Enzmann PJ, Trust TJ. Recombinant infectious hematopoietic necrosis virus and viral hemorrhagic septicemia virus glycoprotein epitopes expressed in *Aeromonas salmonicida* induce protective immunity in rainbow trout *(Oncorhynchus mykiss). Appl Environ Microbiol* 1995; **61**: 3586–91.

257 Michel C. Furunculosis of salmonids: vaccination attempts in rainbow trout *(Salmo gairdneri)* by formalin-killed germs. *Ann Rech Vét* 1979; **10**: 33–40.

258 McCarthy DH, Amend DF, Johnson KA, Bloom JV. *Aeromonas salmonicida*: determination of an antigen associated with protective immunity and evaluation of an experimental bacterin. *J Fish Dis* 1983; **6**: 155–74.

259 Olivier G, Evelyn TPT, Lallier R. Immunogenicity of vaccines from a virulent and an avirulent strain of *Aeromonas salmonicida. J Fish Dis* 1985; **8**: 43–55.

260 Hastings TS, Ellis AE. The humoral immune response of rainbow trout, *Salmo gairdneri* Richardson, and rabbits to *Aeromonas salmonicida* extracellular products. *J Fish Dis* 1988; **11**: 147–60.

261 Lund V, Jorgensen J, Holm KO, Eggset G. Humoral immune response in Atlantic salmon, *Salmon salar* L., to cellular and extracellular antigens of *Aeromonas salmonicida. J Fish Dis* 1991; **14**: 443–52.

262 Olivier G, Evelyn TPT, Lallier R. Immunity to *Aeromonas salmonicida* in coho salmon *(Oncorhynchus kisutch)* induced by modified Freund's complete adjuvant: its non-specific nature and the probable role of macrophages in the phenomenon. *Dev Comp Immunol* 1985; **9**: 419–32.

263 Anderson DP, Siwicki AK. Duration of protection against *Aeromonas salmonicida* in brook trout immunostimulated with glucan or chitosan by

injection or immersion. *Prog Fish Cult* 1994; **56**: 258–61.

264   Nikl L, Albright LJ, Evelyn TPT. Influence of seven immunostimulants on the immune response of coho salmon to *Aeromonas salmonicida. Dis Aquat Org* 1991; **12**: 7–12.

265   Aakre R, Wergeland HI, Aasjerk PM, Endresen C. Enhanced antibody response in Atlantic salmon *(Salmo salar* L.) to *Aeromonas salmonicida* cell wall antigens using a bacterin containing β-1,3-M-glucan as adjuvant. *Fish Shellfish Immunol* 1994; **4**: 47–61.

266   Rodgers CJ. Immersion vaccination for control of fish furunculosis. *Dis Aquat Org* 1990; **8**: 69–72.

267   Tatner F. The quantitative relationship between vaccine dilution, length of immersion time and antigen uptake, using a radiolabelled *Aeromonas salmonicida* bath in direct immersion experiments with rainbow trout, *Salmo gairdneri. Aquaculture* 1987; **62**: 173–85.

268   Embody GC, Hayford CO. The advantage of rearing brook trout fingerlings from selected breeders. *Trans Am Fish Soc* 1925; **55**: 135–42.

269   Cipriano RC. Resistance of salmonids to *Aeromonas salmonicida*: relation between agglutinins and neutralizing activities. *Trans Am Fish Soc* 1983; **112**: 95–9.

270   Stromsheim A, Eide DM, Fjalestad KT, Larsen HJS, Roed KH. Genetic variation in the humoral immune response in Atlantic salmon *(Salmo salar)* against *Aeromonas salmonicida* A-layer. *Vet Immunol Immunopathol* 1994; **41**: 341–52.

271   Lund T, Chiavareesajja J, Larsen JJS, Roed KJ. Antibody response after immunization as a potential indirect marker for improved resistance against furunculosis. *Fish Shellfish Immunol* 1995; **5**: 109–19.

272   Houghton G, Wiergerthes GF, Groeneveld A, Van Muiswinkel WB. Differences in resistance of carp. *Cyprinus carpio* L., to atypical *Aeromonas salmonicida. J Fish Dis* 1991; **14**: 333–41.

273   Smith P, Davey S. Evidence for the competitive exclusion of *Aeromonas salmonicida* from fish with stress-inducible furunculosis by a fluorescent pseudomonad. *J Fish Dis* 1993; **16**: 521–4.

274   Austin B, Stuckey LF, Robertson PAW, Effendi I, Griffith DRW. A probiotic strain of *Vibrio alginolyticus* effective in reducing diseases caused by *Aeromonas salmonicida, Vibrio anguillarum* and *Vibrio ordalii. J Fish Dis* 1995; **18**: 93–6.

9 Pathogenic Mechanisms

PETER J. GOSLING
Department of Health, London, England

9.1 INTRODUCTION

Aeromonas spp. have been found with increasing frequency as opportunistic pathogens in human wounds, often those sustained in aquatic environments[1]. These infections may remain localized and mild, cause extensive muscle necrosis, or lead to septicaemia and infection in other sites, and may be accompanied by skin lesions, fever, chills, and hypotension[2–5]. *Aeromonas* spp. have additionally been incriminated as aetiological agents of human diarrhoeal disease of varying severity, ranging from a mild self-limiting illness to a life-threatening cholera-like disease.

The production of many putative virulence factors by *Aeromonas* species has long been recognized; however, the pathogenesis of *Aeromonas* infections remains incompletely understood. This chapter therefore considers the putative virulence factors of *Aeromonas* spp. and their role in pathogenesis, primarily in humans (the role of *Aeromonas* species in disease of poikilotherms and aspects of their pathogenesis has been considered previously), although much of the information can be considered relevant to other homoiotherms.

9.2 CELL SURFACE LAYERS AND VIRULENCE

Surface antigens of some *Aeromonas* strains have been linked with enhanced virulence[6–8].

9.2.1 OUTER MEMBRANE PROTEINS (OMP)

The OMP of *A. hydrophila* are rather heterogeneous, although most strains produce a 36 kDa protein. In addition, iron-starved *A. hydrophila* cells synthesize new OMP of 68–93 kDa[9]. Mittal and colleagues reported that a group of *A. hydrophila*, virulent in fish, exhibited a number of unique phenotypic properties that were thought to be cell-surface associated, including autoaggregation during growth in static broth culture and resistance to the bactericidal activity of normal serum[10].

The Genus Aeromonas. Edited by B. Austin, M. Altwegg, P.J. Gosling and S. Joseph
© 1996 John Wiley & Sons Ltd

9.2.1.1 S-Layer protein

The surface of such strains has been shown to be completely covered by a regularly arrayed surface layer (S-layer) composed of tetragonally arranged subunits, the major constituent of which is a 52 kDa molecular weight protein. S-layers are present on a number of bacteria pathogenic to humans, including *Campylobacter* spp., as well as on many non-pathogenic bacteria[11,12] S-layers may provide protection for microorganisms in their natural environment; however, at present it is unclear what role (if any) these S-layers play in overt pathogenicity of *Aeromonas* spp. They may provide strains with a selective advantage in causing infection in a susceptible host and be involved in strain dissemination from intestinal body sites by providing a mechanism for microbes to resist phagocytosis and the lytic activity of complement but this needs to be substantiated.

9.2.2   LIPOPOLYSACCHARIDE ENDOTOXIN

The O-antigen structure of virulent strains of *A. hydrophila* has been shown to have many points of similarity with that of the O-antigen of *A. salmonicida*[13]. Such strains possess lipopolysaccharides (LPS) containing O-polysaccharide chains of homogeneous chain length, as has been observed in strains of *A. salmonicida*, where they are considered to play an important role with respect to the ability of virulent strains to produce the surface protein array known to be linked to the overt virulence of this bacterium for fish. LPS endotoxin is an important component of the outer membrane of *A. hydrophila*, which has been shown to enhance red-leg disease in frogs[14]. However, the importance of this toxin in human infections has not yet been elucidated.

9.3   PUTATIVE EXTRACELLULAR VIRULENCE FACTORS

Several biochemical properties of *Aeromonas* spp., i.e. production of haemolysins, enterotoxins, and proteases, have been reported as potential indicators of pathogenicity[15]. The ability of *Aeromonas* strains to produce differing extracellular toxins and enzymes has additionally been associated with virulence[16]. This section therefore considers the role of putative extracellular virulence factors in the pathogenicity of *Aeromonas* infection.

9.3.1   AEROMONAS TOXINS

Our understanding of the role of *Aeromonas* extracellular toxins in the pathogenesis of disease has largely been based upon the moieties resulting from previous attempts to isolate and purify *Aeromonas* haemolysin and enterotoxin. Few studies have successfully separated the haemolytic and enterotoxigenic activities.

Multiple biological activities of cell-free culture filtrates (CFCF) of strains of *A. hydrophila*, including the ability to cause haemolysis of erythrocytes, were reported by Caselitz and Günther[17]. No attempt to separate toxigenic activities was made, however, until Wretlind and colleagues reported the separation of haemolytic activity from other activities by isoelectric focusing. This technique resolved the haemolytic activity into two peaks with pI 4.3 and 5.5, which were interpreted as two forms of the same toxin[18]. The toxin appeared extracellularly during the stationary phase of growth as a result of cell lysis and the yield was higher at 22°C than at 37°C. The haemolytic activity was heat-labile, inactivated by trypsin and shown to be cytotoxic to HeLa cells and diploid lung fibroblasts, dermonecrotic in rabbit skin and lethal to mice.

Partial purification of an extracellular heat-labile haemolysin produced by a strain of *A. hydrophila* was also reported by Bernheimer and Avigad[19]. Purification was carried out on CFCF following 24 h incubation at 37°C by ammonium sulphate precipitation followed by Sephadex G 100 gel chromatography. The resulting toxigenic fractions of partially purified haemolysin contained several different proteins, shown as faintly staining bands by SDS–PAGE, in addition to the haemolysin. The toxin appeared extracellularly during the logarithmic growth phase and its molecular weight was estimated to be 49–53 kDa. The toxin exhibited phospholipase A and C activity, and caused complete lysis of erythrocytes from various animal species, rat erythrocytes being particularly susceptible. This toxin was termed 'aerolysin'. Haemolytic activity was lost upon heating at 50°C for 1 h at pH 7 but, in contrast to the previously reported haemolysin, the toxin was stable to the proteolytic enzymes trypsin and papain. On the basis of the ultraviolet absorption spectrum, molecular weight and heat lability of the toxin, the haemolysin was considered to be a protein.

Studies on the interaction between aerolysin, erythrocytes and erythrocyte membranes demonstrated that erythrocytes from different animal species differ greatly in sensitivity to aerolysin lytic action[20]. The sensitivity to lysis was reported to be a function of the binding efficiency of the haemolysin to the membranes. No investigation of the possible enterotoxigenic activity of the partially purified haemolysin was conducted.

While some investigations recognized only cytotoxic activity in tissue culture cells, others presented evidence that suggested that three activities caused by different toxins were detectable in culture supernatants from strains of *A. hydrophila*: haemolysin, cytotoxic factor and enterotoxin[21,22]. Partially purified toxins were prepared from CFCF prepared after 30 h incubation at 24°C in aerated brain–heart infusion broth by isoelectric focusing. The haemolysin and cytotoxin appeared to be neutralized by heating at 56°C for 10 min, leaving the enterotoxigenic activity still detectable. A further report from the same group of investigators in support of the existence of three different toxins further characterized the toxigenicity of their preparations but no progress was made in purifying the toxins[23]. The pIs of the crude haemolysin, the cytotoxic factor and the enterotoxin were, respectively, 3.4–3.8 and 5.2–5.7, 5.1–5.5 and 3–9. The haemolysin and cytotoxic factor did not produce enterotoxigenic responses in the ligated rabbit ileal loop (RIL) test. In contrast, however, Cumberbatch and colleagues found a correlation between cytotoxic activity and enterotoxigenicity in CFCF and concluded the existence of cytotoxic enterotoxin[24]. No attempt, however, was made in this study to purify the toxigenic moieties from the crude broth filtrates.

Ljungh and colleagues reported further purification of their toxins by gel chromatography of the isoelectric fractions on Sephadex G75 or Biogel P60, and claimed to have fully separated biological activities, but did not present criteria of their purities[25]. It was now apparent that the cytotoxic factor caused incomplete haemolysis and was redesignated α-haemolysin; the haemolysin causing complete haemolysis was designated β-haemolysin. Both the haemolysins gave negative results in the RIL test and only induced a small amount of fluid with high albumin and calcium content, indicative of leakage of intracellular substances due to membrane damage. The toxin designated α-haemolysin appears to correspond to the haemolysin first reported by Wretlind and colleagues[18]. The β-haemolysin appears to correspond to the toxin previously termed 'cytotoxic factor' and possibly to the haemolysin termed 'aerolysin', although β-haemolysin was inactivated by papain whereas aerolysin was not[19,22].

The molecular weights of the purified α-haemolysin, β-haemolysin and enterotoxin were given as 65 ± 5, 50 ± 3 and 15 ± 3 kDa respectively[25]. Although the biological activities were successfully separated, the increases in specific activities and yields were very low: a 12-fold increase in specific activity and a 0.3% yield for α-haemolysin and a 28-fold increase in specific activity and a 0.6% yield for β-haemolysin. The enterotoxin was classed as a cytotonic enterotoxin on the basis of activity in cell culture by the scheme of Keusch and Donta[26]. Only a three-fold increase in specific activity was achieved and the yield was unrecorded. The β-haemolysin, unlike the enterotoxin, did not induce steroidogenesis or increase cAMP production in Y1 adrenal cells, whilst demonstrating a dissimilarity to a cholera-like toxin, and this was later considered by Ljungh to support further the lack of an enteropathogenic effect[27].

A further study reported the purification of an *Aeromonas* β-haemolysin by salt fractionation, gel filtration, ion exchange and hydroxyapatite chromatography[28]. These workers confirmed the molecular weight of the toxin to be approximately 51.5 kDa. Their toxin preparation was homogeneous by SDS–PAGE and consisted of two active components with pIs of 5.39 and 5.46 when separated by isoelectric focusing. It was heat-labile and its activity was inhibited by certain reducing agents, including ferrous iron and cystein. Unfortunately, as with initial studies, no investigation of possible enterotoxigenicity was made that might confirm or refute the observations of Ljungh and co-workers[25].

Three years later, in 1984, a report of a purified *Aeromonas* β-haemolytic enterotoxin emerged from Japan[29]. This toxin, purified by acid precipitation and quaternary aminoethyl–Sephadex chromatography, was shown to have a molecular weight of 48–50 kDa. Homogeneous by SDS–PAGE, the toxin separated into multiple bands by thin-layer isoelectric focusing. The toxin was heat-labile and its enterotoxigenicity was demonstrated in both the infant mouse test and the rabbit ileal loop assay. It was also cytotoxigenic to Vero tissue-culture cells. Unfortunately, no attempt was made to establish if any relationship existed between the non-enterotoxigenic β-haemolysin of Bernheimer and Avigad[19], or Ljungh and colleagues[25], and the enterotoxigenic β-haemolysin[19,25].

Potomski and colleagues[30] reported yet another heat-labile haemolytic enterotoxin purified by affinity chromatography with monoclonal antibodies. This toxin was homogeneous by polyacrylamide gradient gel electrophoresis and had a molecular weight, larger than that of earlier toxins, of 63 kDa with a pI of 6.2. The attempts that were made to establish the immunological relationship of this toxin to cholera toxin (CT) are difficult to interpret since, while the purified toxin did not cross-react with and was not neutralized by anti-CT, a cross-reaction did occur between anti-CT and the crude CFCF. Two years later Bloch and Monteil also purified a heat-labile *Aeromonas* haemolytic enterotoxin of molecular weight 51 kDa, active in the rabbit ileal loop test, by anion-exchange chromatography[31]. This toxin separated into three bands upon isoelectric focusing, of pI 5.50, 5.26 and 5.10.

Further reports from Japan have highlighted the complexity of *Aeromonas* toxins by demonstrating that β-haemolytic enterotoxins purified by SP Sephadex C25 chromatography from different strains of *Aeromonas* differed immunologically and physicochemically[32]. Recently a further β-haemolytic enterotoxin has been reported that also serologically cross-reacts with cholera toxin[33]. This toxin was purified by ammonium sulphate precipitation, hydrophobic interaction chromatography using phenyl–Sepharose, anion-exchange chromatography on DEAE-BIO-Gel A and size-exclusion high-performance liquid chromatography. The toxin had an apparent molecular weight of 52 kDa and was homogeneous by SDS–PAGE.

A recent study to establish the degree of relatedness of four haemolytic enterotoxins produced by different isolates of *Aeromonas* previously used in toxin production by Rose and colleagues[33], Asao and colleagues[32], Chakraborty and co-workers[34], and Buckley and associates[28] has now been conducted[35]. Although the members of the group of toxins displayed structural similarities they also differed from each other at the immunological, biological and genetic levels. As matters stand, it would appear that *Aeromonas* spp. can produce a non-enterotoxigenic β-haemolysin and at least one and possibly several β-haemolytic enterotoxins.

In addition to the controversy regarding the ability of *Aeromonas* β-haemolysin to act as an enterotoxin, reports of cytotonic enterotoxigenic activity in *Aeromonas* preparations have been emerging since Wadström and colleagues[36] originally demonstrated this in the CFCF of *A. hydrophila* and Ljungh and co-workers[25] further characterized it. The enterotoxin they described caused fluid accumulation in the rabbit ileal loop test and, when separated from cytotoxins present in CFCF, did not produce histological damage after 6 h exposure to the enterotoxin in intestinal epithelial cells or haemolysis of erythrocytes. Purification was achieved by isoelectric focusing and gel chromatography. The enterotoxigenic activity appeared between pH intervals of 4 and 6 during isoelectric focusing and could be fully separated from the haemolytic activity by gel filtration, although they found that yield tended to be very low and the activity of the enterotoxin was easily lost. They estimated the molecular weight to be 12–18 kDa. Fluid accumulation was demonstrated in rabbit, rat and mouse intestinal loop assays but in the infant mouse assay the mice died and they regarded the assay as inappropriate. Lethality may have resulted from absorption of contaminating α-haemolysin. Experimental evidence has shown α-haemolysin to be dermonecrotic in the rabbit skin and to be lethal for mice and rabbits[18]. However, although the α-haemolysin is formed in various complex media, yields are higher at lower temperatures (22°C), and repression occurs at 37°C[25]. It is therefore unlikely that significant amounts of α-haemolysin are produced during infection, thus contributing to the pathogenesis of infections in homoiotherms.

Another property of the cytotonic enterotoxin noted by Ljungh and colleagues[37] was that it was labile at 60°C for 20 min. Gel immunodiffusion tests resulted in precipitation lines showing non-identity between the cytotonic enterotoxin and cholera toxin subunits A and B. This was also observed by Gosling and colleagues[36a], who recently purified, by fast protein liquid chromatography (FPLC), an *Aeromonas sobria* cytotonic enterotoxin similar to the active protein described by Ljungh and co-workers. Also in support of this, neutralization studies similarly indicated a lack of antigenic relationship between *Aeromonas* enterotoxin and cholera toxin or *E. coli* LT[37]. In 1984 Chakraborty and colleagues reported genetic cloning of a gene coding for an *Aeromonas* cytotonic enterotoxin devoid of haemolytic or cytotoxic acitivity and which was infant mouse-positive. Genetic probing failed to show homology with *E. coli* LT[38].

Despite reports of immunological cross-reactivity between cholera toxin and crude *Aeromonas* CFCFs, it was not until 1987 that an *Aeromonas* cytotonic enterotoxin was purified by affinity chromatography using monoclonal antibody to cholera toxin[39–42]. This *Aeromonas* toxin was larger than that of Ljungh and colleagues[25]. Four bands were demonstrated by SDS-

PAGE with molecular weights 150, 43.5, 29.5 and 27 kDa. The toxin produced fluid accumulation in the infant mouse and the rabbit ileal loop tests, and these effects could be neutralized by CT antitoxin. The toxin also inhibited ADP-induced platelet aggregation, a phenomenon influenced by compounds that affect intracellular levels of AMP and the basis of the Platelet Aggregation Test for assays of cAMP-mediated enterotoxins[43,44].

More recently a further cytotonic enterotoxin has been reported purified from CFCF by ammonium sulphate precipitation, hydrophobic column chromatography and chromatofocusing[45]. The purified toxin had a molecular weight of 44 kDa and isoelectric points of 4.3 and 5.5. This toxin did not cross-react with CT and was not neutralized by CT antitoxin. The toxin was shown to induce elevation of cAMP levels in Chinese hamster ovary (CHO) tissue-culture cells and although the activity was heat-labile, the structural integrity of the toxin molecules appeared heat-stable. It is of interest that this toxin was isolated from culture filtrates after prolonged incubation periods of 36 h.

In summary, it would appear from the literature that several different *Aeromonas* enterotoxins may exist. However, the wide variation in molecular sizes, particularly among the reported cytotonic enterotoxins with differing degrees of associated cytotoxigenicity and immunological cross-reactivity with cholera toxin, has highlighted the fact that the confusion will not be resolved until each of the claimed toxins has been obtained in highly purified form.

9.3.1.1 Mode of action of haemolysins

It has been demonstrated that *Aeromonas* β-haemolysin is synthesized as a higher-molecular-weight precursor, which crosses the inner bacterial membrane as an inactive protoxin, presumably to protect the bacterium and to increase the chances that it will reach the target eukaryotic cell, which is subsequently activated by the removal of 25 amino acids from the *C*-terminus by a variety of mammalian proteases or by proteases released by the bacterium itself, including trypsin[46,47]. The molecular weight of the pre-trypsinized β-haemolysin isolated by affinity chromatography using monoclonal antibodies has been estimated at 53 kDa and that of post-trypsinized toxin as 49 kDa[48].

The toxin binds with high affinity to the receptor glycophorin on eukaryotic cells. Binding effectively concentrates aerolysin, leading to oligomerization to hexameric structures. Proaerolysin is unable to oligomerize and this accounts for its inactivity. Oligomerization is followed by insertion into the membrane and the formation of 3 nm voltage-gated channels which disrupt the cell. Some of the residues which may participate in oligomerization and channel formation have been identified by chemical modification and site-directed mutagenesis[46,47,49–51].

9.3.1.2   Mode of action of enterotoxins

The confusion regarding the identity of enterotoxigenic moieties from *Aeromonas* spp. has largely precluded obtaining information on their mode of action. Moieties in CFCF of *Aeromonas* species that are immunologically cross-reactive with cholera toxin have been reported and activation of cyclic adenosine 3'5'-monophosphate (cAMP) has been suggested to be involved in the enterotoxigenicity of *Aeromonas* species[25,30,33,45].

9.3.2   ENZYMES

In addition to the toxins discussed in the previous section, *Aeromonas* spp. produce an array of extracellular enzymes during growth, many of which are putative virulence factors (Table 9.1). Production of extracellular proteolytic activity by *A. hydrophila* has been shown to be influenced by temperature, pH and aeration. Conditions which produce maximal growth also result in maximal protease production. Enzyme production appears to be modulated by the inducer catabolite repression system whereby NH_4^+ and glucose repress enzyme production and complex nitrogen and non-glucose carbon energy sources promote it. Under nutritional stress, protease production has been observed to be high, despite poor growth[52,53]. Stresses may induce the cell to produce a series of 'scouting enzymes' (e.g. proteases, amylases, phosphatases, etc.), the purpose of which is to digest potential polymeric nutrient sources to provide readily metabolizable substrates for microbial growth.

Table 9.1. Enzymes produced by motile *Aeromonas* species

Enzyme	Reference
Aminopeptidase	54–56
Chitinase	57–59
Deoxyribonuclease	60
Elastase	61,62
Endopeptidase	54,63,64
Fibrinolysin	65
Lipase	60,66,67
Phospholide acylhydrolase and glycerophospholipid: cholesterol acyltransferase	68,69
Phosphatase	36
Phospholipase A and C	20,60
Proteases	52,53,70–74
Ribonuclease	60

9.3.2.1 Proteases

A variety of different numbers and types of enzymes produced by *A. hydrophila* has been reported. The use of different media and growth conditions, and a variety of assay and purification procedures, makes the literature reports very difficult to interpret and compare. Gross and Coles[75] and Kanai and Wakabayashi[76] identified one protease from isolates of *A. hydrophila*, while Dahle[70] purified two proteases from a strain of *A. hydrophila*. Two and subsequently three proteases were reported from further strains of *A. hydrophila* by Thune and colleagues[74] and Amborski and co-workers[77]. Nieto and Ellis[73] detected four or five zones of proteolytic activity with isoelectric focusing in crude extracts from another strain[73]. Unfortunately, these investigators worked on different strains of *A. hydrophila*, and in each case only a single strain was completely examined.

Leung and Stevenson[72], considering the above reports, conducted their own work to clarify the situation, and concluded that two distinct types of extracellular proteases were produced by strains of *A. hydrophila*: a temperature-stable metallo-protease and a temperature-labile serine-protease. These enzymes were distributed among strains, some producing both, others only one or the other. This situation was also observed among strains of *A. sobria* and *A. caviae*. The functions of the enzymes, including their potential contributions to pathogenicity, remain undefined, and the question of why a strain might produce two forms of protease remains unanswered.

9.4 PATHOGENIC MECHANISMS

9.4.1 THE ROLE OF ATTACHMENT IN PATHOGENESIS

For most microorganisms the ability to adhere to epithelial mucosa is essential in colonization and subsequent development of disease. Many kinds of fimbriae and haemagglutinins of various human pathogens, including enterotoxigenic *E. coli, Neisseria gonorrhoea* and *Bordetella pertussis*, have been shown to be responsible for adhesion and to be colonization factors[78–81]. The surface adhesins, lectins and fimbriae of *Aeromonas* spp. are therefore considered below.

9.4.1.1 Surface adhesins/lectins

A number of recent studies indicate that *A. hydrophila* produces various lectins and/or adhesins which may enable the pathogen to recognize different carbohydrate moieties on red blood cells as well as on other cells. Some of

Table 9.2. Lectins and adhesins in *Aeromonas hydrophila*. Reproduced by kind permission of the University of Lund, Sweden

Lectin/adhesin	Target	Reference
Cell-associated haemagglutinins	Erythrocytes from:	
	Humans	82,83
	Non-human mammals	82–84
	Birds	85
	Fish	86
Extracellular haemagglutinins	Erythrocytes from:	
	Birds	85
	Mammals	85
Cell-associated adhesins	Buccal epithelial cells	82
	Chinook salmon embryo cells	87
	Rabbit intestinal brush borders	88
	Rainbow trout liver cells	87
	Saccharomyces cerevisiae (yeast)	82,84
	Fish mucus	87
Pili agglutinins	Hep-2 cells	89
	INT-407 cells	90

From: Ascencio[91]

them may enable the pathogen to adhere to specific glycoconjugates on epithelial surfaces and in the mucin layers on the gut mucosa, especially early during the colonization process, when maximal adherence ability is needed for the organism to establish an infection successfully.

9.4.1.2   Pili (fimbriae)

At least two distinct morphological types of fimbriae on Hep2-adherent *A. hydrophila* strains recovered from clinical and environmental sources have been reported[89]. One type of pilus is designated as a straight pilus, and the other is a curvilinear and flexible pilus[92,93]. The straight pilus (0.6–2 μm in length) has a subunit molecular weight of 17 kDa to 18 kDa, while the flexible pilus has a molecular weight of 4000. Since the flexible pili are environmentally regulated and maximally expressed at 22°C, it has been proposed that synthesis of flexible pili may be triggered by environmental conditions as a prerequisite for colonization of the mammalian host[92].

9.4.2   IRON ACQUISITION

Aeromonas spp. are able to utilize organic and inorganic forms of iron in their environment, including vertebrate hosts, by means of siderophore-mediated iron acquisition. *A. hydrophila* strains have been shown to express lactoferrin-binding characteristics and to produce novel tryptophan- and phenylalanine-containing phenolate siderophores which enhance the growth of the bacteria in iron-deficient media[94]. It remains to be shown, however, if the siderophores are produced during the development of A. *hydrophila* diseases.

9.4.3 INVASIVENESS

Some strains of A. *hydrophila* are capable of cellular penetration and replication, and this may be an important virulence property of the organism. Plasmid-curing experiments have suggested that this property is probably chromosomally determined. The ability to invade eukaryotic cells may be associated in *Aeromonas* with certain disease syndromes such as dysentery, but there is little evidence to support this hypothesis. However, two studies have reported that 14 and 36% of *Aeromonas* strains recovered from animals and the environment were invasive for Hep2 tissue-culture cells[95,96].

9.4.4 SERUM RESISTANCE

Many clinical isolates of A. *hydrophila* and A. *sobria* are resistant to complement-mediated lysis by fresh human, rabbit and fish sera, as opposed to the more serum-sensitive isolates of A. *caviae*. Serum-susceptible strains activate complement by the classical and alternative pathways, which cause rapid cell lysis, although the nature of the surface components responsible for serum resistance has not been clearly demonstrated[97,98]. This difference in serum sensitivity may potentially explain the greater association of the former species with bacteraemia and invasive disease.

9.5 DISCUSSION

The pathogenicity mechanisms of *Aeromonas* spp. are complex and probably multifactorial. The species possesses a multitude of putative virulence factors to explain its aetiological role in human disease; however, the precise combination of these has yet to be elucidated. In particular the role of these organisms and their toxins in diarrhoeal disease has been the focus of considerable interest. The many reports of enterotoxin and haemolysin isolation have referred to moieties that have not appeared to be related to each other, or very well related to established enterotoxins, including cholera toxin. Full purification of the *Aeromonas* toxins is clearly a prerequisite to assessing their role in diarrhoeal disease. It is suggested that the confusion regarding the nature of *Aeromonas* toxins may be a result of impure preparations.

No experimental animal model for *Aeromonas*-associated diarrhoeal disease has been found, and only a single human volunteer feeding study has been conducted to date[99]. This study failed to demonstrate a clear-cause–effect relationship between the *Aeromonas* strains used and diarrhoea. A total of 57 healthy adult volunteers were split into five groups and orally challenged, following gastric acid neutralization, with one of five viable *Aeromonas* strains in concentrations ranging between approximately 10^4 to greater than 10^{10} organisms/ml. *Aeromonas* strains were recovered from faeces of only two of the groups of volunteers and only two individuals developed mild diarrhoea, the cause–effect being questionable, however, as higher doses of the same strain failed to elicit diarrhoea. This study, however, used five *Aeromonas* strains of questionable suitability to challenge the volunteers. Two of the five strains, although they were faecal isolates from patients with diarrhoea, were reported to be non-enterotoxigenic in both the suckling mouse assay and the rabbit ileal loop test. Of the remaining three strains considered enterotoxigenic by the animal tests, one was a faecal isolate obtained from a healthy individual, which presumably therefore lacked certain necessary virulence factors to cause diarrhoea, and another a wound isolate which, by the lack of demonstrable faecal shedding by volunteers, apparently lacked the necessary attributes to survive the gastrointestinal tract. Only one of the five organisms was a faecal isolate from a patient with diarrhoea which was shown to be enterotoxigenic by the suckling mouse assay and the rabbit ileal loop test. As the pathogenicity of the species has not been elucidated the strains used may well have lacked vital factors necessary to instigate the disease process and it is therefore difficult to draw meaningful conclusions from this study.

Aeromonas spp. possess many putative virulence factors, including proteases, which may contribute to mucus degradation, the β-haemolysin and a variety of cytotonic and cytotoxic enterotoxins. As the immune response to gastrointestinal infection is generally localized at the mucosal surface it is perhaps predictable that no evidence of a systemic response against the enterotoxin in patients with *Aeromonas*-associated diarrhoeal disease has been observed.

If it is postulated that some *Aeromonas* species have the ability to cause diarrhoeal disease in suitable hosts, clearly some aspect of their enteropathogenicity has still to be elucidated. While production of enterotoxigenic moieties may provide strains of *Aeromonas* species with the means to become enteropathogens, it is probable that, as with enterotoxigenic *E. coli*, strains must also possess other criteria such as possession of essential adhesins before they are able to cause disease[100]. The pathways by which diarrhoeal disease may be caused by enteropathogens involving enterotoxins, may be summarized as ingestion of preformed toxins or by ingestion of the organisms, followed by colonization and toxin production in the gastrointestinal tract. *Aeromonas* species may multiply and produce enterotoxins at refrigeration temperatures and may be recovered on several foods at retail. However, the lack of food poisoning outbreaks in which *Aeromonas* species have been incriminated and the heat lability of enterotoxins would suggest that it is unlikely that diarrhoeal disease is brought about by ingestion of preformed toxins.

The colonization/toxin production model of enteropathogenicity is therefore the most probable and a putative model of the pathogenesis of *Aeromonas*-associated non-inflammatory diarrhoeal disease is shown in Figure 9.1.

A recent study recognized two morphologically distinct kinds of fimbriae from a strain of *A. hydrophila* that was shown to adhere to rabbit and human intestines[93]. One type of fimbria appeared rigid, channelled and straight while the other was flexible, wavy and having a helical structure. The rigid fimbriae were purified and characterized and considered not to play a role in the autoaggregation of the organisms but not to be likely colonization factors. These findings raise speculation that the other fimbriae may play an important role in adherence but this remains to be proven.

Approximately 40 different agglutination types are known for *Aeromonas*

INGESTION
(via contaminated drinking water or food products)

↓

COLONIZATION
(probably of the small intestine via cell surface haemagglutins/lectins which provide the organism with a high cell-surface hydrophobicity facilitating the entrapment of organisms in the mucus layer)

↓

MULTIPLICATION
(with production of mucus layer-degrading proteases aiding contact with the mucosal cell surface leading to adherence of the organism by fimbriae or non-fimbrial outer membrane proteins)

↓

TOXIN PRODUCTION
(a variety of cytotonic and cytotoxic enterotoxins and a β-haemolysin. The latter causes some tissue damage which possibly either aids transfer of the enterotoxins into the enterocyte or induces uptake of the organism into the cells, with subsequent enterotoxin production)

↓

NON-INFLAMMATORY DIARRHOEA
(via increased production of cyclic AMP caused by the *Aeromonas* cytotonic enterotoxin)

Figure 9.1. A putative model of the pathogenesis of *Aeromonas*-associated non-inflammatory diarrhoeal disease

on the basis of the animal origin of erythrocytes species, co-agglutination with yeast and the carboxyhydrate inhibition test[84]. Atkinson and colleagues identified a haemagglutinin, A6-type haemagglutinin (A6-HAG), from OMP of *Aeromonas* whose erythrocyte receptor was the H-antigen present on almost all human erythrocytes and expressed on the surface of intestinal cells. Although they did not use the intestine for the adhesion test it seems theoretically possible that A6-HAG may be an important colonizing factor[101].

Other mechanisms may be involved, however, considering the observation of Lawson and co-workers that some strains of *Aeromonas* were capable of invasion of intestinal cells[95]. Recently described mechanisms in other pathogens have included the ability of a surface protein termed 'invasin' in *Yersinia pseudotuberculosis* to induce the uptake of the organism into

Table 9.3. Probable modes of action of *Aeromonas* proteases in enhancing virulence. Reproduced by permission of Raven Press Ltd, New York

Mode	Protease specificity	Advantage to bacterium	Pathological effect
High-level proteolysis	Broad specificity for many tissue proteins or specificity for host proteins present in very high concentrations	Provides large amounts of peptides and free amino acids for use in energy metabolism and biosynthesis	Destruction of tissue proteins with many possible effects, including loss of local tissue integrity and depletion of important protein species
Inhibition of host inflammatory response	Inactivation of key regulatory proteins or peptides that regulate host anti-bacterial responses	Limits host response to ineffective levels	Prevents control of infection at critical early stage
Activation of inappropriate response	Activation of key regulatory proteins or peptides that regulate important host physiological responses	Activated host response may interfere directly or indirectly with normal antibacterial defences	Inappropriate activation of host response without normal physiological regulation may have direct pathogenic effects
Activation of the precursor of β-haemolysin	Removal of 25 amino acids from the *C*-terminus that prevents aggregation of the β-haemolysin	Promotes tissue damage which assists the organism to infect new tissue	Tissue damage

Adapted from Goguen *et al*[106]

intestinal epithelial cells[102]. Induced uptake has also been described with *Listeria monocytogenes* and *Shigella flexneri*[103,104]. The entry of non-invasive *Campylobacter jejuni* into epithelial cells that were co-infected with invasive bacteria such as *S. flexneri* has also been recently described[105]. While speculative, it is possible that *Aeromonas* spp. may also enter cells by such mechanisms, perhaps further aided by membrane alterations brought about by the β-haemolysin, and release their cytotonic enterotoxin from within the intestinal epithelial cells.

Clearly, further study is required to identify the possibly unique colonization factors and possible uptake of organisms involved in the putative enteropathogenic mechanisms of *Aeromonas* spp.

Many species of pathogenic bacteria produce cell-surface or secreted proteases. These enzymes have high potential to enhance bacterial pathogenesis through degradation of critical host proteins and by mimicking the activity of host-regulatory proteases that control important zymogen systems. Although many bacterial proteases have been implicated in virulence, there is currently no system in which both rigorous demonstration of virulence enhancement *in vivo* and convincing identification of the important substrate molecules have been achieved[106]. Bacterial proteases have the potential to enhance the virulence of bacteria and contribute to pathogenesis (Table 9.3).

Elastin is a major structural protein of the lung and comprises a significant proportion of other tissues, such as blood vessel endothelium (arteries) and peridontal ligament. Recent studies have indicated that proteases with elastolytic activity, that is the ability to degrade the native elastin molecule, may be an important virulence factor in a number of pathogenic organisms, including *Aeromonas* spp., in relation to disease-producing potential[107–110].

Most aeromonad wound infections appear to be the result of colonization due to contact with polluted water following traumatic injury. Once established, aeromonad infection can spread throughout and penetrate deeper into the wound site, causing an increase in severity. Elastase from aeromonads, similar to other bacteria and fungi, may play a part in the severity of the infection as one of the factors that can contribute to the virulence of the microorganisms. Hsu and co-authors found that elastase-positive strains of *A. hydrophila* produced lesions and mortality when injected into channel catfish, and concluded that motile aeromonads with positive elastase activity were usually virulent strains[107].

Of 166 *Aeromonas* strains examined for elastase activity only *A. hydrophila* was almost exclusively positive (95%), with high levels of activity[111]. In this study it was therefore suggested that wound infections with *A. hydrophila* may be more severe. However, Trust emphasized that the contribution to bacterial virulence must be confirmed by studies using bacterial isogenic mutants, deficient in the specific exoproducts[112].

Further work is required to clarify the pathogenicity mechanisms of

Aeromonas spp. in systemic disease and conclusively to substantiate or refute its role in diarrhoeal disease. In particular, investigations of adhesins and their receptors, further human volunteer feeding studies involving more completely defined strains of *Aeromonas* and different host populations and a continued search for an animal model for *Aeromonas*-associated diarrhoeal disease, will need to be conducted.

9.6 REFERENCES

1 Deepe GS, Conrad JD. Fulminant wound infection with *Aeromonas hydrophila*. *South Med J* 1980; **73**: 1546–7.
2 Trust TJ, Chipman DC. Clinical involvement of *Aeromonas hydrophila*. *Can Med Assoc J* 1979; **120**: 942–6.
3 Freij BJ. *Aeromonas*: Biology of the organism and diseases in children. *Paediatr Infect Dis*. 1984; **3**: 164–75.
4 Janda JM, Reitano M, Bottone EJ. Biotyping of *Aeromonas* isolates as a correlate to delineating a species-associated disease spectrum. *J Clin Microbiol* 1984; **19**: 44–7.
5 Cahill MM. Virulence factors in motile *Aeromonas* species. *J Appl Bacteriol* 1990; **69**: 1–16.
6 Dooley SG, Lallier R, Trust TJ. Surface antigens of virulent strains of *Aeromonas hydrophila*. *Vet Immunol Immunopathol* 1986; **12**: 339–44.
7 Kokka, RP, Neyland AV, Janda MJ. Electrophoretic analysis of the surface components of autoagglutinating surface array protein-positive and surface array protein-negative *Aeromonas hydrophila* and *Aeromonas sobria*. *J Clin Microbiol* 1990; **28**: 2240–7.
8 Atkinson HM, Trust TJ. Haemagglutination properties and adherence ability of *Aeromonas hydrophila*. Infect Immun 1980; **27**: 938–46.
9 Aoki T, Holland BI. The outer membrane proteins of the fish pathogens *Aeromonas hydrophila, Aeromonas salmonicida*, and *Edwardsiella tarda*. *FEMS Microbiol Lett* 1985; **27**: 299–305.
10 Mittal KR, Lalonde G, Leblanc D, Oliver G, Lallier R. *Aeromonas hydrophila* in rainbow trout: relation between virulence and surface characteristics. *Can J Microbiol* 1980; **26**: 1501–3.
11 Koval SF. Paracrystalline protein surface arrays on bacteria. *Can J Microbiol* 1988; **34**: 407–14.
12 Sleytr UB, Messner P. Crystalline surface layers on bacteria. *Annu Rev Microbiol* 1983; **37**: 311–39.
13 Shaw DH, Squires MJ. O-antigen in a virulent strain of *Aeromonas hydrophila* *FEMS Microbiol Lett* 1984; **24**: 277–80.
14 Rigney M, Zilinsky J, Rouf M. Pathogenicity of *Aeromonas hydrophila* in red leg disease in frogs. *Curr Microbiol* 1978; **1**: 175–81.
15 Santos Y, Toranzo AE, Barja JL, Nieto TP, Villa T. Virulence properties and enterotoxin production of *Aeromonas* strains isolated from fish. *Infect Immun* 1988; **56**: 3285–93.
16 Janda JM, Duffey PS. Mesophilic aeromonads in human disease: current taxonomy, laboratory identification and infectious disease spectrum. *Rev Infect Dis* 1988; **10**: 980–97.
17 Caselitz FH, Günther R. Hamolysinstudien mit Aeromonas-Stammen. *Zentralbl Bakteriol I Abt Orig* 1960; **180**: 30–8.

18 Wretlind B, Möellby R, Wadström T. Separation of two haemolysins from *Aeromonas hydrophila* by isoelectric focusing. *Infect Immun* 1971; **4**: 503–5.
19 Bernheimer AW, Avigad LS. Partial characterization of aerolysin, a lytic exotoxin from *Aeromonas hydrophila*. Infect Immun 1974; **9**: 1016–21.
20 Bernheimer AW, Avigad LS, Avigad G. Interaction between aerolysin, erythrocytes and erythrocyte membranes. *Infect Immun* 1975; **11**: 1312–19.
21 Donta ST, Haddow AD. Cytotoxic activity of *Aeromonas hydrophila Infect Immun* 1978; **11**: 989–93.
22 Wadström T, Ljungh A, Wretlind, B. Enterotoxin, haemolysin and cytotoxic protein in *Aeromonas hydrophila* from human infections. Acta Pathol Microbiol 1976; **84**: 112–14.
23 Ljungh A, Wretlind B, Wadström T. Evidence for enterotoxin and two cytolytic toxins in human isolates of *Aeromonas hydrophila*. In Rosenberg P (ed) *Toxins: Animal, Plant and Microbial* Pergamon Press, Oxford 1978: 947–60.
24 Cumberbatch N, Gurwith MJ, Langston C, Sack RB, Brunton JL. Cytotoxic enterotoxin produced by *Aeromonas hydrophila*: relationship of toxigenic isolates to diarrhoeal disease. *Infect Immun* 1979; **3**: 829–37.
25 Ljungh A, Wretlind B, Möllby R. Separation and characterization of enterotoxin and two haemolysins from *Aeromonas hydrophila*. *Acta Pathol Microbiol Scand* 1981; **89**: 387–97.
26 Keusch GT, Donta ST. Classification of enterotoxins on the basis of activity in cell culture. *J Infect Dis* 1975; **131**: 58–63.
27 Ljungh A. Toxins of *Aeromonas hydrophila*. *Bacterial Protein Toxins* In: Alouf JE, Fehrenbach FJ, Freer JH, Jaljaszewicz J (eds) Academic Press, London, 309–15.
28 Buckley JT, Halasa LN, Lund KD MacIntyre S. Purification and some properties of the haemolytic toxin aerolysin. *Can J Biochem* 1981; **59**: 430–5.
29 Asao T, Konoshita S, Kozaki S, Uemura T, Sakaguchi G. Purification and some properties of *Aeromonas hydrophila* haemolysin. *Infect Immun* 1984; **46**: 122–7.
30 Potomski J, Burke V, Robinson J, Fumarola D, Miragliotta G. *Aeromonas* cytotonic enterotoxin cross reactive with cholera toxin. *J Med Microbiol* 1987; **23**: 179–86.
31 Bloch S, Monteil H. Isolation and properties of beta-haemolysin of *Aeromonas hydrophila* (European strain). *Proc 1st International Workshop on* Aeromonas *and* Plesiomonas, Manchester, UK 1986: 29–30.
32 Asao T, Kozaki S, Kato K, Konishita Y, Otsu K, Uemura T, Sakaguchi G. Purification and characterization of an *Aeromonas hydrophila* haemolysin. *J Clin Microbiol* 1986; **24**: 228–32.
33 Rose JM, Houston CW, Coppenhare DH, Dixon JD, Kurosky A. Purification and chemical characterization of a cholera toxin-cross-reactive cytolytic enterotoxin produced by a human isolate of *Aeromonas hydrophila*. *Infect Immun* 1989; **57**: 1165–9.
34 Chakraborty T, Huhle B, Bergbauer H, Goebel W. Cloning, expression, and mapping of the *Aeromonas hydrophila* aerolysin gene determinant in *Escherichia coli* K-12. *J Bacteriol* 1986; **167**: 368–74.
35 Chopra AK, Houston CW, Kurosky A. Genetic variation in related cytolytic toxins produced by different species of *Aeromonas*. *FEMS Microbiol Lett* 1991; **78**: 231–8.
36 Wadström T, Aust-Kettis T, Habte D, Holmgren J, Meeuwisse G, Möllby R, Söderlind O. Enterotoxin-producing bacteria and parasites in stools of Ethiopian children. *Arch Dis Child* 1976; **51**: 865–70.
36a Gosling PJ, Turnbull PCB, Lightfoot NF, Pether JVS, Lewis RJ. Isolation of

Aeromonas sobria cytotonic enterotoxin and beta-haemolysin by fast protein liquid chromatography. *J Med Microbiol* 1992; **38**: 227–34.

37 Ljungh A, Eneroth P and Wadström T. Cytotonic enterotoxin from *Aeromonas hydrophila*. Toxicon 1982; **20**: 787–94.

38 Chakraborty T, Montenegro MA, Sanyal SC, Helmuth R, Bulling E, Timmis KN. Cloning of enterotoxin gene from *Aeromonas hydrophila* provides conclusive evidence of production of a cytotonic enterotoxin. *Infect Immun* 1984; **46**: 435–44.

39 Shimada T, Sakazaki R, Horigome K, Uesaka Y, Niwano K. Production of cholera-like enterotoxin by *Aeromonas hydrophila*. *Japan J Med Sci Biol* 1984; **37**: 141–4.

40 Campbell JD, Houston CW. Effect of cultural conditions on the presence of a cholera-toxin cross-reactive factor in culture filtrates of *Aeromonas hydrophila*. *Curr Microbiol* 1985; **12**: 101–6.

41 Honda T, Sato M, Nischimura T, Higashitsutsumi M, Fukai K, Miwatani T. Demonstration of cholera toxin-related factor in cultures of *Aeromonas* species by enzyme-linked immunosorbent assay. *Infect Immun* 1985; **50**: 322–3.

42 Potomski J, Burke V, Watson I, Gracey M. Purification of cytotoxic enterotoxin of *Aeromonas sobria* by use of monoclonal antibodies. J Med Microbiol 1987; **23**: 171–7.

43 Salzman EW. Cyclic AMP and platelet function. *New Engl J Med* 1972; **286**: 358–63.

44 Fumarola D, Miragliotta G. Platelet aggregation test and enterotoxins from *Yersinia enterocolitica*. *IRCS Med Sci* 1981; **9**: 1033.

45 Chopra AK, Houston CW, Kurosky A. The purification and partial characterization of a cytotonic enterotoxin produced by *Aeromonas hydrophila*. *Proc 2nd International Workshop on* Aeromonas *and* Plesiomonas, Florida, 1988: 58.

46 Howard SP, Buckley JT. Protein export by a Gram-negative bacterium: production of aerolysin by *Aeromonas hydrophila*. J Bacteriol 1985; **161**: 1118–24.

47 Howard SP, Buckley JT. Activation of the hole-forming toxin aerolysin by extracellular processing. *J Bacteriol* 1985; **161**: 336–40.

48 Kozaki S, Asao T, Kamata Y, Sakaguchi G. Characterization of *Aeromonas sobria* haemolysin by use of monoclonal antibodies against *Aeromonas hydrophila* haemolysins. *J Clin Microbiol* 1989; **27**: 1782–6.

49 Buckley JT. On the mechanism of hole formation by aerolysin. *Abst International Symposium* Aeromonas *and* Plesiomonas, Helsingør, Denmark, 1990: 5.

50 Garland WJ, Buckley JT. The cytolytic toxin aerolysin must aggregate to disrupt erythrocytes, and aggregation is stimulated by human glycophorin. *Infect Immun* 1988; **56**: 1249–53.

51 Howard SP, Garland WJ, Green MJ, Buckley JT. Nucleotide sequence of the gene for the hole-forming toxin aerolysin of *Aeromonas hydrophila*. *J Bacteriol* 1987; **169**: 2869–71.

52 O'Reilly T, Day DF. Effects of cultural conditions on protease production by *Aeromonas hydrophila*. Appl Environ Microbiol 1983; **45**: 1132–5.

53 Pansare AC, Venugopal V, Lewis NF. A note on nutritional influence on extracellular protease synthesis in *Aeromonas hydrophila*. *J Appl Bacteriol* 1985; **58**: 101–4.

54 Prescott JM, Wilkes SH, Wagner FW, Wilson KJ. *Aeromonas* aminopeptidase. Improved isolation and some physical properties. *J Biol Chem* 1971; **246**:

1756–64.

55 Bayliss ME, Prescott JM. Modified activity of *Aeromonas* aminopeptidase: metal ion substitutions and role of substrates. *Biochemistry* 1986; **25**: 8113–17.

56 Kitazono A, Ito K, Yoshimoto T. Prolyl aminopeptidase is not a sulfhydryl enzyme: identification of the active serine residue by site-directed mutagenesis. *J Biochem (Tokyo)* 1994; **116**: 943–5.

57 Ohtakara T. Action pattern of *Aeromonas hydrophila* chitinase on partially *N*-acetylated chitosan. *Agric Biol Chem Tokyo* 1990; **54**: 871–8.

58 Yabuki M, Mizushina K, Amatatsu T, Ando A, Fujii T, Shimada M, Yamashita M. Purification and characterization of chitinase and chitobiase produced by *Aeromonas hydrophila* subsp. *anaerogenes* A52. *J Gen Appl Microbiol* 1986; **32**: 25–38.

59 Sitrit Y, Vorgias CE, Chet I, Oppenheim AB. Cloning and primary structure of the chiA gene from *Aeromonas caviae*. *J Bacteriol* 1995; **177**: 4187–9.

60 Nord CE, Wadström T, Wretlind B. Antibiotic sensitivity of two *Aeromonas* and nine *Pseudomonas* species. *Med Microbiol Immunol* 1975; **161**: 89–97.

61 Hasan JAK, Carnahan AM, Macaluso P, Joseph SW. Elastolytic activity among newly recognised *Aeromonas* spp. using a modified bilayer plate assay. *Abstr Annu Mtg ASM* 1990; 42.

62 Scharmann W. Vorkommen von elastase bei *Pseudomonas* and *Aeromonas*. *Zentralbl Bakteriol Parasitkde. I. Abt. Orig* 1972; **220**: 435–42.

63 Foster BG, Hanna MO. Toxic properties of *Aeromonas proteolytica*. *Can J Microbiol* 1974; **20**: 1403–9.

64 Kanatani A, Yoshimoto T, Kitazono A, Kokubo T, Tsuru, D. Prolyl endopeptidase from *Aeromonas hydrophila*: cloning, sequencing, and expression of the enzyme gene, and characterisation of the expressed enzyme. *J Biochem* 1993; **113**: 790–6.

65 Caselitz FH. Pseudomonas–Aeromonas *und ihre human medizinische Bedeutung*. Jena: VEB Gustav Fischer Verlag, 1966.

66 Caselitz FH, Freitag V, Meyer HJ. Studies of lipase and anti sera on different strains of the genera *Serratia*, *Aeromonas* and *Vibrio*. *Zentral Bakteriol 1 Abt Orig A Med Mikrobiol Infektionskr Parasitol* 1980; **246**: 336–43.

67 Ingham AB, Pemberton JM. A lipase of *Aeromonas hydrophila* showing nonhaemolytic phospholipase C activity. *Curr Microbiol* 1995; **31**: 28–33.

68 McIntyre S, Buckley JT. Presence of glycerophospholipid:cholesterol acyl transferase and phospholipase in culture supernatant of *Aeromonas hydrophila*. *J Bacteriol* 1978; **135**: 402–7.

69 McIntyre S, Trust TJ, Buckley JT. Distribution of glycero-phospho-lipid:cholesterol acyltransferase in selected bacterial species. *J Bacteriol* 1979; **139**: 132–6.

70 Dahle HK. The purification and some properties of two *Aeromonas* proteases. *Acta Pathol Microbiol Scand Sect B* 1971; **79**: 726–38.

71 Alichanidis E. Partial purification and characterization of an extracellular proteinase from *Aeromonas hydrophila* strain A4. *J Dairy Res* 1988; **55**: 97–100.

72 Leung KY, Stevenson RMW. Characteristics and distribution of extracellular proteases from *Aeromonas hydrophila*. *J Gen Microbiol* 1988; **134**: 151–60.

73 Nieto TP, Ellis AE. Characterisation of extracellular metallo- and serin-proteases of *Aeromonas hydrophila* strain B_{51}. *J Gen Microbiol* 1986; **132**: 1975–9.

74 Thune RL, Graham TE, Riddle LM, Amborski RL. Extracellular proteases from *Aeromonas hydrophila*: partial purification and effects on age-0 channel catfish. *Trans Am Fish Soc* 1982; **111**: 749–54.

75 Gross R, Coles NW. A proteinase produced by *Aeromonas hydrophila. Aust J Sci* 1969; **31**: 330–1.
76 Kanai K, Wakabayashi H. Purification and some properties of proteases from *Aeromonas hydrophila. Bull Jap Soc Sci Fish* 1984; **40**: 1367–74.
77 Amborski RL, Borall R, Thune RL. Effects of short-term cold storage on recovery of proteases from extracellular products of *Aeromonas hydrophila. Appl Environ Microbiol* 1984; **48**: 456–8.
78 Evans DG, Evans DJ Jr, Clegg S, Pauley JA. Purification and characterization of the CFA/I antigen of enterotoxigenic *Escherichia coli. Infect Immun* 1979; **25**: 738–48.
79 Knutton S, Lloyd DR, McNeish AS. Identification of a new fimbrial structure in enterotoxigenic *Escherichia coli* (ETEC) serotype 0148:H28 which adheres to human intestinal mucosa: a potentially new human ETEC colonization factor. *Infect Immun* 1987; **55**: 86–92.
80 Punsalang AP Jr, Sawyer WD. Role of pili in the virulence of *Neisseria gonorrhoea. Infect Immun* 1973; **8**: 255–63.
81 Sato Y, Izumiya K, Oda MA, Sato H. Biological significance of *Bordetella pertussis* fimbriae or haemagglutinin: a possible role of the fimbriae or haemagglutinin for pathogenesis and antibacterial immunity. In Manclark CR, Hill JC (eds) *International Symposium on Pertussis.* Washington DC: US Department of Health, Education and Welfare, 1979: 51–7.
82 Atkinson HM, Trust TJ. Haemagglutination properties and adherence ability of *Aeromonas hydrophila. Infect Immun* 1980; **27**: 938–46.
83 Burke VM, Cooper M, Robinson J, Gracey M, Lesmana M, Janda M. Haemagglutination patterns of *Aeromonas* spp. in relation to biotype and source. *J Clin Microbiol* 1984; **19**: 39–43.
84 Adams D, Atkinson HM, Woods WH. *Aeromonas hydrophila* typing scheme based on patterns of agglutination with erythrocyte and yeast cells. *J Clin Microbiol* 1983; **17**: 422–7.
85 Stewart GA, Bundell CS, Burke V. Partial characterisation of a soluble haemagglutinin from human diarrhoeal isolates of *Aeromonas. J Med Microbiol* 1986; **21**: 319–24.
86 Knochel S. *Aeromonas* spp. – Ecology and significance in food and water hygiene. Royal Veterinary and Agricultural University, Denmark *PhD Thesis*, 1989.
87 Krovacek KA, Faris A, Ahne W, Minsson I. Adhesion of *Aeromonas hydrophila* and *Vibrio anguillarum* to fish cells and to mucus-coated glass slides. *FEMS Microbiol Lett* 1987; **42**: 85–9.
88 Levett PN, Daniel RR. Adhesion of vibrios and aeromonads to isolated rabbit brush borders. *J Gen Microbiol* 1981; **125**: 167–72.
89 Carrello A, Silburn KA, Budden JR, Chang BJ. Adhesion of clinical and environmental *Aeromonas* isolates to Hep-2 cells. *J Med Microbiol* 1988; **26**: 19–27.
90 Nishikawa Y, Kimura T, Kishi T. Mannose-resistant adhesion of motile *Aeromonas* to INT 407 cells and the differences among isolates from humans, food and water. *Epidemiol Infect* 1991; **107**: 171–9.
91 Ascencio F. *Aeromonas hydrophila.* A study on cell-associated extracellular matrix and lactoferrin binding-proteins. University of Lund, Sweden, *Thesis*, 1992.
92 Ho ASY, Mietzner TA, Smith AJ, Schoolnik GK. The pili of *Aeromonas hydrophila*: identification of an environmentally regulated 'mini pill'. *J Exp Med* 1990; **172**: 795–806.

93 Honna Y, Nakasone N. Pili of *Aeromonas hydrophila*: purification, characterization and biological role. *Microbiol Immunol* 1990; **34**: 83–98.

94 Barghouthi S, Young R, Olson MOJ. Arceneaux JEL, Clem LW, Bryers BR. Amonabactin, a novel tryptophan- or phenalanine-containing phenolate siderophore in *Aeromonas hydrophila*. J Bacteriol 1989; **171**: 1811–16.

95 Lawson MA, Burke V, Chang BJ. Invasion of Hep-2 cells by faecal isolates of *Aeromonas hydrophila*. *Infect Immun* 1985; **47**: 680–93.

96 Gray SJ, Sticker DJ, Bryant TN. The incidence of virulence factors in mesophilic *Aeromonas* species isolated from farm animals and their environment. *Epidemiol Infect* 1990; **105**: 277–94.

97 Janda JM, Brenden R, Bottone EJ. Differential susceptibility to human serum by *Aeromonas* spp. *Curr Microbiol* 1984; **11**: 325–8.

98 Brenden R, Janda JM. Detection, quantitation and stability of the beta-haemolysin of *Aeromonas* spp. *J Med Microbiol* 1987; **24**: 247–51.

99 Morgan DR, Johnson PC, DuPont HL, Satterwhite TK, Wood LV. Lack of correlation between known virulence properties of *Aeromonas hydrophila* and enteropathogenicity for humans. *Infect Immun* 1985; **50**: 62–5.

100 DuPont HL, Formal SB, Hornick RB. Pathogenesis of *Escherichia coli* diarrhoea. *N Engl J Med* 1971; **285**: 1–9.

101 Atkinson HM, Adams D, Savvas RS, Trust TJ. *Aeromonas* adhesin antigens. *Experimentia* 1987; **43**: 372–4.

102 Isberg RR, Voorhis DL, Falkow S. Identification of invasin: a protein that allows enteric bacteria to penetrate cultured mammalian cells. *Cell* 1987; **50**: 769–78.

103 Gaillard JL, Berche P, Mounier J, Richard S, Sansonetti P. In vitro model of penetration and intracellular growth of *Listeria monocytogenes* in the human enterocyte-like cell line Caco-2. *Infect Immun* 1987; **55**: 2822–9.

104 Clerc P, Sansonetti PJ. Entry of *Shigella flexneri* into HeLa cells: evidence for directed phagocytosis involving actin polymerizal and myosin accumulation. *Infect Immun* 1987; **55**: 2681–8.

105 Bukholm G, Kapperud G. Expression of *Campylobacter jejuni* invasiveness in cell cultures co-infected with other bacteria. *Infect Immun* 1987; **55**: 2816–21.

106 Goguen JD, Hoe NP, Subrahmanyam YVBK. Proteases and bacterial virulence: a view from the trenches. *Infect Agents Dis* 1995; **4**: 47–54.

107 Hsu TC, Waltman WD, Shotts EB. Correlation of extracellular enzymatic activity and biochemical characteristics with regard to virulence of *Aeromonas hydrophila*. *Dev Biol Stand* 1981; **49**: 101–11.

108 Kothary MH, Chase T Jr, MacMillan JD. Correlation of elastase production by some strains of *Aspergillus fumigatus* with ability to cause pulmonary invasive aspergillosis in mice. *Infect Immun* 1984; **43**: 320–5.

109 Janda JM. Elastolytic activity among staphylococci. *J Clin Microbiol* 1986; **24**: 945–6.

110 Morihara K, Tsuzuki H. Production of protease and elastase by *Pseudomonas aeruginosa* strains isolated from patients. *Infect Immun* 1977; **15**: 679–85.

111 Hasan JAK, Macaluso P, Carnahan AM, Joseph SW. Elastolytic activity among *Aeromonas* spp using a modified bilayer plate assay. *Diag Microbiol Infect Dis* 1992; **15**: 201–6.

112 Trust TJ. Pathogenesis of infectious diseases in fish. *Annu Rev Microbiol* 1986; **40**: 479–502.

10 Toxins

S. PETER HOWARD
University of Regina, Regina, Canada

SHEILA MACINTYRE
University of Reading, Reading, England

J. THOMAS BUCKLEY
University of Victoria, Victoria, Canada

Aeromonas spp. produce many products that may be toxic to other cells (Table 10.1). Many of these are released from viable cells in soluble form. Others may remain associated with the cell surface, and still others may be released upon cell death. It seems likely that we have recognized only a few of the factors that may contribute to the pathogenicity of the bacteria *in vivo*, if only because, as with other species, expression depends on growth conditions.

In this chapter, we will concentrate on those products that are released from intact cells. Except for lipopolysaccharide (LPS) or endotoxin, all of those that have been characterized are proteins, and nearly all are enzymes. Many of the genes for these proteins have been cloned and sequenced. They include several proteases (see below), an unusual lipase which is also a glycerophospholipid:cholesterol acyltransferase[1], a second lipase of unusually high molecular weight[2], a chitinase[3], a DNase[4] and two amylases[5,6] as well as the channel-forming protein aerolysin[7]. As far as we know, the primary function of aerolysin is to kill target cells, whereas all of the other proteins are enzymes whose main purpose is likely to be to produce utilizable nutrients by degrading host macromolecules.

Three of the extracellular proteins of *Aeromonas* spp. that have been implicated in pathogenicity have been cloned, sequenced and reasonably well characterized biochemically. These are aerolysin, the lipase/acyltransferase called GCAT (glycerophospholipid:cholesterol acyltransferase), and a serine protease. In this chapter, aerolysin and GCAT will be discussed individually, whereas the serine protease, about which less is known, will be discussed in the section on extracellular proteases.

Most of the proteins found outside the cell, including aerolysin and GCAT as well as at least one of the extracellular proteases, are secreted by a pathway that is commonly referred to as the general secretory pathway. Because of the potential importance of this pathway in determining *Aeromonas* virulence, it will be discussed in a separate section.

The Genus Aeromona]s. Edited by B. Austin, M. Altwegg, P.J. Gosling and S. Joseph
© 1996 John Wiley & Sons Ltd

Table 10.1 *Aeromonas* toxins

Product/toxin*	Size (kDa) a†	b†	Synonym/ Homologue	Source	Role in pathogenesis	Reference
Haemolysins						
Aerolysin	52	47.5	AHH3, AHH4	all aeromonads	Pore-forming cytolysin, broad spectrum, needed for full virulence of *A. hydrophila*.	64
GCAT	37.4	26 + 4.7		*A. hydrophila* *A. salmonicida*	Weakly haemolytic with mammalian cells, strongly haemolytic with fish cells. GCAT–LPS lethal for fish.	84
AHH-1	60	n.d.	ASH4	*A. hydrophila* *A. salmonicida*	n.d.	85
AHH-2	37.7	n.d.	n.d.	*A. hydrophila*	n.d.	
Proteases						
Serine protease	66.8	64.2	'70 kD protease' (caseinase)	*A. salmonicida*	Processes secreted protoxins. Injection results in muscle liquefaction and thrombus formation in salmon.	20
Serine protease	n.d.	22		*A. hydrophila*	Toxic in trout.	48
Metalloprotease	n.d.	19		*A. hydrophila*	n.d.	47
Metalloprotease	n.d.	37	'gelatinase'	*A. salmonicida* *A. hydrophila*	n.d.	49 48
Others						
Acetylcholinesterase	n.d.	15.5		*A. hydrophila*	Toxic for fish.	86
Enterotoxin	n.d.	n.d.	n.d.	*A. hydrophila*	Fluid accumulation in rabbit ileal loop.	62

* Many other biological activities have been reported for partially purified secreted products of *Aeromonas* spp. They have not been included in this table either because their identities have not been established or because it is not clear that they differ from the products that are included.

† a, Size of the encoded gene product; b, apparent size of the active form of the enzyme/toxin. For GCAT there are two fragments (see text).

10.1 THE CHANNEL-FORMING TOXIN AEROLYSIN

If the primary purpose of a toxin is to kill cells, either in order to fend off host defences or so that their contents can become available for bacterial growth, then aerolysin is the only well characterized extracellular product produced by *Aeromonas* spp. that clearly qualifies. This cytoxin, which is responsible for the β-haemolysis exhibited by the great majority of *A. hydrophila* and *A. sobria* strains, is by far the best characterized of the *Aeromonas* cytotoxic proteins[8]. It was first described by Bernheimer and Avigad[9], who reported the partial purification of the protein from culture supernatants of *A. hydrophila* and gave the toxin its name. It was later purified to homogeneity by Buckley and colleagues[10], and Buckley[11] subsequently described the purification of the proform of the *A. hydrophila* protein secreted by *A. salmonicida* containing the cloned *aerA* structural gene. Since then it has been cloned from a number of other aeromonads, including *A. trota*[12] and *A. salmonicida*[13], and Chakraborty and co-authors[14] have presented genetic evidence that the structural gene *aerA* is found in all members of the genus. To our knowledge, the gene has not been found in other species, and at least for the present, it may be considered a distinguishing feature of this genus. Cytolytic toxins from *Pseudomonas aeruginosa* and *Staphylococcus aureus* contain short sequences with some apparent homology to aerolysin, and remarkably, the channel-forming alpha toxin from *Clostridium septicum* shares 20% identity with aerolysin throughout its sequence[15]. It is the most closely related toxin so far described outside the genus *Aeromonas*.

10.1.1 AEROLYSIN AS A VIRULENCE FACTOR

A clear picture of the contribution of aerolysin to the virulence of *Aeromonas* spp. has been difficult to obtain, because in many of the existing studies the experimental procedures that were employed have not been specific enough to allow different toxic substances to be distinguished. Assays based on haemolysis, cytotoxicity or fluid accumulation may or may not detect aerolysin, and many assays cannot distinguish the toxin from other proteins that may be present. Since aerolysin is not the only cytolytic extracellular product that is produced by *Aeromonas* spp., assays that depend on haemolysis can give equivocal results (for example, the lipase will also cause haemolysis). What is more, since aerolysin is completely inactive until it is nicked proteolytically (see below), culture supernatants from *Aeromonas* strains that secrete the protoxin without accompanying proteases might be negative in haemolysis assays, depending on the conditions used.

The clearest, most unequivocal study of the role of aerolysin in systemic infections was carried out by Chakraborty and colleagues[14]. These authors used marker exchange mutagenesis to produce strains of *A. hydrophila* that could not make active aerolysin. These strains were found to be much less toxic to mice than the parent strain. Chakraborty and colleagues[14] were also able to detect anti-aerolysin antibody in mice surviving infection with the wild-type strain, evidence that the toxin was produced during infection. There is also compelling evidence that aerolysin is involved in the virulence of *A. hydrophila* in humans. For example, in a recent study, Vadivelu and co-workers[16] found that the toxin was produced by all 18 strains of *A. hydrophila* that they isolated from patients with bacteraemia. It is worth noting that none of the strains was found to produce enterotoxin.

Although *A. salmonicida* has been reported to contain the aerolysin structural gene[13], it does not appear to secrete active aerolysin when grown in culture, at least under the conditions we have tried. Whether this is because the gene is poorly expressed, or whether the protein is unstable, inactive or not secreted, has not been established. It may be relevant that although the sequence of *A. salmonicida* aerolysin is similar to the sequence of the other aerolysins throughout most of its length, it appears to differ substantially in the region of activation near the carboxy terminus. It is possible that for this reason it is not activated by the proteases that the organism normally produces. Because of the importance of this species as a fish pathogen, further investigation is warranted to establish the possible role of aerolysin in its virulence.

10.1.2   ACTIVATION OF PROAEROLYSIN AND THE PROTOXIN FORM OF GCAT

Both aerolysin and GCAT are released as protoxins. Proaerolysin is completely unable to form channels, and we have shown that activation requires the removal of a *C*-terminal fragment containing approximately 40 amino acids[17,18]. This can be accomplished by a great many proteases, including trypsin and chymotrypsin, as well as by LysC-specific protease and the eukaryotic intracellular proteases related to subtilisin such as furin and KexA, which cleave proteins after basic residues (unpublished observations). Proaerolysin is unable to oligomerize, and since this is a necessary step in channel formation, the protoxin is inactive. It is not known how removal of the *C*-terminal peptide enables oligomerization to occur.

We have found that the serine protease released by *A. salmonicida* (see below) also correctly processes both proaerolysin and GCAT. This protease has a preference for arginines[19], and our preliminary results suggest it cleaves both proaerolysin and GCAT after an arginine doublet. This raises the possibility that the primary function of this protease is to activate extracellular proteins after they are released and this may be the contribution it makes to virulence, although both *A. salmonicida* and *A. hydrophila* release at least one other protease that can activate aerolysin.

The conclusion that the serine protease may process and thereby activate several of the *Aeromonas* products is especially interesting for another reason. This protease is a member of the subtilisin family of enzymes. Not only is it homologous to the enzyme from *Bacillus subtilis*[20] but it shares homology with eukaryotic intracellular proteases such as kexA and furin. Perhaps not coincidentally, many of these enzymes also prefer basic or dibasic sequences (unlike subtilisin itself, which is rather non-specific) and they are involved in the processing of extracellular proteins.

The need for correct processing must always be considered in screens for aerolysin and GCAT that are based on their activity. Since proaerolysin, the form of the toxin released by the cell, is inactive until it is nicked proteolytically[7], a strain of the bacteria that does not secrete protease might appear to be non-haemolytic under some conditions, even if it was producing large amounts of the protoxin.

10.1.3   THE STRUCTURE OF AEROLYSIN

Proaerolysin has been crystallized, and its structure has been solved to 2.8 Å resolution[21]. The protein is a unique fold, unrelated to other structures that have been solved, including that of an aminopeptidase that is also secreted by *A. hydrophila*[22]. Proaerolysin is a dimer in the crystal, and we have shown that it is a dimer in solution[23]. The monomer consists of two lobes, a small *N*-terminal lobe that appears to be required to stabilize the dimer, and a larger lobe that contains regions involved in the activation, binding, oligomerization and insertion of the toxin. The *C. septicum* alpha toxin is homologous to the large lobe of aerolysin, but the clostridia toxin appears to lack the small lobe[24], perhaps a sign that, unlike aerolysin, it is not a dimer.

10.1.4   AEROLYSIN HAS A SPECIFIC RECEPTOR ON SOME CELLS

Aerolysin is one of the most potent of all of the cytolytic toxins. Lower concentrations are required to lyse sensitive cells than have been reported for any other cytolysin. Bernheimer and colleagues[25] were the first to notice that the erythrocytes of different species vary in their sensitivity to aerolysin. Murine species are most sensitive to the toxin. Concentrations as low as 10^{-10} M (5 ng/ml) will lyse rat erythrocytes in 1 h. Human cells are among the least sensitive. Minimum concentrations in the range of 10^{-8} M are needed to lyse these cells.

Both aerolysin and proaerolysin can bind to target cells. Howard and Buckley[26] showed that murine erythrocytes are most sensitive to aerolysin because they contain a glycoprotein receptor that has a high affinity for the toxin and protoxin. Recently, we have described an *in vitro* assay for the receptor and a procedure for its partial purification[27]. It is a 47 kDa glycoprotein with *N*-linked carbohydrate which we have determined is not required for aerolysin binding. In unpublished studies, we have found that the receptor is a glycosylphosphatidylinositol (GPI)-anchored protein that is related to a small family of surface proteins that are involved in ADP-ribosylation reactions. We have also obtained some preliminary data on the aerolysin receptor on the surface of T-lymphocytes. It appears to be the alloantigen RT6, which is also anchored to the plasma membrane by a GPI-anchor. This receptor is thought to be involved in cell signalling; however, the molecular details of its function are unknown. An extraordinary observation about the RT6 gene was made by Haag and co-workers[28]. They found that the human and chimpanzee genes have had three stop codons inserted in their middles so that the protein is not made. The authors speculate that this may have occurred to reduce the sensitivity of their T-lymphocytes to attack by a specific pathogen. Of course, with the knowledge that RT6 may be a receptor for aerolysin, *Aeromonas* spp. are obvious candidates.

10.1.5   CHANNEL FORMATION IS RESPONSIBLE FOR CELL DEATH

Although cells that present a specific receptor are most sensitive to aerolysin, our experience with a number of different cell lines from a variety of species has shown that all cells that have an exposed plasma membrane are

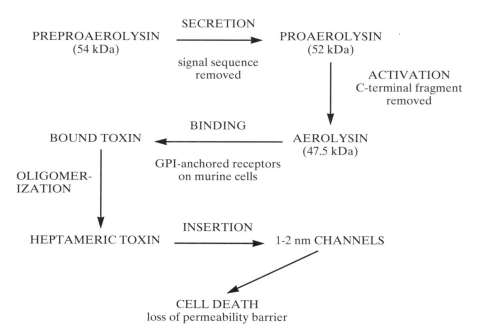

Figure 10.1.   Suggested mechanism of action of aerolysin

susceptible to the toxin. This is because aerolysin is able to insert directly into the membrane lipid bilayer and form channels. Cells that are surrounded by a glycocalyx, such as bacteria and intracellular parasites like trypanosomes, are insensitive to the toxin. We have taken advantage of the refractory nature of the trypanosomal cell wall to develop a purification procedure for the intracellular form of this parasite in blood samples[29].

Proaerolysin that is bound to the cell surface can be converted to aerolysin by endogenous proteases[17,30], or it can be internalized and activated inside the cell (unpublished results). Binding of aerolysin to the receptor serves to concentrate the toxin and this promotes oligomerization, which is the step that converts the toxin to an insertion-competent form. The oligomer then enters the membrane and forms a discrete and very stable channel[31]. The process of channel formation is depicted in Figure 10.1. It is the channels that destroy erythrocytes, by disrupting their permeability barrier and causing osmotic lysis. Presumably other cells are killed in the same way; however, this has not been established. The toxin is not known to have any enzymatic activity, nor is it structurally related to other toxins that are known to be enzymes, such as lipases and the ADP-ribosylating toxins.

10.2   THE LIPASE/ACYLTRANSFERASE GCAT

The lipase produced by *Aeromonas* and *Vibrio* spp. was named GCAT, an acronym for glycerophospholipid:cholesterol acyltransferase, based on the reaction first shown to be catalysed by the enzyme[32]. The unusual ability to produce cholesteryl esters is a distinguishing feature of GCAT. No other bacterial lipases have been shown to carry out the same reaction. In fact, the only enzyme with a similar reaction mechanism is the important mammalian plasma enzyme lecithin:cholesterol acyltransferase. The reaction has only been described in members of the *Vibrio* group and in *Staphylococcus aureus*[33]. Subsequent studies have shown that GCAT is capable of catalysing the hydrolysis of a number of lipids, including diacylglycerols and all of the common glycerophospholipids[34,35]. Interestingly, in the absence of detergent it is unable to hydrolyse triacylglycerols. However, in some detergents it can degrade even cholesteryl esters which are extremely hydrophobic[36]. In contrast to aerolysin, GCAT is secreted in late log or early stationary phase by *A. hydrophila*.

Because GCAT can act as a phospholipase, it can be regarded as a degradative enzyme, and it is not surprising that it can destroy cells by digesting their plasma membranes. For this reason the enzyme will cause erythrocyte haemolysis and it therefore qualifies as a haemolysin. It is at least partly responsible for the weak haemolysis of many strains of *A. salmonicida*, which do not appear to produce active aerolysin when grown in culture.

10.2.1 GCAT AS A VIRULENCE FACTOR

The importance of GCAT as a virulence factor has only been investigated in any depth with respect to its role in the fish disease furunculosis, caused by *A. salmonicida*. Both GCAT and the extracellular 70 kDa serine protease have been implicated as important factors in this disease process. Interestingly, fish erythrocytes are much more sensitive to GCAT-induced haemolysis than mammalian erythrocytes, perhaps because their membranes contain higher proportions of phosphatidylcholine and correspondingly lower proportions of sphingomyelin, which is not a substrate for the enzyme[34]. Injection of GCAT into salmonid fish causes extensive loss of erythrocytes and reproduces many of the pathological signs typical of furunculosis, including eventual death, but only when large amounts are administered (5–10 μg GCAT/g bodyweight[37]; 0.33 μg GCAT/g bodyweight[38]).

It has been reported that GCAT will readily associate with LPS[39]. The relevance of this to the disease process *in vivo* has not been established, but with an LD_{50} of 45 ng protein/g bodyweight of fish, a purified GCAT–LPS preparation is 10–100 times more toxic than the purified protein itself[38], and the complex is currently considered the major lethal toxin produced by *A. salmonicida*[40]. The LPS itself exhibits negligible toxicity. As well as enhancing targeting of GCAT to eukaryotic cells, the association with LPS also protects GCAT from proteolytic degradation[38], potentially prolonging the lifetime of the enzyme in fish. Histopathological changes observed on intraperitoneal injection of GCAT include haemorrhaging, decrease in circulating erythrocytes, stimulation of blood coagulation and dramatic infiltration and degranulation of eosinophilic granulocytes in the gills of fish[37,40,41].

To establish the importance of GCAT in furunculosis, marker exchange mutagenesis to remove the GCAT gene has recently been performed on two virulent isolates of *A. salmonicida*. No difference in LD_{50} was detected between the two mutants and their respective parents when each was injected intraperitoneally into Atlantic salmon (MacIntyre and colleagues, unpublished results). Thus we must conclude that GCAT itself is not absolutely essential for death of Atlantic salmon caused by *A. salmonicida*. However, the relative importance of GCAT to the disease furunculosis may well depend on the model used. A less invasive model, such as cohabitation, may yield quite different results. Once *A. salmonicida* is introduced into the body by intraperitoneal injection, it may be able to grow and cause disease without GCAT. The enzyme could be more important in establishing an early phase of infection. Alternatively, although GCAT is by far the most toxic extracellular protein identified in this organism, other factors may compensate for its loss. It should be noted that the serine protease also leads to thrombosis, resulting in circulatory failure[41], and other attributes such as a maximum gross rate of bacterial growth may be more important in *A. salmonicida* pathogenesis. Of course, the possibility exists that an as yet unidentified extracellular product plays an important role in *A. salmonicida* infection.

10.2.2   PROPERTIES OF GCAT

The structural gene for GCAT has been cloned and sequenced, first from *A. hydrophila*[1], and later from *A. salmonicida* (Nerland, unpublished results). It does not appear to be closely related to any of the other lipases that have been sequenced. In a way this is not surprising as lipases constitute a very large and diverse family of enzymes, and the ability of GCAT to carry out acyl transfer to cholesterol distinguishes it from all of them, except the mammalian enzyme LCAT. In contrast to virtually all of the serine esterases (both lipases and proteases), GCAT is unusual in that it does not contain the consensus sequence G.X.S.X.G. Instead the *Aeromonas* enzyme has the sequence G.X.S.X.S. We have recently identified a group of proteins that also have this sequence and which also share four other sequence blocks with GCAT[42]. One of these is the heat-labile haemolysin of *Vibrio parahaemolyticus*, which is known to be a lipase. Another is a lipase produced by *Xenorhabdus luminescens*, and a third is an uncharacterized gene product from *Pseudomonas putida*. Interestingly, most of the others are uncharacterized plant proteins, one of which we have recently shown is also a lipase (unpublished results).

10.2.3   ACTIVATION OF GCAT

Unlike aerolysin, GCAT is not completely inactive in the form in which it is released. It is secreted as a soluble monomer of 35 kDa which is capable of hydrolysing lipids or carrying out acyl transfer in artificial lipid preparations; however, it is nearly inactive against lipids in artificial or membrane bilayers. This is because the surface pressure of the bilayers prevents the enzyme from reaching the ester linkages. Proteolytic nicking between the two cysteines near the *C*-terminus results in a change in the conformation of the protein so that it becomes a dimer[43,44]. The processed form of the enzyme has a much higher surface activity, and it is able to degrade lipids in bilayers. Whether proteolytic processing plays any role in the toxicity of the lipase has not been established. It seems reasonable to suppose that this unusual and novel need for 'activation' to enable attack on natural membranes protects the bacteria from toxic effects of the enzyme during its secretion.

Activation of GCAT has not been as thoroughly studied as aerolysin activation, however it is known to be caused by trypsin as well as by other proteases including the 70 kDa serine protease secreted by *A. salmonicida*[45], which we have found also correctly processes proaerolysin, as noted above. The possible importance of this protease in GCAT activation is supported by the observation that an *A. salmonicida aspA*::kan mutant is unable to process proGCAT *in vivo* (Vipond and MacIntyre, unpublished observations). As pointed out above for aerolysin, this raises the possibility that the primary function and contribution to pathogenicity of this protease is to activate extracellular proteins after they are released. This seems unlikely, however, as under laboratory conditions the serine protease is released at a much later stage in the *A. salmonicida* growth cycle than GCAT, and what is more, both *A. salmonicida* and *A. hydrophila* release at least one other protease that can activate aerolysin, although not GCAT.

10.3   *AEROMONAS* PROTEASES

A number of proteases have been described in the culture supernatants of *Aeromonas* spp. and correlations have been made between extracellular protease activity and virulence. Some of these proteases have attracted the attention of biochemists interested in reaction mechanisms and structure–function relationships. For most of the proteases, there is little or no evidence of direct involvement in pathogenicity. As pointed out by Leung and Stevenson[46], at least some of them may aid the organism in overcoming initial host defences and they probably provide amino acids for cell proliferation. Not surprisingly, as proteases are degradative enzymes, they induce tissue damage and they can cause cell death. However, it seems reasonable to presume that (in contrast to aerolysin) this is not their primary purpose.

10.3.1   METALLOPROTEASES

Leung and Stevenson[46] found that strains of *A. hydrophila* made deficient in the production of metalloprotease by transposon Tn5 mutagenesis had reduced virulence for fish. The best characterized metalloprotease in *A. hydrophila* cultures is a small (19 kDa) zinc protease with unusual substrate specificity. It appears to be homologous to the LasA enzyme from *P. aeruginosa*[47]. Its toxicity has not been investigated. Rodriguez and colleagues[48] reported another extracellular metalloprotease of 38 kDa which was lethal when injected into fish. Although the enzyme was said to be pure, it exhibited 11 bands in narrow-range isoelectric focusing, a sign that it had been partially degraded, or that it was associated with lipopolysaccharide. A similar protease of 37 kDa from *A. salmonicida* has been reported by Arnesen and colleagues[49]. The substrate specificity of this enzyme is not known, nor is any sequence information available. One possibility is that it is related to the 43 kDa *A. proteolytica* protease described below.

The best characterized of all the *Aeromonas* proteases is the aminopeptidase secreted by *A. proteolytica* (now *Vibrio proteolytica*), which has been extensively studied by Prescott and colleagues[50]. This enzyme is released as a 43 kDa protein that is rapidly processed to 32 or 30 kDa. It is a metalloenzyme with remarkable heat stability. The gene for the protein has recently been cloned and sequenced[51] and the structure of the protein has been solved by X-ray crystallography[22]. It has not been implicated in *Aeromonas* virulence.

Prescott described a second metalloprotease in *A. proteolytica*, with a molecular weight of 35 kDa, which in contrast to the aminopeptidase, is heat-labile and is an endopeptidase preferring phenylalanine residues[52]. Nothing is known of its role in the virulence of this organism. It may be identical to the 37–38 kDa metalloprotease from *A. hydrophila* and *A. salmonicida*, discussed above.

10.3.2 SERINE PROTEASES

A. hydrophila and *A. salmonicida* produce a serine protease of unusually high molecular weight. This endoproteolytic enzyme is thought to be one of the two most important virulence factors of *A. salmonicida*, although with an LD_{50} of 2.4 μg/g fish[38], it is much less toxic than GCAT complexed with LPS. When purified preparations are injected intramuscularly into salmonids, they produce lesions similar to the elongated swellings of furunculosis, and classic furuncles are formed when it is injected in combination with GCAT[40]. The enzyme has also been shown to stimulate coagulation and thrombus formation in fish blood[41]. Contradictory results regarding the importance of serine protease to virulence of *A. salmonicida* have been obtained using chemically induced serine protease-deficient mutants. Sakai[53] found mutants to be essentially non-virulent, while Drinan and Smith[54] observed only a tenfold reduction in virulence, which they considered to be negligible. Clearly, the role of this enzyme will only be resolved with trials using well defined deletion mutants. The serine protease has been cloned from both *A. hydrophila*[55] and *A. salmonicida*, and the sequence of the gene from the latter organism is available[56]. At 64 kDa, the protein is unusually large for a protease. It is a member of the subtilisin family of proteases, and it is sensitive to serine protease inhibitors such as phenylmethylsulphonylfluoride. This protease is capable of activating both proaerolysin and GCAT and, as indicate above, this may be one of its most important contributions to virulence.

A second serine protease has been reported by Rodriguez and co-workers[48], which they found to be 22 kDa and heat-stable. It was lethal when injected into rainbow trout. As with the metalloprotease described in the same article by these authors, the purified serine esterase contained several bands on isoelectric focusing, suggesting that it had been partially degraded, or that it was contaminated with LPS.

10.4 ENTEROTOXIN

In addition to those discussed above, a great many other *Aeromonas* extracellular proteins which could be considered toxins have been reported in the literature. However, in most cases they have not been studied further, and we therefore know little about their structure and function. Among the most frequently reported 'activities' found in *Aeromonas* culture supernatants is that of an enterotoxin. However, in some cases, the nature of the protein is unclear. Whether or not *Aeromonas* spp. produce a toxin analogous to the heat-labile enterotoxins of *V. cholerae* and *E. coli* has not been convincingly established. Rose and colleagues[57,58] described the toxin they have cloned as an enterotoxin related to cholera toxin, but it is clear that this toxin is, in fact, aerolysin as it has the same *N*-terminal sequence as aerolysin, and cross-reacts with an anti-aerolysin antibody. There is no reason to believe that aerolysin is a true enterotoxin. Unlike the enterotoxins, aerolysin is cytolytic rather than cytotonic and it has no sequence homology with the *Vibrio* or *Escherichia* enterotoxins. What is more, aerolysin is a homodimer in solution while the others are heterohexamers. Finally, no enzyme activity has been associated with aerolysin, and there is no evidence that it can directly cause a change in intracellular cyclic AMP levels. Unfortunately, the presence of aerolysin in the culture supernatants of the majority of *Aeromonas* strains is likely to continue to cause confusion in studies of the presence and/or properties of *Aeromonas* cytotonic enterotoxins.

There is some evidence that a small fraction of *A. hydrophila* isolates may be capable of enterotoxin production. Ljungh and colleagues[59] obtained a partially purified preparation from *A. hydrophila* that was free of cytolytic activity, and which caused intestinal fluid secretion without mucosal damage in rabbit and rat ileal loops, and gave a positive response in the rabbit skin test. This preparation caused cytotonic alterations in adrenal Y1 cells, accompanied by increased cAMP content. In related studies, Honda and co-authors[60] and Potomski and colleagues[61] found that 5–20% of the *A. hydrophila* strains produced a factor that cross-reacted with cholera toxin antibody. It caused fluid accumulation in mice and elongation of Chinese hamster ovary (CHO) cells. It is worth noting that Rose and co-workers[58] reported that their enterotoxin (that is, aerolysin) also cross-reacted with a cholera toxin antibody. Since these proteins are clearly completely unrelated, doubt is cast on the significance of these cross-reactivity observations.

In an effort to establish firmly the presence of a true enterotoxin in *Aeromonas* strains, Chakraborty and colleagues[62] cloned enterotoxic, cytotoxic and haemolytic activities from a strain of *A. hydrophila* as a means of convincingly separating the activities for analysis. It was found that the enterotoxin activity caused elongation of CHO cells as opposed to the rounding caused by haemolytic or cytotoxic toxins, and caused fluid accumulation in both rabbit ileal loop and suckling mouse assays that was not caused by the cytotoxin or haemolysin clones. It was also demonstrated, however, that the toxin was resistant to heat denaturation at 56°C but not at 100°C, which, in addition to differences in the kinetics of fluid accumulation, suggested that it is unrelated to either the heat-labile or heat-stable enterotoxins of *E. coli*. Given these results, it is unfortunate that no further studies with this clone have been reported.

Finally, in the molecular age it is perhaps telling that there are no reports of the demonstration of an enterotoxin gene in *Aeromonas* spp. using oligonucleotide probes based on the sequence of the *Vibrio* or *Escherichia* genes, that no gene with sequence homology to the classical enterotoxins has been cloned from *Aeromonas*, and that no amino acid sequence corresponding to these enterotoxins has been obtained for any *Aeromonas* protein.

10.5   TOXIN SECRETION BY *AEROMONAS* SPP.

Molecular studies on extracellular, as opposed to extracytoplasmic, secretion of proteins by *Aeromonas* spp. have used the same extracellular proteins studied as toxins as secretion models. One of the first such studies examined aerolysin secretion. A number of mutants that could not export aerolysin, a phospholipase, or proteolytic activity were isolated using chemical mutagenesis[63]. Pulse chase analysis showed that the mutants, like the wild type, synthesized and translocated aerolysin across the inner membrane in an energy-dependent manner. The labelled protein could be osmotically shocked from the mutants, but could not be digested by proteases added to a cell suspension. Finally, the cloned aerolysin gene revealed the presence of a 'standard' amino terminal signal sequence, and when synthesized in *E. coli*, aerolysin was translocated only across the inner membrane and could be recovered in the periplasmic fraction released by osmotic shock[64].

These results led to the hypothesis that aerolysin is translocated across the envelope in two steps, the first involving signal sequence and *sec*-dependent translocation across the inner membrane, and the second, again dependent on extragenic factors, involving specific transfer across the outer membrane. Most, if not all, other *A. hydrophila* extracellular proteins are probably secreted in a similar manner, given that the secretion mutants are pleiotropically defective in extracellular export[63,65], and that each of the extracellular proteins that has been cloned also contains a normal amino-terminal signal sequence[1,64].

There are now a number of research groups interested specifically in Gram-negative extracellular export. Three major pathways for the specific transport of proteins across the outer membrane have been identified, and they are represented in many different bacteria. The diversity of these pathways has recently been reviewed[66].

Perhaps the most widespread of the three extracellular secretion systems includes the *A. hydrophila exe* genes and is termed the general secretory pathway or the Type II secretion system[66]. This system, for which the *pul* genes of *Klebsiella oxytoca* represent the first isolated and best studied example, is *sec* gene and signal sequence-dependent and in fact can be viewed as an extension of the inner membrane translocation pathway utilized by all bacteria[67,68]. The genes that encode this pathway in *Aeromonas*, as well as the other bacteria in which it is found, are chromosomally encoded[63]. The system is composed of (at least) a large operon of 12 genes, *exeC–exeN* in *A. hydrophila*[69] and it has also been cloned from *A. salmonicida*[70]. All of the encoded proteins are either predicted or have been shown to be localized to the envelope, but only one of them (ExeD and its homologues) contains a signal sequence. In addition, two further genes, *exeA* and *exeB*, are required in *A. hydrophila*[71] but have not yet been demonstrated to play a role in other bacteria.

Although it has been well established that the *pul* genes and their homologues are required for extracellular secretion in the bacteria that contain them, little is yet known about how they function. In what may be a clue to the function of the large operon, the most detailed information concerns a group of the Exe proteins and their homologues which bear strong homology to proteins involved in the assembly and the structure of the adhesive type IV pili which many bacteria, including *Aeromonas*, elaborate[69,72]. Each of the operons encodes a number of proteins that contain consensus prepilin cleavage sequences at their amino termini, although the remainder of the proteins do not bear homology to the type IV pilin structural protein. Amongst the *A. hydrophila* Exe proteins, five of them (ExeG–K) contain consensus prepilin cleavage sequences at their amino termini and the ExeG protein was shown to be processed by the *P. aeruginosa* prepilin peptidase[69]. As might be expected from the presence of these prepilin sequences in the type II secretion operons, it has been shown that the prepilin peptidase itself is required for the functional assembly of this secretion pathway. In *A. hydrophila*, the gene for this peptidase has recently been cloned, and, as is found in *P. aeruginosa*, was shown to form part of a type IV prepilin assembly operon[73-75].

The ExeE protein and its homologues contain a consensus ATP-binding site, which has been shown to be required for the function of the protein in the secretion process[76,77]. Since it had been shown that aerolysin secretion across the outer membrane of *A. salmonicida* could be inhibited by proton ionophores[78], it was proposed that ExeE might act to transduce the energy derived from ATP hydrolysis to other Exe proteins in the membranes which were responsible for the actual translocation across the outer membrane[79]. It has now been shown that EpsE, the ExeE homologue of *V. cholerae*, is an autokinase, and furthermore that it associates with the inner membrane, but only if EpsL is also being produced by the bacteria[77].

An additional phenotype of the strains containing mutations in the *exeC–N* genes is a radical decrease in the quantities of major outer membrane proteins, especially the OmpF homologue Protein II[80]. Cells lysed during osmotic shock procedures done to determine the localization of aerolysin entrapped in the cells of the non-secretory mutant L1.97 (*exeE::Tn5–751*) indicating that the cells were also osmotically fragile, an effect probably caused by the altered outer membrane structure[79]. In a recent article it was shown that these *exe* cells appear to be specifically downregulated with respect to Protein II synthesis, since it was shown that *exeE* mutants were able to induce and properly assemble both the LamB homologue and the PhoE homologue of this bacterium under the appropriate inducing conditions[81]. The fact that mutations in the *exeC–N* genes cause the downregulation of Protein II, whereas mutations in the *exeAB* genes do not indicates clearly that the two loci affect different aspects or stages of the Type II secretion pathway[71].

The attributes of the secreted proteins that are important for translocation across the outer membrane have not yet been determined, although a number of interesting findings which bear on this question have been reported. Wong and Buckley[78] showed that the alteration of a single Trp residue at position 227 of mature aerolysin to a Gly or Leu caused the accumulation of the active (and therefore not radically altered in structure) protein in the outer membrane of cells producing it. In addition, it was recently reported that aerolysin is exported from the periplasm as the mature dimer, which could be crosslinked by disulphide bonds[82]. These results suggest that aerolysin has folded into its final, quaternary structure before passing across the outer membrane. This would imply that the secretion 'signal' that presumably differentiates secretory from true periplasmic proteins is contained in all or part of the three-dimensional structure of the protein rather than in a linear sequence of amino acids. It will be of great interest to learn what such a signal might consist of, especially since it has also been shown that *A. hydrophila* can secrete the normally periplasmic alkaline phosphatase of *E. coli*, suggesting that it too contains this signal structure[83].

10.6 REFERENCES

1   Thornton J, Howard SP, Buckley JT. Molecular cloning of a phospholipid–cholesterol acyltransferase from *Aeromonas hydrophila*. Sequence homologies with lecithin–cholesterol acyltransferase and other lipases. *Biochim Biophys Acta* 1988; **959**: 153–9.

2   Anguita J, Rodriguez LB, Naharro G. Purification, gene cloning, amino acid sequence analysis, and expression of an extracellular lipase from an *Aeromonas hydrophila* human isolate. *Appl Environ Microbiol* 1993; **59**: 2411–17.

3   Ueda M, Kawaguchi T, Arai M. Molecular cloning and nucleotide sequence of the gene encoding chitinase II from *Aeromonas* sp. NO.1OS-24. *J Ferment Bioeng* 1994; **78**: 205–21.

4   Chang MC, Chang SY, Chen SL, Chuang SM. Cloning and expression in *Escherichia coli* of the gene encoding an extracellular deoxyribonuclease (Dnase) from *Aeromonas hydrophila*. *Gene* 1992; **122**: 175–80.

5   Chang MC, Chang JC, Chen JP. Cloning and nucleotide sequence of an extracellular α-amylase gene from *Aeromonas hydrophila* MCC-1. *J Gen Microbiol* 1993; **139**: 3215–23.

6   Pemberton JM, Gobius KS. Molecular cloning, characterization, and nucleotide sequence of an extracellular amylase gene from *Aermonas hydrophila*. *J Bacteriol* 1988; **170**: 1325–32.

7   Howard SP, Garland WJ, Green MJ, Buckley JT. Nucleotide sequence of the gene for the hole-forming toxin of *Aeromonas hydrophila*. *J Bacteriol*, 1987; **169**: 2869–71.

8   Sanyal SC, Singh DV. Production of haemolysis and its correlation with enterotoxicity in *Aeromonas* spp. *J Med Microbiol* 1992; **37**: 262–7.

9   Bernheimer AW, Avigad LS. Partial characterization of aerolysin, a lytic exotoxin from *Aeromonas hydrophila*. *Infect Immun* 1974; **9**: 1016–21.

10 Buckley JT, Halasa LN, Lund KD, MacIntyre S. Purification and some properties of the haemolytic toxin aerolysin. *Can J Biochem* 1981; **59**: 430–5.

11 Buckley JT. Purification of cloned proaerolysin released by a low protease mutant of *Aeromonas salmonicida*. *Biochem Cell Biol* 1990; **68**: 221–4.

12 Husslein V, Huhle B, Jarchau T, Lurz R, Goebel W, Chakraborty T. Nucleotide sequence and translational analysis of the *aerCaerA* region of *Aeromonas sobria* encoding aerolysin and its regulatory region. *Mol Microbiol* 1988; **2**: 507–17.

13 Hirono I, Aoki T. Cloning and characterization of three hemolysin genes from *Aeromonas salmonicida*. *Microb Pathogen* 1993; **15**: 269–82.

14 Chakraborty T, Huhle B, Hof H, Bergbauer H, Goebel W. Marker exchange mutagenesis of the aerolysin determinant in *Aeromonas hydrophila* demonstrates the role of aerolysin in *A. hydrophila*-associated systemic infections. *Infect Immun* 1987; **55**: 2274–80.

15 Parker MW, van der Goot FG, Buckley JT. Aerolysin — the ins and outs of a model channel-forming toxin. *Molec Microbiol* 1995; **19**: 205–12.

16 Vadivelu J, Puthucheary SD, Phipps M, Che, YW. Possible virulence factors involved in bacteraemia caused by *Aeromonas hydrophila*. *J Med Microbiol* 1995; **42**: 171–4.

17 Howard SP, Buckley JT. Activation of the hole-forming toxin aerolysin by extracellular processing. *J Bacteriol* 1985; **163**: 336–40.

18 van der Goot FG, Lakey J, Pattus F, Kay CM, Sorokine O, Van Dorsselaer A, Buckley JT. Spectroscopic study of the activation and oligomerization of the channel-forming toxin aerolysin: identification of the site of proteolytic activation. *Biochemistry* 1992; **31**: 8566–70.

19 Price NC, Banks RM, Campbell CM, Duncan D, Stevens L. The specificity of the major (70 kDa) protease secreted by *Aeromonas salmonicida*. *J Fish Dis* 1990; **13**: 49–58.

20 Coleman G, Landon M, Whitby PW. The cloning and nucleotide sequence of the serine protease gene (*aspA*) of *Aeromonas salmonicida* spp. *salmonicida FEMS Microbiol Lett* 1992; **99**: 65–72.

21 Parker MW, Buckley JT, Postma JPM, Tucker AD, Leonard K, Pattus F, Tsernoglou D. Structure of the *Aeromonas* toxin proaerolysin in its water-soluble and membrane-channel states. *Nature (London)* 1994; **367**: 292–5.

22 Chevrier B, Schalk C, D'Orrchymont H, Rondeau JM, Moras D, Tarnus C.

Crystal structure of *Aeromonas proteolytica* aminopeptidase: a prototypical member of the co-catalytic zine enzyme family. *Structure* 1994; **2**: 283–93.

23 van der Goot FG, Ausio J, Wong KR, Pattus F, Buckley JT. Dimerization stabilizes the pore-forming toxin aerolysin in solution. *J Biol Chem* 1993; **268**: 18272–9.

24 Ballard J, Crabtree J, Roe BA, Tweten RK. The primary structure of *Clostridium septicum* alpha-toxin exhibits similarity with that of *Aeromonas hydrophila* aerolysin. *Infect Immun* 1995; **63**: 340–4.

25 Bernheimer AW, Avigad LS, Avigad G. Interactions between aerolysin, erythrocytes and erythrocyte membranes. *Infect Immun* 1975; **11**: 1312–19.

26 Howard SP, Buckley JT. Membrane glycoprotein receptor and hole-forming properties of a cytolytic protein toxin. *Biochemistry* 1982; **21**: 1662–7.

27 Gruber HJ, Wilmsen HU, Cowell S, Schindler H, Buckley JT. Partial purification of the rat erythrocyte receptor for the channel-forming toxin aerolysin and reconstitution into planar lipid bilayers. *Mol Microbiol* 1994; **14**: 1093–1011.

28 Haag F, Koch-Nolte F, Kühl M, Lorenzen S, Thiele H-G. Premature stop codons inactivate the RT6 genes of the human and chimpanzee species. *J Mol Biol* 1994; **243**: 537–46.

29 Pearson MW, Saya LE, Howard SP, Buckley JT. The use of aerolysin toxin as an aid for visualization of low numbers of African trypanosomes in whole blood. *Acta Tropica* 1982; **39**: 73–7.

30 Garland WJ, Buckley JT. The cytolytic toxin aerolysin must aggregate to disrupt erythrocytes, and aggregation is stimulated by human glycophorin. *Infect Immun* 1988; **56**: 1249–53.

31 Wilmsen HU, Pattus F, Buckley JT. Aerolysin, a hemolysin from *Aeromonas hydrophila*, forms voltage-gated channels in planar lipid bilayers. *J Membr Biol* 1990; **115**: 71–81.

32 MacIntyre S, Buckley JT. Presence of glycerophospholipid–cholesterol acyltransferase and phospholipase in culture supernatant of *Aeromonas hydrophila. J Bacteriol* 1978; **135**: 402–7.

33 MacIntrye S, Trust TJ, Buckley JT. Distribution of glycerophospholipid–cholesterol acyltransferase in selected bacterial species. *J Bacteriol* 1979; **139**: 132–6.

34 Buckley JT. Substrate specificity of bacterial glycerophospholipid:cholesterol acyltransferase. *Biochemistry* 1982; **21**: 6699–703.

35 Buckley JT. On the mechanism of action of bacterial glycerophospholipid:cholesterol acyltransferase. *Biochemistry* 1983; **22**: 5490–4.

36 Buckley JT, McLeod R, Frohlich J. Action of a microbial glycerophospholipid:cholesterol acyltransferase on plasma from normal and LCAT-deficient subjects. *J Lipid Res* 1984; **25**: 913–18.

37 Huntly PJ, Coleman G, Munro ALS. The nature of the lethal effect on Atlantic salmon, *Salmo salar* L., of a lipopolysaccharide-free phospholipase activity isolated from the extracellular products of *Aeromonas salmonicida. J Fish Dis* 1992; **15**: 99–102.

38 Lee KK, Ellis AE. Glycerophospholipid:cholesterol acyltransferase complexed with lipopolysaccharide (LPS) is a major lethal exotoxin and cytolysin of *Aeromonas salmonicida*: LPS stabilizes and enhances toxicity of the enzyme. *J Bacteriol* 1990; **172**: 5382–93.

39 McIntyre S, Trust TJ, Buckley JT. Identification and characterization of outer membrane fragments released by *Aeromonas* sp. *Can J Biochem* 1980; **58**: 1018–25.

40 Ellis AE. An appraisal of the extracellular toxins of *Aeromonas salmonicida* ssp. *salmonicida. J Fish Dis* 1991; **14**: 265–77.

41 Salte RK, Norberg K, Arnesen JA, Odegaard OR, Eggset G. Serine protease and glycerophospholipid:cholesterol acyltransferase of *Aeromonas salmonicida* work in concert in thrombus formation; *in vitro* the process is counteracted by plasma antithrombin and alpha 2-macroglobulin. *J Fish Dis* 1992; **15**: 215–27.

42 Upton C, Buckley JT. A new family of lipolytic enzymes? *Trends Biochem Sci* 1995; **233**: 178–9.

43 Ausio J, van der Goot FG, Buckley JT. Physical and chemical characterization of the oligomerization state of the *Aeromonas hydrophila* lipase/acyltransferase. *FEBS Lett* 1993; **333**: 296–300.

44 Hilton S, McCubbin WD, Kay CM, Buckley JT. Purification and spectral study of a microbial fatty acyltransferase: activation by limited proteolysis. *Biochemistry* 1990; **29**: 9072–8.

45 Eggset G, Bjornsdottir R, Leifson RM, Arnesen JA, Coucheron DH, Jorgensen TO. Extracellular glycerophospholipid:cholesterol acyltransferase from *Aeromonas salmonicida*: activation by serine protease. *J Fish Dis* 1994; **17**: 17–29.

46 Leung KY, Stevenson RMW. Tn-5 induced protease-deficient strains of *Aeromonas hydrophila* with reduced virulence for fish. *Infect Immun* 1988; **56**: 2639–44.

47 Loewy AG, Santer UV, Wieczorek M, Blodgett JK, Jones SW, Cheronis JC. Purification and characterization of a novel zinc-proteinase from cultures of *Aeromonas hydrophila*. *J Biol Chem* 1993; **268**: 9071–8.

48 Rodriguez LA, Ellis AE, Nieto TP. Purification and characterization of an extracellular metalloprotease, serine protease and haemolysin of *Aeromonas hydrophila* strain B$_{32}$: all are lethal for fish. *Microb Pathogen* 1992; **13**: 17–24.

49 Arnesen JA, Eggset G, Jorgensen TO. Partial purification and characterisation of extracellular metalloproteases from *Aeromonas salmonicida* ssp. *salmonicida*. *J Fish Dis* 1995; **18**: 283–95.

50 Prescott JM, Wilkes SH, Wagner FW, Wilson KJ. *Aeromonas* aminopeptidase: Improved isolation and some physical properties. *J Biol Chem* 1971; **246**: 1756–64.

51 Guenet C, Lepage P, Harris BA. Isolation of the leucine aminopeptidase gene from *Aeromonas proteolytica*. *J Biol Chem* 1992; **267**: 8390–5.

52 Bayliss ME, Wilkes SH, Prescott JM. *Aeromonas* neutral protease: specificity toward extended substrates. *Arch Bioch Biophys* 1980; **204**: 214–19.

53 Sakai DK. Loss of virulence in a protease-deficient mutant of *Aeromonas salmonicida*. *Infect Immun* 1985; **48**: 146–52.

54 Drinan EM, Smith PR. Histopathology of a mutant *Aeromonas salmonicida* infection in Atlantic Salmon *(Salmo salar)*. In Ellis AE (ed.) *Fish and Shellfish Pathology*. Academic Press, London: 1985: 79–83.

55 Rivero O, Anguita J, Mateos D, Paniagua C, Naharro G. Cloning and characterization of an extracellular temperature-labile serine protease gene from *Aeromonas hydrophila*. *FEMS Microbiol Lett* 1991; **81**: 1–8.

56 Whitby PW, Landon M, Coleman G. The cloning and nucleotide sequence of the serine protease gene (*aspA*) of *Aeromonas salmonicida* ssp. *salmonicida FEMS Microbiol Lett* 1992; **99**: 65–72.

57 Rose JM, Houston CW, Kurosky A. Bioactivity and immunological characterization of a cholera toxin-cross-reactive cytolytic enterotoxin from *Aermonas hydrophila*. *Infect Immun* 1989; **57**: 1170–6.

58 Rose JM, Houston CW, Coppenhaver DH, Dixon JD, Kurosky A. Purification and chemical characterization of a cholera toxin-cross-reactive cytolytic enterotoxin produced by a human isolate of *Aeromonas hydrophila*. *Infect Immun* 1989; **57**: 1165–9.

59 Ljungh Å, Wretlind B, Möllby R. Separation and characterization of enterotoxin

and two haemolysins from *Aermonas hydrophila. Acta Pathol Scand Sect B* 1981; **89**: 387–97.

60 Honda T, Sato M, Nishimura T, Higashitsutsumi M, Fukai K, Miwatani T. Demonstration of cholera toxin-related factor in cultures of *Aeromonas* species by enzyme-linked immunosorbent assay. *Infect Immun* 1985; **50**: 322–3.

61 Potomski J, Burke V, Robinson J, Fumarola D, Miragliotta G. *Aeromonas* cytotonic enterotoxin cross-reactive with cholera toxin. *J Med Microbiol* 1987; **23**: 179–86.

62 Chakraborty T, Montenegro MA, Sanyal SC, Helmuth R, Bulling E, Timmis KN. Cloning of enterotoxin gene from *Aeromonas hydrophila* provides conclusive evidence of production of a cytotoxic enterotoxin. *Infect Immun* 1984; **46**: 435–41.

63 Howard SP, Buckley JT. Intracellular accumulation of extracellular proteins by peliotropic export mutants of *Aeromonas hydrophila. J Bacteriol* 1983; **154**: 413–18.

64 Howard SP, Buckley JT. Molecular cloning and expression in *Escherichia coli* of the structural gene for the haemolytic toxin aerolysin from *Aeromonas hydrophila. Mol Gen Genet* 1986; **204**: 289–95.

65 Jiang B, Howard SP. Mutagenesis and isolation of *Aeromonas hydrophila* genes which are required for extracellular secretion. *J Bacteriol* 1991; **173**: 1241–9.

66 Salmond GPC, Reeves PJ. Membrane traffic wardens and protein secretion in Gram-negative bacteria. *Trends Biochem Sci* 1993; **18**: 7–12.

67 Pugsley AP. The complete general secretory pathway in Gram-negative bacteria. *Microbiol Rev* 1993; **57**: 50–108.

68 Schatz PJ, Beckwith J. Genetic analysis of protein export in *Escherichia coli.* Annu Rev Genet 1990; **24**: 215–48.

69 Howard SP, Critch J, Bedi A. Isolation and analysis of eight *exe* genes and their involvement in extracellular protein secretion and outer membrane assembly in *Aeromonas hydrophila. J Bacteriol* 1993; **175**: 6695–703.

70 Karlyshev AV, MacIntyre S. Cloning and study of the genetic organization of the exe gene cluster of *Aeromonas salmonicida. Gene* 1995; **158**: 77–82.

71 Jahagirdar R, Howard SP. Isolation and characterization of a 2nd exe operon required for extracellular protein secretion in *Aeromonas hydrophila. J Bacteriol* 1994; **176**: 6819–26.

72 Hokama A, Iwanaga M. Purification and characterization of *Aeromonas sobria* pili, a possible colonization factor. *Infect Immun* 1991; **59**: 3478–83.

73 Bally M, Ball G, Badere A, Lazdunski A. Protein secretion in *Pseudomonas aeruginosa*: The *xcpA* gene encodes an integral inner membrane protein homologous to *Klebsiella pneumoniae* secretion function protein PulO *J Bacteriol* 1991; **173**: 479–86.

74 Nunn D, Bergman S, Lory S. Products of three accessory genes, *pilB, pilC* and *pilD*, are required for biogenesis of *Pseudomonas aeruginosa* pili. *J Bacteriol* 1990; **17**: 2911–19.

75 Pepe CM, Eklund MW, Strom MS. Cloning of an *Aeromonas hydrophila* type IV pilus biogenesis gene cluster: complementation of pilus assembly functions and characterization of a type IV leaderpeptidase/*N*-methyltransferase required for extracellular protein secretion. *Mol Microbiol* 1996; in press.

76 Possot O, PugsleyAP. Molecular characterization of pule, a protein required for pullulanase secretion. *Mol Microbiol* 1994; **12**: 287–99.

77 Sandkvist M, Bagdasarian M, Howard SP, DiRita VJ. Interaction between the autokinase EpsE and EpsL in the cytoplasmic membrane is required for extracellular secretion in *Vibrio cholerae. EMBO J* 1995; in press.

78 Wong KR, Buckley JT. Proton motive force involved in protein transport across the outer membrane of *Aeromonas salmonicida. Science* 1989; **246**: 654–6.

79 Jiang B, Howard SP. The *Aeromonas hydrophila* exeE gene, required both for protein secretion and normal outer membrane biogenesis, is a member of a general secretion pathway. *Mol Microbiol* 1992; **6**: 1351–61.

80 Jeanteur D, Gletsu N, Pattus F, Buckley JT. Purification of *Aeromonas hydrophila* major outer-membrane proteins: *N*-terminal sequence analysis and channel-forming properties. *Mol Microbiol* 1992; **6**: 3355–63.

81 Howard SP, Meiklejohn HG. Effect of mutations in the general secretory pathway on outer membrane protein and surface layer assembly in *Aeromonas spp. Can J Microbiol* 1995; **41**: 525–32.

82 Hardie KR, Schulze A, Parker MW, Buckley JT. *Vibrio* sp. secrete proaerolysin as a folded dimer without the need for disulphide bond formation. *Mol Microbiol* 1995; **17**: 1035–44.

83 Wong KR, Buckley JT. *Aeromonas* spp. can secrete *Escherichia coli* alkaline phosphatase into the culture supernatant, and its release requires a functional general secretion pathway. *Mol Microbiol* 1993; **9**: 955–63.

84 Buckley JT, Halasa LN, MacIntyre S. Purification and partial characterization of a bacterial phospholipid–cholesterol acyltransferase. *J Biol Chem* 1982; **59**: 430–6.

85 Aoki T, Hirono I. Clonding and characterization of the haemolysin determinants from *Aeromonas hydrophila. J Fish Dis* 1991; **14**: 303–12.

86 Nieto TP, Santos Y, Rodriguez LA, Ellis AE. An extracellular acetycholinesterase produced by *Aeromonas hydrophila* is a major lethal toxin for fish. *Microb Pathog* 1991; **11**: 101–10.

11 The *Aeromonas hydrophila* Group in Food

SAMUEL A. PALUMBO
Eastern Regional Research Center, Wyndmoor, Pennsylvania, USA

Mention of brand or firm names does not constitute an endorsement by the USDA, over others of a similar nature not mentioned.

Members of the *Aeromonas hydrophila* group occur widely in the aquatic environment, including freshwater, estuaries and marine[1–9]. This close association with water is reflected in the name of the group, *hydrophila*, i.e. water-loving. Factors controlling their numbers include temperature and the trophic state of the water[10–14]. The presence of the *A. hydrophila* group in foods undoubtedly reflects contact of that food with water.

11.1 FOOD

Organisms of the *A. hydrophila* group have been isolated from a wide range of both animal and plant food products, including raw red meats (Tables 11.1 and 11.2), poultry (Table 11.3), finned fish (Table 11.4), seafood (Table 11.5),

Table 11.1. Isolation and incidence of the *A. hydrophila* group in red meats stored under various modified atmospheres

Product	Organism isolated	Type of study; conditions	Reference
Pork loins	*Aeromonas, Ah*	CO_2 storage; 0 and 4°C	15
Pork loins	*Ah*	N_2 storage; 4°C	16
Pork loins	*Aeromonas*	Vacuum and N_2 packed storage; 0°C and above	17
Fresh pork	*Ah*	Vacuum packed storage; 5°C	18
Lamb carcasses	*Aeromonas* spp.	N_2-storage; 0°C	19
Boxed beef	*Aeromonas* spp.	Vacuum packed storage; 4°C	20
Top sirloin	*Aeromonas*	Vacuum packed storage; 7°C	21

Ah, A. hydrophila.

The Genus Aeromonas. Edited by B. Austin, M. Altwegg, P.J. Gosling and S. Joseph
© 1996 John Wiley & Sons Ltd

Table 11.2. Isolation and incidence of the *A. hydrophila* group in raw red meat

Product	Organism isolated	Type of study; conditions	Reference
Ground pork, beef, lamb, veal	*Ah, As, Ac, Aeromonas* spp.	Survey and storage; 5°C	22,23,24, 25,26,27
Sliced beef; wholesale and retail cuts of beef	*Aeromonas*	Survey and storage; refrigerated, 0–10°C	28,29,30
Pork, beef, lamb	*Ah, As, Ac, Aeromonas*	Survey	31,32,33, 34,35
Ground meat	*Ah, Ac*	Survey	36
Raw pork	*Ah, Ac, As, A. jandaei, Aeromonas* spp.	Survey	37
Lamb cuts and carcasses	*Ah, As, Ac*	Survey	38
Ground antelope	*Aeromonas* spp.	5°C storage	26

Ah, A. hydrophila; As, A. sobria; Ac, A. caviae.

Table 11.3. Isolation and incidence of the *A. hydrophila* group in poultry

Product	Organism isolated	Type of study; conditions	Reference
Chicken	*Ah*	Survey and storage; 5°C	22
Chicken	*Aeromonas*	Storage; 4.4°C	40
Chicken	*Aeromonas* spp; *Ah, As, Ac*	Survey	23, 24, 25 31, 32, 33
Raw chicken	*Ah, As, Ac, A. jandaei, Aeromonas* spp., *A. veronii* biotype sobria	Survey	37
Broiler carcasses	*Ah*	Survey	41
Turkey	*Ah*	Survey	31

Ah, A. hydrophila; As, A. sobria; Ac, A. caviae.

dairy products (Table 11.6), vegetables (Table 11.7), and miscellaneous foods (Tables 11.8 and 11.9). As can be seen from Tables 11.1 to 11.9, this group can be isolated from a very wide range of food products, both fresh and processed (cooked, cured, fermented). Organisms were detected both in

Table 11.4.   Isolation and incidence of the *A. hydrophila* group in finned fish

Product	Organism isolated	Type of study; conditions	Reference
Various fish – freshwater and seawater	*Ah, As, Ac*	Survey	27, 23, 32, 30, 42
Fish	*Aeromonas*	Survey	24
River fish	*Ah*	Survey	9
Carp	*Ah, Ac, As, A. jandaei, Aeromonas* spp.	Survey	36
Irradiated fillets	*Aeromonas*	Storage at 10 –12°C	43
Herring fillets	*Ah, Aeromonas* spp.	Air and CO_2 storage; 10 –12°C	44

Ah, A. hydrophila; As, A. sobria; Ac, A. caviae.

Table 11.5.   Isolation and incidence of the *A. hydrophila* group in seafood

Product	Organism isolated	Type of study; conditions	Reference
Cockles	*Ah*	Outbreak	45
Shrimp	*Aeromonas* spp.	Outbreak	46
Scallops	*Ah*	Survey, storage; 5°C	22
Crab	*Aeromonas, Ah*	Survey	47, 48
Shellfish	*Ah, Ac*	Survey	30
Oysters	*Ah*	Survey, storage; 4.4°C	49
Oysters, clams, cockles	*Aeromonas, Ah*	Survey	50, 51, 52
Oysters, crabs, shrimp, hard clams, freshwater clams	*Ah, As, Ac*	Survey	53

Ah, A. hydrophila; As, A. sobria; Ac, A. caviae.

surveys of products purchased at retail and from food products held at storage temperatures ranging from 0 to 12°C. Their presence in heat-processed foods such as pasteurized milk and ice-cream (Table 11.6) and ham and cooked meats (Table 11.9) undoubtedly reflects post-processing recontamination because, as will be shown below, these bacteria are relatively heat-sensitive and are destroyed by the usual pasteurization treatments given to meat and dairy products.

Table 11.6.   Isolation and incidence of the *A. hydrophila* group in dairy products

Product	Organism isolated	Type of study; conditions	Reference
Raw milk	*Ah*	5°C; storage/survey	22
Raw milk	*Ah, As, Ac*	Survey	33, 54
Ice-cream	*Ac*	Survey	55
Pasteurized milk	*Aeromonas* spp.	7°C storage; 12°C storage	56, 57, 58 54,
Pasteurized milk	*Ah, Ac, A. schubertii*	Survey	59
White cheese	*Ah, Ac, As*	Survey	59
Whipped cream	*Aeromonas* spp.	Survey	6

Ah, A. hydrophila; As, A. sobria; Ac, A. caviae.

Table 11.7.   Isolation and incidence of the *A. hydrophila* group in vegetables and vegetable products

Product	Organism isolated	Type of study; conditions	Reference
Vegetables	*Ah, As, Ac* *Aeromonas* spp.	Survey; storage at 5°C	60
Vegetables	*Aeromonas* spp.	Survey	35
Vegetables	*Ah, As, Ac*	Survey	23
Spices and herbs	*Aeromonas* spp.	Survey	61

Ah, A. hydrophila; As, A. sobria; Ac, A. caviae.

Until the last 10 years, the bacteria isolated were often identified simply as *Aeromonas* spp. As the taxonomy and nomenclature of this group became clarified[64], *A. hydrophila*, *A. sobria* and *A. caviae* were identified based on biochemical characteristics. Most recently, taxonomy based on DNA hybridization groups was developed for the *A. hydrophila* group and additional genotypes were identified and these have been isolated from foods, including *A. schubertii* from pasteurized milk[59], *A. jandaei* from raw pork and chicken and from carp[37] and *A. veronii* biotype sobria from raw chicken[37].

As indicated in Tables 11.1–11.9, members of the *A. hydrophila* group as well as the newly defined genospecies have been isolated from various fresh and processed foods. Three approaches can be used to determine the

Table 11.8. Isolation and incidence of the *A. hydrophila* group in miscellaneous products

Product	Organism isolated	Type of study; conditions	Reference
Pre-prepared salads	*Ah, As, Ac*	Survey	32
Mayonnaise-based salads	*Aeromonas* spp.	Survey	6
Meat/seafood deli	*Aeromonas* spp.	Survey	6

Table 11.9. Isolation and incidence of the *A. hydrophila* group in miscellaneous meat products

Product	Organism isolated	Type of study; conditions	Reference
Offal	*Ah, As, Ac*	Survey	37
Offal	*Ah, As, Ac*	Survey	32
Cooked meats	*Ah, As, Ac*	Survey	32
Cooked tripe	*Aeromonas* spp.	Survey	62
Cooked sliced ham	*Ah, Ac, As*	Survey	63
Mettwurst	*Ah*	Survey	63
Smoked cooked sausage	*Ah, Ac*	Survey	63
Fresh pork sausage	*Ah, As, Ac*	Survey	24
Fresh pork sausage	*Ah*	Survey; storage at 5°C	22
Chicken liver	*Ah*	Survey; storage at 5°C	22
Calf liver	*Ah*	Survey; storage at 5°C	22

Ah, A. hydrophila; As, A. sobria; Ac, A. caviae.

incidence of specific bacteria and the bacterial microbiota of foods: (1) a general, all-purpose plating medium, followed by identification and verification of representative colony types; (2) a selective plating medium specific for the bacteria/bacterial group desired, which may or may not be followed by verification of the isolates; or (3) an enrichment procedure, either selective or non-selective, when low numbers are anticipated or if injured or damaged cells may be encountered. Bacteria of the *A. hydrophila* group have been readily isolated from foods using any of the three approaches.

Many of the bacteriological surveys of refrigerated foods, especially raw red meats[15,16,19–21,28–30], poultry[39,40], fish[43,44], vegetables[35], crabs[47] and oysters[51] have employed the first approach. Besides being very labour-intensive, this approach often yields presence/absence information and not an actual number of bacteria per gram/millilitre of food.

The usefulness of the second approach depends on having a selective medium specific for the bacterium or group of bacteria desired. Many plating media have been developed for the isolation of the *A. hydrophila* group in the clinical laboratory (see Refs 65 and 66 for a description of some of these media). Several additional media have been described for use on clinical specimens. As in the case of other bacterial foodborne pathogens, media developed for use on clinical specimens do not peform very well when used for food samples, primarily because of the need for quantitative recovery of bacteria from food and the presence of competing microflora in the food.

Many recent food surveys have employed one or more selective media. Some of the *Aeromonas*-specific media used to isolate bacteria from food products include: Rimler–Shotts agar for oysters[50]; mA agar of Rippey and Cabelli for poultry[41]; starch ampicillin agar (SAA) of Palumbo and colleagues[22] for a wide variety of foods of animal and plant origin by several investigators[6,24,25,34,38,39,60,67,68], bile salts brilliant green agar[23] and bile salts brilliant green starch agar (developed because of an apparent problem with swarming of *Proteus* when certain samples are plated on SAA[69] for a variety of human, food and environmental specimens). Other selective, non-specific media have also been used on food products, including thiosulphate–citrate–bile salts–sucrose agar (a medium typically used for Vibrios)[49] and seafood and aquaculture foods[53]. In addition, MacConkey agar and xylose deoxycholate citrate agar, general purpose Gram-negative media, have also been used in a limited number of studies[33,52,55]. Stern and co-workers[25] attempted to use cefsulodin–irgasan–novobiocin (CIN) agar to isolate *Aeromonas* spp. from livestock faeces and ground beef, but observed a much lower quantitative recovery on CIN agar than on SAA. In addition to yielding quantitative recovery of *Aeromonas* spp. from food samples, the usefulness of SAA can also be evaluated by the number (percentage) of colonies of the *A. hydrophila* group bacteria that can be verified. Several investigators[22,23,34,37,40,60,68] have determined that essentially all the presumptive colonies on SAA were verified as *A. hydrophila* group.

The third approach, use of an enrichment broth (selective or non-selective), is needed when low numbers of *Aeromonas* are anticipated, there are large numbers of competing background microflora or there are injured cells that must be allowed to recover/repair. The general-purpose, non-selective alkaline peptone water (10 g/l peptone, 10 g/l NaCl, pH 8.5; BAM) has been favoured by many investigators, including Nishikawa and Kishi[23] for a variety of food and water samples, Wong and colleagues[53] for fish and seafood, Krovacek and colleagues[32] for fish, raw meat, poultry and vegetables, Hood and associates[49] for oysters, Hunter and Burge[55] for ice-cream, Majeed and co-workers[68] for lamb faecal samples and carcasses, and Fricker and Tompsett[33] for fish, raw and cooked meats and pre-prepared salads. Other enrichment broths have also been employed. Pathak and co-authors[9] used Rimler–Shotts medium for fish, and tryptic soy broth plus

10–30 mg/l ampicillin (TSBA) has been used for raw meat and poultry[24] and frozen oysters[50]. Abeyta and colleagues[50] have suggested that TSBA was also useful if freeze-injured cells were present in the food. Using TSBA, they were able to recover *A. hydrophila* from oysters frozen for 18 months.

In most instances, these bacteria are not considered part of the normal gut flora of either meat animals or humans. Stern and co-workers[25] found a very low incidence (four of 99 samples) of *Aeromonas* spp. in the faeces of red meat animals and poultry. Although various investigators have isolated *A. hydrophila* from human faeces, they have not considered it part of the normal intestinal flora[70,71]. Thus, water probably represents the major if not sole source of these bacteria in foods[72].

11.2   CONTROL

11.2.1   GROWTH

The growth of foodborne bacteria (pathogens and spoilage bacteria) can be controlled by various intrinsic and extrinsic parameters of the food[73]. The intrinsic parameters of a food which control the growth of bacteria and other microorganisms include pH, moisture content, oxidation–reduction potential (Eh), nutrient content, presence of antimicrobial compounds, and biological structures (skins of meat animals and the peel of fruits and vegetables). The extrinsic parameters include storage temperature, relative humidity of the environment, and presence and concentration of gases in the environment.

11.2.1.1   Temperature

Food microbiologists have traditionally relied on refrigeration (low-temperature storage, ≤5°C) to prevent the growth of pathogens in foods, and thus maintain the safety of foods. However, over the past 30 or more years, a group of pathogens has been recognized that are capable of growing in foods held at ≤5°C[75]. *A. hydrophila* is included in this group. As is readily evident from the information in Tables 11.1–11.9, strains of the *A. hydrophila* group are readily isolated from a wide range of food products held refrigerated under different conditions. Thus, food microbiologists must depend on other factors individually or a combination of refrigeration and other factors to control the growth of the *A. hydrophila* group in foods. The combined influence of temperature with other factors such as pH, NaCl and sodium nitrite level, and atmosphere has been studied and forms the basis of the polynomial models developed for *A. hydrophila*[76,77].

While the *A. hydrophila* group are often also called mesophilic aeromonads[78], the survey information shown in Tables 11.1–11.9 indicates that these bacteria are readily capable of growth in culture and various foods at 0–5°C (Table 11.10). Also shown in Table 11.10 are the optimum and maximum temperatures for *A. hydrophila*. This group should perhaps be described as psychrotrophic. As will be discussed below, whether the *A. hydrophila* group bacteria grow at a given low temperature in a food is a function of multiple factors.

11.2.1.2 pH

The minimum pH values permitting growth are a function of both culture conditions (NaCl concentration and temperature) and acidulant used (Table 11.11). Acetic acid is inhibitory at higher pH values compared to other acids, which appears directly related to its higher pK_as (Freese and colleagues[82]). The sensitivity to acetic and lactic acids may account for the generally low

Table 11.10. Temperature parameters for growth of the *A. hydrophila* group in culture and food systems

Range/condition (°C)	References
4/5–42	79
28, optimum	78
28, shortest generation time	Palumbo, unpublished observation for *A. hydrophila* K144
37, shortest lag time	Palumbo, unpublished observation for *A. hydrophila* K144
43, in culture	80
0, in lamb carcass purge	19
+1, in culture	81

Table 11.11.   pH values and limiting conditions for *A. hydrophila*

Acid	pH				
	Growth at	Conditions	No growth at	Conditions	Reference
HCl	4.5 (2/10)*	28°C, 0.5% NaCl	n.d.	n.d.	83
HCl	5.5 (4/10)	5°C, 0.5% NaCl	4.5	5°C, 0.5% NaCl	83
HCl	5.0	19°C, 2% NaCl	5.5	5°C, 0.5% NaCl	86
Sulphuric	5.0	19°C, 2% NaCl	5.5	5°C, 0.5% NaCl	86
Acetic	6.0	19°C, 0.5% NaCl	6.0	5°C, 0.5% NaCl	86
Lactic	5.5	19°C, 0.5% NaCl	5.5	5°C, 0.5% NaCl	86
Citric	5.5	19°C, 0.5% NaCl	5.5	5°C, 0.5% NaCl	86
Tartaric	5.5	19°C, 0.5% NaCl	5.5	5°C, 0.5% NaCl	86

* Strains positive/strains tested.

incidence and level of the bacterium in fermented or acidified foods. Aytac and Ozbas[84,85] observed that *A. hydrophila* inoculated into pasteurized milk used for the manufacture of yoghurt died off during storage of the product at 4°C. The decline observed in the product was similar to that observed by Palumbo and Williams[86] in a culture system. High or maximum pH values are usually not of interest to food microbiologists since there are relatively few foods with high pH values. However, Palumbo and colleagues[87] observed that *A. hydrophila* grew equally well in BHI broth at pH 8.2 and 2% (w/v) NaCl as the bacterium did at pH 7.3.

11.2.1.3   NaCl

As with the effect of pH, the inhibitory activity of NaCl can be increased by low temperature and pH. For example, Palumbo and co-workers[83] reported that, at 28°C, some strains of the bacterium will grow at 5% (w/v) NaCl, while at 5°C the strains will only grow at 3% (w/v) NaCl. Santos and colleagues[88] verified the temperature dependency of NaCl inhibition for other strains of *A. hydrophila*.

NaCl exerts its inhibitory action against bacteria and other microorganisms by lowering the water activity (a_w, amount of free water for growth; Sperber[89]) of the growth medium or food. The humectant (substance employed to lower a_w) used can also have an independent inhibitory effect, i.e. it can take larger amounts (lower a_ws) of certain substances to achieve the same inhibitory activity. Santos and colleagues[88] observed that, for strains of *A. hydrophila* at 28°C, the minimum a_w for growth was 0.971 when the a_w of the broth was adjusted with NaCl (4.5%), 0.940 when the a_w was adjusted with glycerol (18%), and 0.967 when adjusted with polyethylene glycol (30%).

11.2.1.4   Nitrite

The preservation of meats by curing (combination of salt and nitrite) almost predates recorded history. Early procedures depended on the chance contamination of the salt used with small amounts of nitrate, which was converted to nitrite, the actual curing agent (responsible for the typical pink colour of cured meats as well as its recognized preservative action), by an appropriate bacterial flora. Current procedures simply add nitrite to meat products directly.

There are only a limited number of studies on the effect of nitrite on the growth of the *A. hydrophila* group bacteria (Palumbo and co-workers[76,77]), and these are generally on the effect of nitrite in combination with other factors. However, examination of the growth responses of *A. hydrophila* K144 under both aerobic and anaerobic conditions indicated that this bacterium is similar to many other bacteria in that it is inhibited by nitrite only at pH values below 6.0[90]. *A. hydrophila* has a generally low incidence in cured meat products and, when it occurs, it is at fairly low levels[63]. The low incidence and level could also reflect sensitivity to NaCl and the thermal processing given to these products.

11.2.1.5   Atmosphere

Popoff[78] described bacteria of the *A. hydrophila* group as facultative anaerobes (a facultative anaerobe is a bacterium that can be grown in the presence or absence of oxygen[91]). The isolation of the *A. hydrophila* group from red meats stored refrigerated under vacuum-, nitrogen-, and CO_2-packaging (Table 11.1) as well as the ready growth of the bacterium in anaerobic culture (broth under a nitrogen atmosphere[77]) support this description. It appears that the primary effect that atmosphere has on the *A. hydrophila* group is on the microflora that develop in the food held under various atmospheres

11.2.1.6   Miscellaneous inhibitors

There have been few studies on the effect of miscellaneous food components and inhibitors on *A. hydrophila*. Stecchini and colleagues[92] observed that the organism was inhibited by the essential oils of clove, coriander, nutmeg and pepper in a model system and by coriander and clove in non-cured cooked pork. Sofos and co-workers[93] determined that *A. hydrophila* was inhibited by liquid smoke isolated from several different species of wood including Douglas fir, birch, southern yellow pine and aspen. Moir and Eyles[94] studied the response of *A. hydrophila* to the food preservatives methyl-*p*-hydroxybenzoate and potassium sorbate; they observed that the minimum inhibitory concentration (MIC) of potassium sorbate was, as anticipated, pH-dependent (MIC of 50 mg/l at pH 5 compared to MIC of 550 mg/l at pH 6). Of the four foodborne psychrotrophic bacteria studied *(Listeria monocytogenes, Yersinia enterocolitica, Pseudomonas putida* and *A. hydrophila)*, *A. hydrophila* was the most sensitive to these two preservatives. Tea polyphenols, crude catechin extracts from green tea and crude theaflavins from black tea were active against a number of foodborne pathogens including *A. sobria*, but were not active against *A. hydrophila*[95]. Venugopal and colleagues[96] studied the effect of the food preservatives butylated hydroxyanisole, propylhydroxy parabenzoate and sodium tripolyphosphate on growth and protease production by *A. hydrophila*, with protease secretion inhibited at lower concentration than growth. Palumbo and co-workers[87] investigated the effect of four polyphosphates (Sodaphos, Hexaphos, sodium pyrophosphate and sodium tripolyphosphate) in conjunction with increased NaCl and lower temperatures on the growth of *A. hydrophila* in a model system and in a food system. In the model system, the polyphosphates individually caused small decreases in generation and lag times; however, 2% polyphosphate combined with 3.5% (w/v) NaCl inactivated the organism. In ground pork, the combination of high NaCl and high polyphosphate restricted the outgrowth of *A. hydrophila* when held at 5°C. As can be seen from the above, most of the compounds offer relatively little protection against the hazard from *A. hydrophila* by themselves; however, their use may be combined with NaCl (see Palumbo and colleagues[87] mentioned above) and/or low pH to provide additional hurdles to growth of this organism. Since many of the compounds are already present in the food or in use, further studies are encouraged.

11.2.2   INACTIVATION

There have been relatively few studies on the behaviour of *A. hydrophila* group bacteria in foods under food processing conditions[97–99]. Heat, acidification (fermentation) and irradiation are most commonly used to destroy bacteria during food preparation.

11.2.2.1   Heat

Palumbo and colleagues[99] reported that the thermal resistance of *A. hydrophila* is similar to that of other Gram-negative bacteria found in foods, and that any heating protocol designed to eliminate *Salmonella* would destroy *A. hydrophila*. When heated in egg yolk or hamburger, Nishikawa and colleagues[98] observed that *A. hydrophila* was more heat-sensitive than *Salmonella typhimurium*, *Staphylococcus aureus*, or *Escherichia coli* 0157:H7; in hamburger, no viable *A. hydrophila* were detected when the internal temperature reached 60°C. Based on the D-values (decimal reduction times) given by Condon and co-workers[97] and Palumbo and colleagues[99], heat-processed foods such as pasteurized dairy products and cooked meats (e.g. luncheon meats) should be free of the *A. hydrophila* group bacteria and they generally are, particularly immediately after processing. However, low numbers of these bacteria can be detected after extended refrigerated storage (Tables 11.6 and 11.9); this undoubtedly reflects recontamination after the heating step.

11.2.2.2   Irradiation

Although currently approved for just a few foods, irradiation represents a promising technology for the destruction of pathogenic and spoilage microorganisms in foods. Palumbo and co-workers[100] studied the radiation resistance of *A. hydrophila* in ground beef and blue fish. They calculated D-values at 2°C of 14–19 kRad. Comparison of these values to those for other Gram-negative pathogens[101] indicates that *A. hydrophila* has radiation resistance similar to other Gram-negative pathogens and any dose used to reduce/eliminate them should reduce/eliminate this bacterium.

11.2.2.3   Acidification

Pathogens can also be inactivated by acid, either added directly or produced by fermentation. As part of a study of the influence of pH on the growth of *A. hydrophila* in broth, the effect of pH adjustment with different acids was investigated[86]. In addition to changes in lag and generation times related to both pH and acidulant. Palumbo and Williams[86] observed that certain combinations of salt, pH and acidulant not only altered the growth kinetics of the bacterium, but also inactivated it (Figure 11.1). The inactivation kinetics appear related to the pK_a of the acid: those with higher pK_as, lactic and acetic, are more lethal than H_2SO_4 or HCl. The lethality of acetic and lactic acids towards *A. hydrophila* has been observed under product conditions[84,85,98]. Nishikawa and colleagues[98] added *A. hydrophila* (starting count of *ca.* 10^6 CFU/ml) to a cucumber–seaweed salad acidified with vinegar; the count of *A. hydrophila* declined to undetectable levels after 7 h at room temperature. Aytac and Ozbas[84,85] added *A. hydrophila* to the

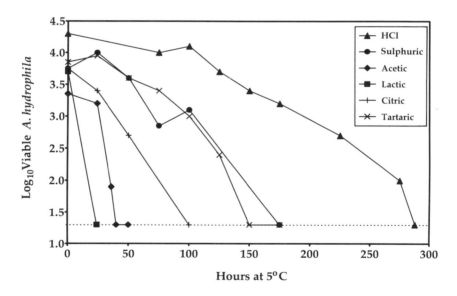

Figure 11.1   Effect of different acids (HCl, sulphuric, tartaric, citric, acetic, and lactic) on the decline in viable count of *A. hydrophila* K144 in brain heart infusion (BHI) broth at pH 5.0 and 5°C (0.5% NaCl). [dashed line = lower limit of detection (Log_{10} = 1.33)]. From *Journal of Food Science* (reference 86). Reproduced with permission.)

pasteurized milk used to prepare yoghurt and acidophilus yoghurt; they observed a dramatic decline during the fermentation (pH 4.65) and then the numbers declined to undetectable levels by day 5 of storage at 4°C (pH 4.50). The organism is generally not identified in, or isolated from, ripened cheese or other fermented dairy products, but it often is not looked for. Freitas and colleagues[59] isolated *Aeromonas* spp. from approximately 1/3 of the white cheese samples (pH 5.1–5.6) in Brazil. The percentage positive isolations were the same as those from pasteurized milk, suggesting that the cheese may become recontaminated during processing. Although the survey data in Table 11.8 indicate that *A. hydrophila* can be isolated from pre-prepared and mayonnaise-based salads (>20% positive), often a description of the product (pH value) is lacking and on many occasions investigators will simply indicate the presence or absence of the bacterium.

11.2.2.4 Chlorine and disinfectants

The response of the *A. hydrophila* group to chlorine seems to represent a paradox. On the one hand, these bacteria are more susceptible to chlorination than *E. coli*[102]; on the other hand, they are isolated from chlorinated drinking water[3] and from waters with negative coliform tests. Several explanations can be offered for this paradox: (a) the bacterium is injured by chlorine and is thus not recovered by the selective media used in the procedure[103]; (b) high organic loading of the water may result in inadequate chlorination[72]; (c) the ability of *A. hydrophila* to grow on very low levels of various organic substrates such as simple sugars, amino acids and free fatty acids as well as proteins and polysaccharides[104,105]; and (d) post-treatment recontamination. Since water, whether for washing/cleaning the foods before processing or cleaning of the processing/handling equipment, is generally the source of the *A. hydrophila* group in foods, the presence of these bacteria in the water supply should be closely monitored.

A. hydrophila can also be inactived by chlorine dioxide treatment of water[106]. According to these authors, ClO_2 has high stability in the distribution system, but its use is limited to waters treated with activated carbon and/or low direct organic carbon.

11.2.3 INJURY

Food microbiologists are interested in both qualitative (incidence) as well as quantitative recovery of various bacterial groups from foods. Quantitation is important because the risk to the consumer is proportional to the number of a particular bacterium present and because the number present at any time reflects the history of the food, i.e. temperature of holding or heating or the effect of various food parameters, for example, pH or salt level, on the response of the organism. Food microbiologists use media containing selective agents to recover bacteria from food. However, research has shown that bacteria stressed or injured by food-processing operations such as heating, freezing, drying or sanitizing cannot be recovered on selective media, that is, stressed or injured cells have become sensitive to the selective agents[107]. There have been few studies on injury in *A. hydrophila*[66]. Cattabiani and Brindani[103] studied the recovery of *A. hydrophila, A. sobria* and *A. caviae* treated with the disinfectants chlorine, iodine, and a quaternary ammonium compound; they determined that significantly fewer cells were recovered on selective media (RS-agar or McConkey agar) compared with the non-selective tryptone soya agar. The amount of injury was proportional to both contact time and concentration of disinfectant. Studies have shown that this organism can also be injured by the food preservatives methyl-*p*-hydroxybenzoate and potassium sorbate[92] and the combination of high polyphosphate and high NaCl[87]. Palumbo and colleagues[22] determined that even non-injured *A. hydrophila* was sensitive to the selective agents bile salts and sodium desoxycholate. The use of selective media should be carefully assessed when recovering *A. hydrophila* from any food, particularly one which was processed, or from any environment in which the organism might have been stressed, if quantitative recovery is desired.

11.2.4   BACTERIOCINS

Today, consumers are demanding foods with no or fewer additives or with natural preservatives. Bacteriocins (protein substances produced by lactic acid bacteria [LAB]) or bacteriocin-producing cultures have seen increased use as natural preservatives for foods. Bacteriocins with activity against specific pathogens or spoilage flora have been described. Lewus *et al.*[108] described the isolation of several LAB cultures from meat which had activity against *A. hydrophila* as well as *Listeria monocytogenes* and *Staphylococcus aureus*. This inhibition was demonstrated in a model system; it remains to be investigated whether this inhibition will occur in an actual food system held at refrigeration temperatures. Kalchayanand and co-workers[109] observed that *A. hydrophila* was resistant to the action of the bacteriocins nisin and pediocin; however, when the cells were given a preliminary treatment of (injured by) either heat or freezing, *A. hydrophila* became sensitive to these bacteriocins.

11.2.4   VIRULENCE OF FOODBORNE ISOLATES

The data in Tables 11.1–11.9 indicate a widespread occurrence of the *A. hydrophila* group bacteria in foods. Further, Palumbo and colleagues[22] among others have reported that they often occur at high levels. However, the link between consumption of *A. hydrophila*-containing food and gastrointestinal illness (diarrhoea) has not been conclusively made. In the feeding study of Morgan and co-authors[79], diarrhoea was observed in only a few volunteers and only at high dose levels. In certain instances[46,50] *A. hydrophila* has been the only pathogen isolated from the implicated food. To complicate the situation further, there is not a specific set of traits associated with virulence in bacteria of the *A. hydrophila* group.

Isolates can be characterized either by certain biochemical reactions (including those which have been observed to correlate with enterotoxin production) or by possession of various virulence-associated factors (Table 11.12). Based on examination of a series of clinical and environmental isolates, Stelma and colleagues[116] proposed that the major if not sole virulence determinant in *A. hydrophila* was β-haemolysin, and observed that β-haemolytic activity correlated with fluid accumulation in the rabbit ileal loop test (enterotoxin). Kaper and colleagues[5] observed cytotoxin (to Y-1 adrenal cells) correlated with lysine decarboxylase and the Voges–Proskauer reaction (VP). Burke and colleagues[117] and Mascher and co-workers[8] both observed a correlation between VP and enterotoxin. However, multiple factors may actually be involved in virulence in this organism. Cahill[118] and Janda[64] proposed that virulence in *A. hydrophila* may be related to the organism's structures such as pili, S-layer, lipopolysaccharide (LPS; endotoxin), outer membrane proteins (OMPs) and flagella, extracellular factors including haemolysins, enterotoxins, proteases and siderophores, and cell-associated factors such as invasins, serum resistance, plasmids and adherence. There have been several studies that have examined the virulence factors of *A. hydrophila* group isolated from food and water (Table 11.12). These investigators observed that most food and water isolates possessed the same putative virulence factors as clinical isolates. Gray and colleagues[119] determined the phenotype and virulence characteristics (cytotoxin, cell elongation factor and haemagglutinins) of 61 *Aeromonas* strains isolated from the faeces of pigs, cows and a variety of farm environments. In general, they observed good correlation between the three virulence traits and phenotype (species): most *A. hydrophila* isolates were positive for the three factors, few *A. caviae* were, and *A. sobria* isolates were intermediate. Esteve and associates[120] observed in their study of motile *Aeromonas* isolated from eels that virulence factors for eels such as elastases and haemolysins resided mainly in the strains phenotyped as *A. hydrophila* and *A. jandaei*, while isolates identified as *A. caviae*, *A. sobria* and *A. allosaccharophila* were non-pathogenic.

11.3   CONCLUSIONS

The information presented in this chapter further supports the enigma that has been *A. hydrophila*. Members of the *A. hydrophila* group occur widely in both fresh and processed foods, often in very high numbers (Tables

Table 11.12. Comparison of virulence factors of clinical and environmental (food) isolates of the *A. hydrophila* group

Investigator(s) (reference)	Organisms isolated	Source	Putative virulence factors	Comments
Abeyta et al. (50)	*Ah*	Oysters	Haemolysin, suckling mouse, cytotoxin (Y-1), rabbit ileal loop	23/28 strains positive for at least one virulence factor
Abeyta et al. (52)	*Aeromonas* spp.	Water, shellfish	Haemolysin, Y-1 adrenal cells	91% positive for haemolysin; 54% produced heat-lable cytotoxin
Baloda et al. (110)	*Ah, As*	Water, fish, food, seafood	Protease, cytotoxin (HeLa and McCoy cells), haemolysin	57/60 positive for protease; 58/60 produced cytotoxin; 52/60 were β-haemolytic
Burke et al. (111)	*Ah*	Faeces, water	Enterotoxin (suckling mouse), haemolysin, haemagglutination	Many clinical and water strains possessed identical properties
Kaper et al. (5)	*Ah*	Water	Y1 Adrenal cells (cytotoxin), rabbit ileal loop	83/116 produced cytotoxin; 8/11 positive in rabbit ileal loop
Krovacek et al. (7, 32)	*Ah, As*	Water	Haemolysin, protease, CHO-K1 cells, cytotoxic, cytotonic	49/61 haemolysin-positive; 52/61 protease-positive; 55/61 cytotoxin-positive; 19/61 cytotonic enterotoxin-positive
Mascher et al. (8)	*Aeromonas* spp.	Water	Haemolysin	All produced haemolysin, with 40/46 giving titre of >32

Neves *et al.* (112)	*Av*	Water	Haemolysin, enterotoxin (suckling mouse), Verotoxin	Most strains positive for enterotoxin and haemolysin; Verotoxin production corresponded with haemolysin
Okrend *et al.* (24)	*Ah, As, Ac*	Poultry, pork, beef	Haemolysin, CHO cells, Y-1 adrenal cells	At least half of all strains positive for the three factors
Palumbo *et al.* (113)	*Ah*	Fish and seafood, poultry, raw milk, red meat, clinical	Serum resistance, cytotoxin (Y-1 adrenal cells), elastase, protease, haemagglutination, β-haemolysin	Food isolates similar to the clinical isolates
Pin *et al.* (114)	*Ah, Ac, As, Aeromonas* spp.	Faeces, food	Autoagglutination, haemolysin, haemagglutination, self-pelleting, precipitate after boiling	89% of *Aeromonas* isolated from food and faeces produced putative virulence factors
Rahim and Aziz (115)	*Ah, Ac, As*	Freshwater prawns	Rabbit ileal loop (enterotoxin); haemolysin	11/21 were enterotoxin-positive; enterotoxin correlated with α-haemolysin

Ah, A. hydrophila; As, A. sobria; Ac, A. caviae.

11.1–11.9). They can grow or at least survive in many of these foods, even when held at what is considered adequate refrigeration, i.e. 5°C; Tables 11.1–11.10. Food isolates have been shown to possess the virulence attributes necessary to cause disease (Table 11.11), yet the definitive link between food and disease has eluded researchers. Palumbo[121] suggested some possible explanations for these apparent contradictions:

(1) Temperature of growth of the cultures. Temperature can have a major influence on expression of virulence-associated factors, with increased expression at low temperatures (room temperature down to refrigeration temperature).
(2) Clinical isolates (those capable of causing disease) may represent a unique subset of the *A. hydrophila* group and newer epidemiological techniques such as fatty acid methyl esters (FAMES), multilocus enzyme electrophoresis (MLEE) and ribotyping may provide the necessary differentiation among isolates and thus allow designation of disease-causing strains.
(3) Additional virulence-associated factors such as pili may need to be investigated.

Food and clinical microbiologists need to cooperate and determine what questions to ask and thus ultimately provide the link between *A. hydrophila* in food and disease.

11.4 REFERENCES

1 Abeyta C, Wekell MM. Potential sources of *Aeromonas hydrophila*. *J Food Safety* 1988; **9**: 11–22.
2 Araujo RM, Arribas RM, Pares R. Distribution of *Aeromonas* species in waters with different levels of pollution. *J Appl Bacteriol* 1991; **71**: 182–6.
3 Burke V, Robinson J, Gracey M, Peterson D, Partridge K. Isolation of *Aeromonas hydrophila* from a metropolitan water supply: seasonal correlation with clinical isolates. *Appl Environ Microbiol* 1984; **48**: 361–6.
4 Hazen TC, Fliermans, CB, Hirsch RP, Esch GW. Prevalence and distribution of *Aeromonas hydrophila* in the United States. *Appl Environ Microbiol* 1978; **36**: 731–8.
5 Kaper JB, Lockman H, Colwell RR. *Aeromonas hydrophila*: ecology and toxigenicity of isolates from an estuary. *J Appl Bacteriol* 1981; **50**: 359–77.
6 Knochel S, Jeppesen C. Distribution and characterization of *Aeromonas* in food and drinking water in Denmark. *Int J Food Microbiol* 1990; **10**: 317–22.
7 Krovacek K, Faris A, Baloda SJ, Lindberg T, Peterz M, Mansson I. Isolation and virulence profiles of *Aeromonas* spp. from different municipal drinking water supplies in Sweden. *Food Microbiol* 1992; **9**: 215–22.
8 Mascher F, Reinthaller FF, Stunzner D, Lamberger B. *Aeromonas* species in a municipal water supply of a central European city: biotyping of strains and detection of toxins. *Zentralbl Bakteriol B* 1988; **186**: 333–7.
9 Pathak SP, Bhattacherjee JW, Kalra N, Chandra S. Seasonal distribution of *Aeromonas hydrophila* in river water and isolation from fish. *J Appl Bacteriol* 1988; **65**: 347–52.
10 Rhodes MW, Kator H. Seasonal occurrence of mesophilic *Aeromonas* as a function of biotype and water quality in temperate freshwater lakes. *Water Res* 1994; **28**: 2241–51.

11 Rippey SR, Cabelli VJ. Growth characteristics of *Aeromonas hydrophila* in limnetic waters of varying trophic state. *Arch Hydrobiol* 1985; **104**: 311–19.

12 Rippey SR, Cabelli VJ. Use of the thermotolerant *Aermonas* group for the trophic state classification of fresh waters. *Water Res* 1989; **23**: 1107–14.

13 Rippey SR, Troy MA, Cabelli VJ. Growth kinetics of *Aeromonas hydrophila* in freshwaters supplemented with various organic and inorganic nutrients. *World J Microbiol Technol* 1994; **10**: 159–64.

14 Williams A, LaRock PA. Temporal occurrence of *Vibrio* and *Aeromonas hydrophila* in estuarine sediments. *Appl Environ Microbiol* 1985; **50**: 1490–5.

15 Blickstad E, Molin G. Carbon dioxide as a controller of the spoilage flora of pork, with special reference to temperature and sodium chloride. *J Food Protect* 1983; **46**: 756–63.

16 Enfors S-O, Molin G, Ternstrom A. Effect of packaging under carbon dioxide, nitrogen or air on the microbial flora of pork stored at 4°C. *J Appl Bacteriol* 1979; **47**: 197–208.

17 Lee BH, Simard RE, Laleye LC, Holley RA. Effects of temperature and storage duration on the microflora, physiochemical and sensory changes of vacuum- or nitrogen-packed pork. *Meat Sci* 1985; **13**: 99–112.

18 Myers BR, Marshall RT, Edmondson JE, Stringer WC. Isolation of pectinolytic *Aeromonas hydrophila* and *Yersinia enterocolitica* from vacuum-packaged pork. *J Food Protect* 1982; **45**: 33–7.

19 Grau FH, Eustace IJ, Bill BA. Microbial flora of lamb carcasses stored at 0°C in packs flushed with nitrogen or filled with carbon dioxide. *J Food Sci* 1985; **50**: 482–5.

20 Simard RE, Zee J, L'Heureux L. Microbial growth in carcasses and boxed beef during storage. *J Food Protect* 1984; **47**: 773–7.

21 Beebe SD, Vanderzant C, Hanna MO, Carpenter ZL, Smith GC. Effect of initial internal temperature and storage temperature on the microbial flora of vacuum packaged beef. *J Milk Food Technol* 1976; **39**: 600–5.

22 Palumbo SA, Maxino F, Williams AC, Buchanan RL, Thayer DW. Starch-ampicillin agar for the quantitative detection of *Aeromonas hydrophila*. *Appl Environ Microbiol* 1985; **50**: 1027–30.

23 Nishikawa Y, Kishi T. Isolation and characterization of motile *Aeromonas* from human, food and environmental specimens. *Epidemiol Infect* 1988; **101**: 213–23.

24 Okrend AJG, Rose BE, Bennett B. Incidence and toxigenicity of *Aeromonas* species in retail poultry, beef and pork. *J Food Protect* 1987; **50**: 509–13.

25 Stern NJ, Drazek ES, Joseph SW. Low incidence of *Aeromonas* sp. in livestock faeces. *J Food Protect* 1987; **50**: 66–9.

26 Smith FC, Adams JC, Field RA. Predominant psychrotrophic bacteria on fresh and aged ground beef and antelope. *J Milk Food Technol* 1975; **38**: 516–17.

27 Cattabiani F. Tossigenicita di *Aeromonas hydrophila*, *caviae* e *sobria* isolate da varie fonti. *Arch Vet Ital* 1985; **36**: 25–33.

28 Ayres JC. Temperature relationships and some other characteristics of the microbial flora developing on refrigerated beef. *Food Res* 1960; **25**: 1–18.

29 Jay JM. Nature, characteristics, and proteolytic properties of beef spoilage bacteria at low and high temperatures. *Appl Microbiol* 1967; **15**: 943–4.

30 Seideman SC, Vanderzant C, Hanna MO, Carpenter ZL, Smith GC. Effect of various types of vacuum packages and length of storage on the microbial flora of wholesale and retail cuts of beef. *J Milk Food Technol* 1976; **39**: 745–53.

31 Hudson JA, Mott SJ, Delacy KM, Eldridge AL. Incidence and coincidence of *Listeria* spp., motile aeromonads and *Yersinia enterocolitica* on ready-to-eat fleshfoods. *Int J Food Microbiol* 1992; **16**: 99–108.

32 Krovacek K, Faris A, Baloda SJ, Peterz M, Lindberg T, Mansson I. Prevalence

and characterization of *Aeromonas* spp. isolated from foods in Uppsala, Sweden. *Food Microbiol* 1992; **9**: 29–36.

33 Fricker CR, Tompsett S. *Aeromonas* spp. in foods: a significant cause of food poisoning. *Int J Food Microbiol* 1989; **9**: 17–23.

34 Ibrahim A, Mac Rae IC. Incidence of *Aeromonas* and *Listeria* spp. in red meat and milk samples in Brisbane, Australia. *Int J Food Microbiol* 1991; **12**: 263–70.

35 Comi G, d'Aubert S, Cantoni C. *Aeromonas* enterotossici in alimenti. *Ind Alimenti* 1984; **23**: 597–600.

36 Hanninen M-L. Occurrence of *Aeromonas* spp. in samples of ground meat and chicken. *Int J Food Microbiol* 1993; **18**: 339–42.

37 Yamamoto K, Nagamine S, Kamiya T, Murakami R, Kikuchi J, Sato H. Identification to the genospecies of *Aeromonas* strains from raw meat and carp, and evaluation of their biological activities. *J Food Hyg Soc Japan* 1994; **35**: 187–94.

38 Majeed KN, Egan AF, Mac Rae IC. Enterotoxigenic aeromonads on retail lamb meat and offal. *J Appl Bacteriol* 1989; **67**: 165–70.

39 Kirov SM, Anderson MJ, McMeekin TA. A note on *Aeromonas* spp. from chickens as possible food-borne pathogens. *J Appl Bacteriol* 1990; **68**: 327–4.

40 Nagel CW, Simpson KL, Ng H, Vaughn RH, Stewart CF. Microorganisms associated with spoilage of refrigerated poultry. *Food Technol* 1960; **21**: 21–3.

41 Barnhart HM, Pancorbo OC, Dreesen DW, Shotts EB Jr. Recovery of *Aeromonas hydrophila* from carcasses and processing water in a broiler processing operation. *J Food Protect* 1989; **52**: 646–9.

42 Santos Y, Toranzo AE, Barja JL, Nieto TP, Villa TG. Virulence properties and enterotoxin production of *Aeromonas* strains isolated from fish. *Infect Immun* 1988; **56**: 3285–93.

43 Mavinkurve SS, Gangal SV, Sawant PL, Kumta US. Bacterial studies on irradiated tropical fish-Bombay Duck (*Harpondon nehereus*). *J Food Sci* 1967; **32**: 711–16.

44 Molin G, Stenstrom I-M. Effect of temperature on the microbial flora of herring fillets stored in air or carbon dioxide. *J Appl Bacteriol* 1984; **56**: 275–82.

45 Bernardeschi P, Bonnechi I, Cavallini G. *Aeromonas hydrophila* infection after cockles ingestion. *Haematologia* 1988; **73**: 545.

46 Altwegg M, Martinetti-Lucchini G, Lüthy-Hottenstein J, Rohrbach M. *Aeromonas*-associated gastroenteritis after consumption of contamination shrimp. *Eur J Clin Microbiol Infect Dis* 1991; **10**: 44–5.

47 Faghri MA, Pennington CL, Cronholm LS, Atlas RM. Bacteria associated with crabs from cold waters with emphasis on the occurrence of potential human pathogens. *Appl Environ Microbiol* 1984; **47**: 1054–61.

48 Flynn TJ, Knepp IG. Seafood shucking as an etiology for *Aeromonas hydrophila* infection. *Arch Intern Med* 1987; **147**: 1816–17.

49 Hood MA, Baker RM, Singleton FL. Effect of processing and storing oyster meats on concentrations of indicator bacteria, Vibrios, and *Aeromonas hydrophila*. *J Food Protect* 1984; **47**: 598–601.

50 Abeyta C Jr, Kaysner CA, Wekell MM, Sullivan JJ, Stelma GN. Recovery of *Aeromonas hydrophila* from oysters implicated in an outbreak of foodborne illness. *J Food Protect* 1986; **49**: 643–6.

51 Kueh CSW, Chan K-Y. Bacteria in bivalve shellfish with special reference to the oyster. *J Appl Bacteriol* 1985; **59**: 41–7.

52 Abeyta C Jr, Kaysner CA, Wekell MM, Stott RF. Incidence of motile aeromonads from United States West Coast shellfish growing estuaries. *J Food Protect* 1990; **53**: 849–55.

53 Wong H-C, Ting S-H, Shieh W-R. Incidence of toxigenic vibrios in foods available in Taiwan. *J Appl Bacteriol* 1992; **73**: 197–202.

54 Kirov SM, Hui DS, Hayward LJ. Milk as a potential source of *Aeromonas* gastrointestinal infection. *J Food Protect* 1993; **56**: 306–12.

55 Hunter PR, Burge SH. Isolation of *Aeromonas caviae* from ice-cream. *Lett Appl Microbiol* 1987; **4**: 45–6.

56 Craven HM, Macauley BJ. Microorganisms in pasteurised milk after refrigerated storage. 1. Identification of types. *Aust J Dairy Sci* 1992; **47**: 38–45.

57 Reinheimer JA, Suarez VB, Haye MA. Microbial and chemical changes in refrigerated pasteurised milk processed in the Santa Fe area (Argentina). *Aust J Dairy Technol* 1993; **48**: 5–9.

58 Ternstrom A, Lindberg A-M, Molin G. Classification of the spoilage of raw and pasteurized bovine milk, with special reference to *Pseudomonas* and *Bacillus*. *J Appl Bacteriol* 1993; **75**: 25–34.

59 Freitas AC, Nunes MP, Milhomem AM, Ricciardi ID. Occurrence and characterization of *Aeromonas* species in pasteurized milk and white cheese in Rio De Janeiro, Brazil. *J Food Protect* 1993; **56**: 62–5.

60 Callister SM, Agger WA. Enumeration and characterization of *Aeromonas hydrophila* and *Aeromonas caviae* isolated from grocery store produce. *Appl Environ Microbiol* 1987; **53**: 249–53.

61 Kneifel W, Berger E. Microbiological criteria of random samples of spices and herbs retailed on the Austrian market. *J Food Protect* 1994; **57**: 893–901.

62 Hunter PR, Cooper-Poole B, Hornby H. Isolation of *Aeromonas hydrophila* from cooked tripe. *Lett Appl Microbiol* 1992; **15**: 222–3.

63 Gobat P-F, Jemmi T. Distribution of mesophilic *Aeromonas* species in raw and ready-to-eat fish and meat products in Switzerland. *Int J Food Microbiol* 1993; **20**: 117–20.

64 Janda JM. Recent advances in the study of the taxonomy, pathogenicity, and infectious syndromes associated with the genus *Aeromonas*. *Clin Microbiol Rev* 1991; **4**: 397–410.

65 von Graevenitz A, Bucher C. Evaluation of differential and selective media for isolation of *Aeromonas* and *Plesiomonas* spp. from human faeces. *J Clin Microbiol* 1983; **17**: 16–21.

66 Palumbo SA. A review of methods for detection of the psychrotrophic foodborne pathogens *Listeria monocytogenes* and *Aeromonas hydrophila*. *J Food Safety* 1991; **11**: 105–22.

67 Knøchel S. Effect of temperature on hemolysin production in *Aeromonas* spp. isolated from warm and cold environments. *Int J Food Microbiol* 1989; **9**: 225–35.

68 Majeed KN, Egan AF, Mac Rae IC. Incidence of aeromonads in samples from an abattoir processing lambs. *J Appl Bacteriol* 1989; **67**: 597–604.

69 Nishikawa Y, Kishi T. A modification of bile salts brilliant green agar for isolation of motile *Aeromonas* from foods and environmental specimens. *Epidemiol Infect* 1987; **98**: 331–6.

70 Catsaras M, Buttiaux R. Les *Aeromonas* dans les matières fecales humaines. *Ann Inst Pasteur (Lille)* 1965; **16**: 85–8.

71 Lautrop H. *Aeromonas hydrophila* isolated from human faeces and its possible pathological significance. *Acta Pathol Microbiol Scand* 1961; **51**: 299–301.

72 Schubert RHW. Aeromonads and their significance as potential pathogens in water. *J Appl Bacteriol* 1991; **70**: 131–5S.

73 Jay JM. *Modern Food Microbiology*, 4th Edn. New York: Van Nostrand Reinhold, 1992; 38–62.

74 Palumbo SA. The growth of *Aeromonas hydrophila* K144 in ground pork at 5°C. *Int J Food Microbiol* 1988; **7**: 41–8.

75 Palumbo SA. Is refrigeration enough to restrain foodborne pathogens? *J Food Protect* 1986; **49**: 1003–9.

76 Palumbo SA, Williams AC, Buchanan RI, Phillips JG. Model for the aerobic growth of *Aeromonas hydrophila* K144. *J Food Protect* 1991; **54**: 429–35.

77 Palumbo SA, Williams AC, Buchanan RI, Phillips JG. Model for the anaerobic growth of *Aeromonas hydrophila* K144. *J Food Protect* 1992; **55**: 260–5.

78 Popoff M. Genus III. *Aeromonas* Kluyver and Van Niel 1936, 398[AL]. Krieg NR, Holt JG. (eds) *Bergey's Manual of Systematic Bacteriology*, Vol. 1. Baltimore, MD: Williams and Wilkins, 1984: 545–8.

79 Morgan DR, Johnson PC, DuPont HL, Satterwhite JK, Wood LV. Lack of correlation between known virulence properties of *Aeromonas hydrophila* and enteropathogenicity for humans. *Infect Immun* 1985; **50**: 62–5.

80 Kirov SM, Rees B, Wellock RC, Goldsmid JM, Van Galen AD. Virulence characteristics of *Aeromonas* spp. in relation to source and biotype. *J Clin Microbiol* 1986; **24**: 827–34.

81 Eddy BP. Cephalotrichous, fermentative gram-negative bacteria: the genus *Aeromonas*. *J Appl Bacteriol* 1960; **25**: 216–49.

82 Freese E, Sheu CW, Galliers E. Function of lipophilic acids as antimicrobial food additives. *Nature* 1973; **241**: 321–5.

83 Palumbo SA, Morgan DR, Buchanan RL. Influence of temperature, NaCl, and pH on the growth of *Aeromonas hydrophila*. *J Food Sci* 1985; **50**: 1417–21.

84 Aytac SA, Ozbas ZY. Survey of the growth and survival of *Yersinia enterocolitica* and *Aeromonas hydrophila* in yoghurt. *Milchwissenschaft* 1994; **49**: 322–5.

85 Aytac SA, Ozbas ZY. Growth of *Yersinia enterocolitica* and *Aeromonas hydrophila* in acidophilus yoghurt. *Aust J Dairy Technol* 1994; **49**: 90–2.

86 Palumbo SA, Williams, AC. Growth of *Aeromonas hydrophila* K144 as affected by organic acids. *J Food Sci* 1992; **57**: 233–5.

87 Palumbo SA, Call JE, Cooke PH, Williams AC. Effect of polyphosphates and NaCl on *Aeromonas hydrophila* K144. *J Food Safety* 1995; **15**: 77–87.

88 Santos J, Lopez-Diaz T-M, Garcia-Lopez ML, Garcia-Fernandez M-C, Otero A. Minimum water activity for the growth of *Aeromonas hydrophila* as affected by strain, temperature and humectant. *Lett Appl Microbiol* 1994; **19**: 76–8.

89 Sperber WH. Influence of water activity on foodborne bacteria — a review. *J Food Protect* 1983; **46**: 142–50.

90 Castellani AG, Niven CF Jr. Factors affecting the bacteriostatic action of sodium nitrite. *Appl Microbiol* 1955; **3**: 154–9.

91 Gould WA. *Glossary for the Food Industries*. Baltimore, MD: CTI Publications, 1990.

92 Stecchini ML, Sarais I, Giavedoni P. Effect of essential oils on *Aeromonas hydrophila* in a culture medium and in cooked pork. *J Food Protect* 1993; **56**: 406–9.

93 Sofos JN, Maga JA, Boyle DL. Effect of ether extracts from condensed wood smokes on the growth of *Aeromonas hydrophila* and *Staphylococcus aureus*. *J Food Sci* 1988; **53**: 1840–3.

94 Moir CJ, Eyles MJ. Inhibition, injury, and inactivation of four psychrotrophic foodborne bacteria by the preservatives methyl-p-hydroxybenzoate and potassium sorbate. *J Food Protect* 1992; **55**: 360–6.

95 Hara Y, Ishigami T. Antibacterial activities of tea polyphenols against foodborne pathogenic bacteria. *Nip Shokuhin Kogyo Gakkasihi* 1989; **36**: 996–9.

96 Venugopal V, Pansare AC, Lewis FN. Inhibitory effect of food preservatives on protease secretion by *Aeromonas hydrophila*. *J Food Sci* 1984; **49**: 1078–81.

97 Condon S, Garcia ML, Otero A, Sala FJ. Effect of culture age, pre-incubation at

low temperature and pH on the thermal resistance of *Aeromonas hydrophila. J Appl Bacteriol* 1992; **72**: 322–6.

98 Nishikawa Y, Ogasawara J, Kimura T. Heat and acid sensitivity of motile *Aeromonas*: a comparison with other food-poisoning bacteria. *Int J Food Microbiol* 1993; **18**: 271–8.

99 Palumbo SA, Williams AC, Buchanan RL, Phillips JG. Thermal resistance of *Aeromonas hydrophila. J Food Protect* 1987; **50**: 761–4.

100 Palumbo SA, Jenkins RK, Buchanan RL, Thayer DW. Determination of irradiation D-values for *Aeromonas hydrophila. J Food Protect* 1986; **49**: 189–91.

101 Monk JD, Beuchat LR, Doyle MP. Irradiation inactivation of food-borne microorganisms. *J Food Protect* 1995; **58**: 197–208.

102 Knøchel S. Chlorine resistance of motile *Aeromonas* spp. *Water Sci Technol* 1991; **24**: 327–30.

103 Cattabiani F, Brindani F. Valutazione del danneggiamento di tipo structurale da disinfettanti in *Aeromonas, Vibrio* e *Plesiomonas. Arch Vet Ital* 1988; **39**: 245–53.

104 van der Kooij D. Nutritional requirements of aeromonads and their multiplication in drinking water. *Experientia* 1991; **47**: 444–6.

105 van der Kooij D, Hijnen WAM. Nutritional versatility and growth kinetics of an *Aeromonas hydrophila* strain isolated from drinking water. *Appl Environ Microbiol* 1989; **54**: 2842–51.

106 Medema GJ, Wondergem E, Dijk-Looyaard AM, Havelaar AH. Effectivity of chlorine dioxide to control *Aeromonas* in drinking water distribution systems. *Water Sci Technol* 1991; **24**: 325–6.

107 Busta FF. Practical implications of injured microorganisms in food. *J Milk Food Technol* 1976; **39**: 138–45.

108 Lewus CB, Kaiser A, Montville TJ. Inhibition of food-borne bacterial pathogens by bacteriocins from lactic acid bacteria isolated from meat. *Appl Environ Microbiol* 1991; **57**: 1683–8.

109 Kalchayanand N, Hanlin MB, Ray B. Sublethal injury makes Gram-negative and resistant Gram-positive bacteria sensitive to the bacteriocins, pediocin AcH and nisin. *Lett Appl Microbiol* 1992; **15**: 239–43.

110 Baloda SB, Krovacek K, Eriksson L, Linne T, Mansson I. Detection of aerolysin gene in *Aeromonas* strains from drinking water, fish and foods by the polymerase chain reaction. *Comp Immun Microbiol Dis* 1995; **18**: 17–26.

111 Burke V, Robinson J, Cooper M, Beaman J, Partridge K, Peterson D, Gracey M. Biotyping and virulence factors in clinical and environmental isolates of *Aeromonas* species. *Appl Environ Microbiol* 1984; **47**: 1146–9.

112 Neves MS, Nunes MP, Milhomem AM, Ricciardi ID. Production of enterotoxin and cytotoxin in *Aeromonas veronii. Braz J Med Biol Res* 1990; **23**: 437–40.

113 Palumbo SA, Bencivengo MM, Del Corral F, Williams AC, Buchanan RL. Characterization of the *Aeromonas hydrophila* group isolated from retail foods of animal origin. *J Clin Microbiol* 1989; **27**: 854–9.

114 Pin C, Marin ML, Selgas MD, Garcia ML, Tormo J, Casas C. Virulence factors in clinical and food isolates of *Aeromonas* species. *Folia Microbiol* 1994; **39**: 331–6.

115 Rahim Z, Aziz KMS. Enterotoxigenicity, hemolytic activity and antibiotic resistance of *Aeromonas* spp. isolated from freshwater prawn marketed in Dhaka, Bangladesh. *Microbiol Immunol* 1994; **38**: 773–8.

116 Stelma GN, Johnson CH, Spaulding P. Evidence for the direct involvement of β-hemolysin in *Aeromonas hydrophila* enteropathogenicity. *Curr Microbiol* 1986; **14**: 71–7.

117 Burke V, Robinson J, Max H, Gracey M. Biochemical characteristics of enterotoxigenic *Aeromonas* spp. *J Clin Microbiol* 1988; **15**: 48–52.

118 Cahill MM. Virulence factors in motile *Aeromonas* species. *J Appl Bacteriol* 1990; **69**: 1–16.
119 Gray SJ, Stickler DJ, Bryant TN. The incidence of virulence factors in mesophilic *Aeromonas* species isolated from farm animals and their environment. *Epidemiol Infect* 1990; **105**: 277–94.
120 Esteve C, Amaro C, Garay E, Santos Y, Toranzo AE. Pathogenicity of live bacteria and extracellular products of motile *Aeromonas* isolated from eels. *J Appl Bacteriol* 1995; **78**: 555–62.
121 Palumbo SA. The occurrence and significance of organisms of the *Aeromonas hydrophila* group in food and water. *Med Microbiol Lett* 1993; **2**: 339–46.

12 *Aeromonas* Gastrointestinal Disease: A Case Study in Causation?

S. W. JOSEPH
University of Maryland, College Park, USA

All observations must be for or against some view, if it is to be of any service

Charles Darwin

12.1 INTRODUCTION

Over the past two decades, the genus *Aeromonas* has probably been studied more intensively than at any other time since its designation by Kluyver and Van Niel in 1936[1]. During this time it was suspected that *Aeromonas*, contrary to earlier belief, might in fact be a primary pathogen of both intestinal and extra-intestinal diseases. Consequently, numerous studies, including clinical, virulence, epidemiological, taxonomic, structural and ecological aspects, were performed. While suggesting that *Aeromonas* was a very intriguing organism, the information gathered in all of these areas enhanced the knowledge base on *Aeromonas*, particularly regarding its taxonomy[2–10]. Thus, when the proper taxonomic designations began to be used appropriately in subsequent studies, the information gathered was enormously more valuable than that obtained in earlier studies.

In the absence of correct taxonomic designations of *Aeromonas* isolates in many of the earlier studies, it is rather arduous, if not impossible, to attempt to construct a clear picture of the significance of individual *Aeromonas* species in disease causation over the years. Obviously, it is difficult to assess the pathogenic potential of a specific aeromonad retrospectively, when there were problems with accurately designating the organism to its correct species in the original study.

Despite the increase in information gathered from these studies, a clearly defined image of this organism as an aetiologic agent of diarrhoeal disease, i.e. primary cause, has not been clearly established in the eyes and minds of some investigators. While there are instances of extra-intestinal disease caused in otherwise healthy individuals and in a small population resulting from intestinal cases. However, there is still doubt expressed that *Aeromonas*

The Genus Aeromonas. Edited by B. Austin, M. Altwegg, P.J. Gosling and S. Joseph
© 1996 John Wiley & Sons Ltd

can be regarded as an intestinal pathogen in the same fashion that one might, for example, regard *Vibrio parahaemolyticus*.

The following sections will be devoted to exploring the potential causation of *Aeromonas* in diarrhoeal diseases, its present status and future considerations.

12.2   HISTORICAL PERSPECTIVE

For many years, there was little proof and hardly any acceptance that aquatic bacteria were involved in human disease. Only *Vibrio cholerae* might have qualified as such, except that conventional opinion, at the time, considered it to be a human pathogen which occasionally contaminated aquatic environments, leading to other human outbreaks. Only recently has that view changed, so that *V. cholerae* is now regarded as an aquatic organism capable of causing disease in epidemic/pandemic proportions[11,12].

Aeromonas is an aquatic organism that was first described in 1890 by Zimmerman[13], who named it *Bacillus punctatus* because of the nature of the colonies on gelatin plates. During the next few decades, organisms resembling *Aeromonas* morphologically and biochemically, at that time, were isolated from both cold- and warm-blooded animals, including fish[14], frogs[15], snakes[16], domestic animals[17], and birds[18]. These isolates were variously classified into many different genera including, among others, *Aerobacter, Pseudomonas* and *Vibrio*[2].

While there were early cases of suggested *Aeromonas*-caused disease in humans in the 1930s[19,20], a case reported by Hill and colleagues[21] in 1954 is considered to be the first verified instance of human disease caused by *Aeromonas*, although, at that time, it was considered a new member of the family Pseudomonadaceae. The organism was isolated from the blood of a woman who was suffering from a fulminant septicaemia with metastatic myositis. This organism was later termed *Vibrio jamaicensis* by Caselitz in 1955[22].

There were numerous isolations and studies of strains isolated from humans, other animals, and the environment between 1890 and 1957 that were probably *Aeromonas* but were designated otherwise. Others followed, suggesting that *Aeromonas* could cause intestinal or extra-intestinal disease[23–25]. Apparently, Caselitz[26] in 1958 was the first to describe the enteropathogenicity of *Aeromonas* after repeatedly isolating a strain of *Aeromonas* from a child with enteritis in Jamaica[26]. He was actually able to cause agglutination of the organism with the child's homologous antiserum[26].

The genus *Aeromonas* was first proposed by Kluyver and van Niel[1] in 1936, but the suggestion was not formally accepted until the seventh edition of *Bergey's Manual of Determinative Bacteriology*, published in 1957[27]. For many years, *Aeromonas* was thought to cause disease only in humans, who were somehow compromised or predisposed to infection[28,29].

In 1968, Von Graevenitz and Mensch reported 27 cases of infection or intestinal colonization associated with *Aeromonas*, although a small percentage of healthy individuals were found to be colonized[28]. This seminal study was one of the first to show a strong relationship between the presence of large numbers of aeromonads and human diarrhoeal disease. Many studies on the relationship of *Aeromonas* spp. and diarrhoeal disease have been subsequently reported and vary from single-case reports to multiple cases. Some of the reports simply described the isolation of aeromonads over a particular time period in a clinical laboratory while others were based on attempts to establish prospective or retrospective studies.

A number of these reports will be reviewed herein to determine if, in fact, we now have sufficient evidence to reach a conclusion regarding the possibility of causality by *Aeromonas* spp. in diarrhoeal disease.

12.2.1 SINGLE-CASE STUDIES

Single-case studies are generally reported because of their unusual nature in relation to the cases ordinarily observed in the clinical environment. One of the earliest treated cases of diarrhoeal disease attributed to *Aeromonas* was described by Rosner in 1964[24].

Some 30 years later there was a case reported of a man who had returned to Finland from Turkey, and had 15 positive *A. caviae* cultures over a 17-month period as the sole enteropathogen, and was treated successfully with ciprofloxacin only to become culture-positive once again six months later with a novel strain of *A. media*. The individual responded to another course of ciprofloxacin, became asymptomatic and subsequent cultures were negative[30].

Another case involving repeated cultures of *A. hydrophila* from a diarrhoeal male, who had travelled to India, was reported in 1980. In this case, the initial cultures were not considered significant, although no other viral or bacterial pathogens could be demonstrated. Metronidazole treatment did not eliminate the *A. hydrophila*, which was subsequently cultured on several occasions. Eventual treatment with trimethoprim-sulphamethoxazole led to remission, discontinuation of symptoms and negative cultures for *Aeromonas*[31].

Cholera-like disease was reported in the case of an individual infected with *A. sobria* having a cytotonic toxin. This was one of the two reported cases of *A. sobria* (now *A. veronii* bv sobria) causing disease in humans prior to 1983[32]. The other was a soft tissue infection in a SCUBA diver[33].

Another interesting case was that of a male child, 3 months old, suffering from Kwashiorkor and acute diarrhoeal disease, who had been fed tap water prior to 1 month of age. He had 3–10 watery stools per day, abdominal cramping and variable vomiting. Both *A. sobria* and *A. caviae* were isolated from the child's faeces. After amoxyacillin therapy for recurrent otitis media, only *A. caviae* could be recovered during subsequent acute diarrhoeal relapses. When the child was returned home, both foster parents contracted *A. caviae* gastroenteritis, suggesting intrafamilial transmission. Subsequently, the child and parents were treated with trimethoprim-sulphamethoxazole, which was successful judging by the resolution of symptoms and negative stool cultures[34].

Two other unique cases bear consideration. One involved a research microbiologist, who was working with a well-characterized strain of *Aeromonas*. While inadvertently mouth-pipetting a broth culture of the organism, he swallowed a large amount of the suspension (approximately 1.0 ml) and contracted an acute but self-limiting (within 3 days) case of diarrhoea. Subsequent stool cultures revealed the presence of the same organism on several occasions[35].

The other case also involved a microbiologist working in a large teaching hospital, who consumed a large amount of egg salad and contracted acute diarrhoea. Stool cultures were positive, and culture of the egg salad revealed an enormous growth of *A. hydrophila*. Investigation into the nature of the contamination led to the discovery that the boiled eggs, used in the egg salad, were being shelled and placed into a sink prior to preparation of the salad. Culture of the sink also revealed a heavy growth of *A. hydrophila*. Similar to the above case, the individual's diarrhoea was self-limiting and remission occurred without antibiotic therapy. There were a total of four employees in the hospital, who consumed egg salad from the hospital cafeteria, became ill and were found to be culture-positive for *A. hydrophila*[36,37].

In the 1960s an accidental laboratory ingestion of a suspension of *V. parahaemolyticus* led to the general acceptance that this organism was an intestinal pathogen[38]. Subsequent large-scale outbreaks of diarrhoeal disease confirmed this opinion[39]. The laboratory incidents described above with *Aeromonas* are analogous to the situation with *V. parahaemolyticus*. However, there have been no similarly confirmed and repeated outbreaks of *Aeromonas* diarrhoeal disease in large populations. Conversely, such information does stand in stark contrast to the fact that the only previous human volunteer study with *Aeromonas* was essentially unsuccessful in proving that *Aeromonas* was a cause of intestinal disease[40].

There have been case reports of complications of *Aeromonas* intestinal infection including: severe cholera-like disease[32,41]; chronic colitis post-*Aeromonas* gastrointestinal infection[42]; association with pancreatitis[43]; prolonged diarrhoea[44]; fatality[45] and small bowel obstruction[46].

When viewing the nature of these individual cases, the most obvious question to be asked then and now is whether or not there is a causal relationship between *Aeromonas* and the signs and symptoms of the affected individual. While no definitive answer is possible based on so few cases, it is tempting to believe that such symptomatology suggests that *Aeromonas* is an intestinal pathogen. Without larger numbers of cases, an absolute decision cannot be reached. Thus, the next section will review reports of large populations of individuals with diarrhoea and associated *Aeromonas* spp. identified through surveillance studies.

12.2.2 SURVEILLANCE STUDIES

There have been many reports resulting from the compilation of data obtained over various time periods to assess the relevant number of pathogens isolated over the total numbers of stool samples analysed. There is a general value in these numbers because they tend to show trends in the population at risk and suggest whether or not particular organisms are increasing or decreasing, relatively.

12.2.2.1 Outbreaks

Infectious diarrhoeal disease can be a serious problem in day care or long-term care settings. In a long-term care institution, there were 17 patients involved in an acute diarrhoeal episode in which 13 had two or more loose bowel movements lasting less than 48 h. Four of 11 patients who were cultured were positive for *A. hydrophila*. The condition was fatal to one patient who suffered profuse diarrhoea and concomitant complications leading to death. Presumably, previous, predisposing infirmities were associated with this patient's complications and eventual demise. The rapid occurrence of illness over a short period of time (3 days) strongly suggested the possibility of a foodborne or waterborne point-source outbreak[36].

An outbreak was reported from Sweden where 22 of 27 people enjoying a Swedish '*landgang*' (*smorgasbord*) became ill within 20–34 h of consumption and suffered the general signs and symptoms of acute diarrhoeal illness, that is severe acute diarrhoea, abdominal pain, headache, fever and vomiting. Two others became ill a day later. The symptoms were short in duration (48 h). *A. hydrophila* was isolated from all of the meat samples tested and quantitatively showed 10^6–10^7 cfu/g of sample. Other organisms including, faecal *Streptococcus*, coliforms, *Staphylococcus aureus*, fungi and yeast were present in low numbers[47].

Other outbreaks have been associated with seafoods. A direct link between the patient and the incriminated source of infection was suspected in a patient with severe *Aeromonas* infection after shucking shellfish[48]. Similarly, a case of moderate diarrhoea was reported after consumption of ready-to-eat shrimp cocktail[49].

Several recent studies have similarly implicated *Aeromonas* in outbreaks or in population studies in various locations including Scotland[50]; an industrial camp in British Columbia[51]; three nursing homes in Pennsylvania in the USA[52]; and in Palm Island and Townsville, Australia[53].

12.2.2.2   Field studies

In a study of US Navy personnel on deployment in Egypt, there were 183/4500 troops who suffered acute diarrhoeal episodes in a two-month period. A possible aetiologic agent was identified in 49% of the cases. Only one individual was associated with *A. hydrophila*[54]. In another study of US military on deployment[55], 36 individuals with diarrhoeal disease were cultured and none were found to have *Aeromonas* in their stools.

In a temporary clinic established in Dos Palos, Peru, faecal specimens from 20 diarrhoeic children and 10 non-diarrhoeic controls were tested for common agents of diarrhoea. The largest number of cases was attributed to enterotoxigenic *Escherichia coli* (ETEC) (30%), while *Aeromonas* was found in 20% of the cases. Other agents were enteropathogenic *E. coli* (15%) and *Campylobacter* (15%). Isolates from control specimens were *Aeromonas* (10%) and *Campylobacter* (10%). In Lima, diarrhoeic children less than 3 years old had an *Aeromonas* prevalence of 21–42%. Isolations from healthy children were 4–7% for *Aeromonas* and 9–13% for *Campylobacter*[56].

In a study of an adult population in Lima, Pazzaglia and colleagues observed a frequency of 9.2% in 655 *Aeromonas*-associated diarrhoeic cases and 3.5% in 287 non-diarrhoeic individuals. This compared closely with the finding for *Shigella* of 9.8% and 0%, respectively[57].

A survey was conducted in Djibouti to examine the causes of diarrhoea in that country. Surprisingly, in a population of 209 diarrhoeal cases and 100 controls, most of the common diarrhoeal agents showed similar frequencies in both groups. The exceptions were *A. hydrophila* group (3.3%) and *Shigella* spp. (7.7%). Neither organism was found in any of the control samples[58]. This finding in Djibouti was similar to the results obtained in a study earlier in nearby Mogadishu, Somalia, which concluded that *Aeromonas* spp. should be considered enteric pathogens, whenever isolated from diarrhoeal stools in the region[59].

Echeverria and co-workers conducted a prospective study in 35 Peace Corps volunteers during their first five weeks in Thailand and found that 20 (57%) developed the travellers' diarrhoea syndrome. Recognized bacterial enteropathogens were isolated in 47% of 39 diarrhoeal episodes. *A. hydrophila* was isolated from 31% of the 39 diarrhoeal episodes, and was the most common organism associated with traveller's diarrhoea. After recovery, only three of 35 volunteers exhibited *A. hydrophila* in their stools. The organism was isolated from 14% of stools from volunteers with traveller's diarrhoea and 9% of stools from individuals without diarrhoea. *A. hydrophila* was the only potential enteric pathogen isolated from two individuals within 24 h after the onset of traveller's diarrhoea[60].

In a corollary study in Thailand, Pitarangsi and colleagues studied another group of 33 Peace Corps volunteers and three populations of Thais. The results with the second group of volunteers were similar to those described above. In seven episodes of travellers' diarrhoea, *A. hydrophila* was the only possible enteric pathogen isolated. In the Thai populations, there was virtually no difference between experimental and control populations. However, in the 0–2-year-old group in the Soongnern district, there was a 20% rate in the experimental population, with an 8% rate in the control group. In a similar group seen in a children's hospital, the rates were 9% for both groups. Perusal of the data for Soongnern district also shows that 3–20-year-old individuals showed little difference between experimental and control groups. However, adults (>20 years old) showed a 34% rate in the experimental group and 27% in the control group, a significant difference. This type of activity is not uncommon. Others have shown that effects of *Aeromonas* are exhibited more in the very young and in adult populations. This observation perhaps reflects a reduced antibody level to the organism in the more susceptible age groups[61].

In a later Thai study comparing incidence in a hospital-based population with that in household contacts and neighbours, *Aeromonas* was isolated in 9% of cases with diarrhoea versus 2% without diarrhoea. *Aeromonas* was isolated from 5 (8%) of 65 household contacts and only one (1%) of 70 neighbours of persons with *Aeromonas*-associated diarrhoea, and from 28 (2%) of 1471 persons who lived in the same rural area, but had not been in contact with persons with diarrhoea from whom *Aeromonas* was isolated[62].

In a follow-up study in Peace Corps volunteers in Thailand, Taylor and co-workers found an incidence of 10% in symptomatic cases from whom *Aeromonas* was isolated and only one (3%) in an asymptomatic case[63,64].

While there has been diminished emphasis lately over infection by *Aeromonas* occurring because of the immunocompromised nature of an individual, acquired immunodeficiency syndrome in HIV-positive individuals still requires attention. In a study of HIV-positive patients *Aeromonas* was found in 9/43 cases (21%). The incidence of *Aeromonas* was higher in these patients than in an otherwise healthy cohort also exhibiting diarrhoea[65].

Thus far, California has been the only state in the USA to require reporting of *Aeromonas* infections. A report of the data collected showed that the gastrointestinal tract was the most common site for isolation of *Aeromonas* (81%). There were no common source outbreaks but the high rate supported evidence from studies, such as those described above, that *Aeromonas* is an enteric pathogen. Because *Aeromonas* was thought to be largely non-preventable, it was suggested that public health surveillance was not necessary and mandatory reporting was discontinued[66].

In a retrospective study, Moyer compiled information on 56/107 patients, who had *Aeromonas* recovered from their stool specimens. *A. caviae* was the most frequently isolated species, with seasonality for all species occurring in the warmer months of June, July and August. There was a clear association between drinking and swimming in untreated water and *A. caviae* isolation[67]. Similarly, Holmberg and colleagues found a distinct relationship between the drinking of untreated water and the occurrence of acute gastroenteritis in children and enteritis in adults[68].

In a study of 978 Finnish tourists travelling to Morocco, 16 travellers with diarrhoea were compared with 39 diarrhoeal individuals without recent history of travelling to a foreign country. The isolation rates for *Aeromonas* were 8.7% and 1.4%, respectively, with *Aeromonas* as the sole pathogen in 5.5% of the cases. Interestingly, in the Moroccan travel group, *A. veronii* bv sobria and *A. caviae* were the most common isolates, while in the untravelled group, *A. hydrophila* and *A. caviae* were most common[69].

12.2.2.3   Laboratory-based studies

Numerous reports have been published on the rate of isolation of *Aeromonas* from stool specimens based on compilation and analysis of laboratory data. Various procedures have been used for isolation and identification, and some reports show attempts to identify to species, others to major phenotypes and others only to the genus *Aeromonas*. In varying numbers, the frequency reported is based on *Aeromonas* as the sole isolate and in others, in combination with other enteropathogens. Some reports differentiate between the two categories. When these reports are reviewed from a geographical point of view, they offer some interesting comparisons and significant insights into the occurrence of *Aeromonas*-related gastrointestinal disease.

In assessing the frequency of isolation of *Aeromonas* in Table 12.1, there is a broad range from 80% in Canada to 0.18% in Denmark. Other countries with high rates were Peru (52.4%), Japan (11.1%), Australia (11%), Thailand (18% and 31%) and England (11%). Interestingly, a second study in Manitoba, Canada revealed that *A. hydrophila* was the fourth-ranked enteropathogen in frequency of occurrence. A profile of the study is given in Table 12.2. Only *A. hydrophila* was identified. The largest number of cases were <5 and >16 years old. The signs and symptoms of disease were rather typical of *Aeromonas*-related diarrhoea[70].

The study by Gluskin and colleagues[80] in Israel provides further details regarding *Aeromonas* intestinal disease. They found *Aeromonas* to constitute only 4% of all enteropathogens; 94% of all cases were <3 years old; 78% were <1 year old; peak incidence of the disease occurred in children, ages 2–6 months, in whom diarrhoea was most severe with dehydration, vomiting, acidaemia and azotaemia; bloody diarrhoea in 7% of the children; and almost all strains were resistant to ampicillin. The latter observation suggests that these strains were either *A. caviae* or *A. trota*.

In US Navy overseas research, studies in Jakarta, Indonesia revealed an incidence of 11% *Aeromonas* in patients with diarrhoea and 4% in controls,

Table 12.1 Association of *Aeromonas* with diarrhoea

Location	Diarrhoea (%)	Non-diarrhoeic (%)	P < 0.05	Year	Reference
Canada	32/40 (80)	9/22 (41)	*	1979	71
Australia	105/975 (11)	7 (975)	*	1983	72
India	45/2480 (1.8)	0/512 (0)	*	1991	73
India	170/12, 605 (13.5) 1978–81	ND		1990	74
	830/12, 605 (5.0) 1986–87	ND			
Bangladesh	875/2, 654 (33)	ND		1986	75
Bangladesh	67/271 (24)	ND		1988	76
Peru	205/391 (52.4)	12/138 (8.7)	*	1991	56
Peru	40/343 (11) Peruvian adults	ND			77
	50/378 (12) Peruvian children	ND			77
	8/156 (5) American adults	ND			77
Saudi Arabia	58/15, 548 (0.4)	0/1368 (0)	*	1991	78
Israel	17/1005 (1.7)	0/500 (0)	*	1990	79
Israel	146/13, 820 (1.1)	ND		1992	80
Nigeria	53/2350 (2.6)	2/500 (0.4)	*	1990	81
Djibouti	7/209 (3.3)	0/100 (0)	*	1990	58
Thailand (Travellers)	18/76 (24)	5/105 (5)	*	1981	60
Thailand (Thais)	37/207 (18)	44/367 (12)	*	1982	61
Thailand (Travellers)	12/39 (31)	3/35 (9)	*	1982	61
Indonesia	186/1695 (11)	14/338 (4)	*	1987	77
Japan	29/262 (11.1)	202/9104 (2.2)	*	1988	82
England	13/123 (11)	28/846 (3)	*	1983	83
England	38/445 (8.5)	5/402 (1.3)	*		84
Netherlands	208/34 311 (0.61)	ND		1989	7
Denmark		(0.18)		1961	23
France	30/4426 (0.67)	–		1965	85
Finland	249/13, 027 (1.9)	0/343 (0)	*	1995	30
Italy	21/561 (4)	12/576 (2)		1986	86
USA	20/1797 (1)	0/553 (0)	*	1985	87
USA	14/321 (3.2)	2/380 (0.5)	*	1988	88
USA	15/246 (6)	11/155 (7)		1988	89
USA	112/5461 (2.1)	ND		1965	90
USA	243/3334 (7.3)	ND		1987	91

Adapted and modified from References 92 and 93

Table 12.2   Profile of *A. hydrophila*-associated
diarrhoeal disease in Manitoba, Canada[70]

- *A. hydrophila* ranked fourth in frequency
- 392 positive stools for *A. hydrophila*
- 284 cases could be followed up
- 211 had diarrhoea
- 21 had a co-pathogen
- 182 had *A. hydrophila* only
- Most were from rural communities
- 66.3% <5 years old
- 3.1% 6–15 years old
- 20.4% >16 years old
- Mean diarrhoea duration – 6.9 days
- XS mucus – 21%
- Bloody diarrhoea – 27%
- Vomiting – 35%
- Abdominal pain – 40%

and in a second study of secretory diarrhoea, 34/196 individuals had *Aeromonas* isolated. Seven had *Aeromonas* as the sole pathogen[77]. In another study in Manila, Philippines, there was an incidence of 2% in diarrhoeal patients in 1983 and 1% in 1984. Controls were 0.5% and 0.3% respectively[77].

Most of the studies listed in Table 12.1 involve populations of all ages. In general, these studies showed that *A. caviae* was the major species isolated, usually followed by *A. hydrophila* then *A. veronii* bv veronii or bv sobria.

A major exception to this finding is in the report from Bangladesh[75], where *A. veronii* bv sobria was the major species isolated (43%), followed by *A. hydrophila* (30%), and then *A. caviae* (27%). These numbers are similar to those of Janda and colleagues[94] who isolated the three species in approximately equal numbers from faecal samples, and Carnahan and co-workers, who found a similar distribution, except for one year when there was an increase in *A. caviae* from the very young and very old patients[95]. Another contrast in the Bangladesh study was that the species ratio did not vary among population age groups, while, as a group, the patients tended to be older.

In Hawaii, *Aeromonas* was found to be the third most common pathogen. In most cases, the disease was a self-limiting enteritis. The stool isolation rate for *Aeromonas* was 2.9% versus 6.4% for *Campylobacter jejuni* and 3.5% for *Salmonella* spp. It was found retrospectively that patients with *Aeromonas* isolations had more enduring symptoms than diarrhoeal patients from whom no enteropathogen was isolated[96].

In a 3-year study of *Aeromonas* in a community-based hospital of the Annapolis area near the Chesapeake Bay in the US, Carnahan and colleagues[95] found that most of their *A. caviae* isolates were from faecal samples (75%) and that *A. caviae* was predominant in one of the three years of study. Otherwise, the distribution was equal among the three phenotypes. The age distribution was fairly uniform, with the highest number of cases associated with *A. caviae* occurring in children under 1 year old and adults over 70 years old. The enteric pathogen isolation rate for 1990–91 showed *Aeromonas* to be the second most commonly isolated enteric pathogen (1.7%) following *Salmonella* (4.3%).

Also of interest were repeated studies from the same locations, e.g. India[74] and Thailand[60,61], where frequencies actually increased in later studies presumably because of either greater awareness, better methods of isolation, or actual increases in the prevalence of *Aeromonas*.

In the USA, frequency varied from 1.0 to 7.3%. In the study by Brake and Osterhout[90] at Johns Hopkins University in Baltimore, Maryland, *Aeromonas* ranked as high as second among enteropathogens, whereas in other US studies it ranked as low as number five. Their findings were consistent with those of Carnahan and colleagues[95], whose studies were done in a proximal location to Baltimore.

12.2.3   *A. CAVIAE* AS A PAEDIATRIC PATHOGEN

Several of the studies mentioned above have either alluded to or directly stated that *A. caviae* is the most prevalent of the three major phenospecies, the other two being *A. hydrophila* and *A. veronii* biovars sobria and veronii. The unanswered question concerned the ability of *A. caviae* actually to cause disease during colonization. Since many investigators have addressed this question more closely in young children, perhaps the answer lies in their investigations.

It appears that *Aeromonas* can colonize very early in life. Pazzaglia and colleagues[97] obtained daily swabs from Caesarean newborns for approximately one week and recovered the organism from 12 of 52 of the infants. None were clinically ill, but a few intermittent watery stools were noted. Hospital water seemed to be the source of the organism. Some of those who were culture-positive had different phenotypes on repeated culture and most were negative at 15- and 30-day follow-ups. Four of the eight positive neonates exhibited *A. caviae*, suggesting that *A. caviae* was not an enteric pathogen.

Aeromonas as a paediatric pathogen has been studied virtually worldwide. In the USA Namdari and Bottone[98] found *A. caviae* to be predominant mostly in infants who were <1 year old. None were breast fed. Fourteen of 17 were positive for *Aeromonas* solely, and all had a stool pH of >7.5 They postulated that formula-fed infants have an intestinal shift from an acidic to an alkaline environment, which favours survival of *A. caviae*. They later demonstrated cytotoxicity of *A. caviae* clinical strains as well as adhesion to

HEp-2 cells[98]. Karunakaran and Sur[99] similarly found a β-haemolytic toxin, which was significantly reduced in the presence of iron chelators. Cohen[100] reported that others have found prevalence of *Aeromonas* in children <3 years old. Burke and colleagues[72] reported an incidence of 10% diarrhoeal disease in children caused by *Aeromonas*, and found the organism in only 0.6% of the control population. Also, Figura and colleagues[86], found prevalence of *A. caviae*, but no difference in isolation rates from paediatric patients and controls in Italy, which was similar to the results obtained by Kotlof and co-workers in Baltimore, MD, although they did not identify their *Aeromonas* isolates to species[89].

In two studies in the USA, Philadelphia, Pennsylvania[101] and Shreveport, Louisiana[102], there were prevalences of 4% and 14% observed, respectively. In the Louisiana study a cohort population showed only 1.4% positive isolations for *Aeromonas*. The species were not given. In another study in Tallahassee, Florida, the prevalence was less than 0.1%[103], while in New York, in 1984, an 18% prevalence was noted. As many as 80% of the cases were in children ≤16 months[104].

In a prospective study of children <24 months in day-care centres, there were two major outbreaks, with *Aeromonas* isolated in 24% of the children in outbreak 1, and 21% in outbreak 2. In 37 sporadic cases of children with diarrhoea, only 2.4% were positive for *Aeromonas*. There were no other enteropathogens in the second outbreak[105].

In Europe, the results approximated those reported in the USA. Hunt and colleagues[106] in England found an isolation rate of 4.4%, while Wilcox and co-workers[107] observed 2.5%. In the prior study, *A. caviae* was the predominant species isolated (80%). In Germany, symptomatic patients were positive for *Aeromonas* at a 2.5% rate as compared with 0.3% in the non-diarrhoeic group[108]. In France, the rate was similar (2.3%), with *A. caviae* predominating[109]. A retrospective study in Spain[110] showed 3.7% of infectious diarrhoeal patients <14 years old to have *Aeromonas*, 56% of which were the sole pathogen. *A. caviae* predominated, especially in children who were being or had been fed with artificial lactation, an interesting observation in view of the conclusions by Namdari and Bottone[98].

On the South American continent two studies, one in Peru, and the other in Chile, presented an interesting contrast. Pazzaglia and colleagues[111] in Peru, when studying infants <18 months, found 46% of the diarrhoeic patients to be positive compared with 9% in matched controls. In 47% of the isolates, *Aeromonas* was the only enteropathogen isolated.

Conversely, in Chile, the isolation rate in the same age group was 7.1% and 5.2% in paired asymptomatic infants. However, carriage of enteropathogenic *E. coli*, enterotoxigenic *E. coli* and *Campylobacter* in the asymptomatic group was also high. The researchers found that diarrhoea was longer lasting in cases where the *Aeromonas* isolates were invasive and/or toxigenic[112].

In two studies in the middle East, the observations were again similar in

Israel, with an incidence of 4%[80]. However, in Egyptian children, *Aeromonas* was isolated from 88% of diarrhoeic and 45% of non-diarrhoeic children. Of interest, studies of the water supplies in Cairo showed nine of 10 samples to be positive for *Aeromonas*[113].

Gracey and colleagues[114] also found 10% of hospitalized children in Jakarta, Indonesia to have *Aeromonas*, with fewer than 0.4% in an age-matched group without diarrhoea. Santoso and colleagues[115] in a study in Denpasar, Bali, Indonesia found rates of 15.5% and 12.7% respectively in diarrhoeic and non-diarrhoeic children. However, when they considered only the strains which produced cytotoxic enterotoxin, the rates were 7.1% and 2.8% respectively.

A frequency of 3.7% was observed in a single study in Varanasi, India[116], while no *Aeromonas* spp. were isolated in Seoul, Korea during a 14-month period[117].

The rate of isolation from paediatric populations in the above studies varies with geographic location, ranging from 0% in Korea to almost 90% in Egypt. A number of other variables could be responsible for the variations seen among these studies as noted above. However, the premise of Namdari and Bottone[98], regarding favourable alteration of the pH of the intestinal tract for *A. caviae*, seems to be valid thus far. Further, in every study where identification to species was performed, *A. caviae* was the predominant species.

The presence of *A. caviae* in the control populations of some of the studies begs the question as to whether all *A. caviae* strains are enteropathogens. If all hosts are considered equal, then the answer should be 'no'. However, if there are host immune or physiologic factors involved, then the answer could be 'yes'. The remaining possibility is that all of the *A. caviae* strains are not equal, which suggests the possibility of virulent biotypes. If this is the case, we probably do not yet have the tools to verify or disprove such a hypothesis. The question of causality is further considered below.

12.3   IMMUNE RESPONSE

Several studies using the agglutination test reached different conclusions regarding its value in measuring antibody response to acute diarrhoea presumably caused by *Aeromonas* spp. Some found it to be valuable and showed significant results, whereas others found no differences between tests and control samples[118–121].

Kuijper and colleagues[122] used three different serum assays including bacterial agglutination, enzyme-linked immunosorbent assay (ELISA) and toxin-neutralization. Agglutination was not considered very useful, suggesting that the above application studies were of questionable value. The sensitivity of the ELISA was only 30% in patients with acute or chronic *Aeromonas*-related diarrhoea, and specificity was 74%. The results correlated IgM and IgG responses to lipopolysaccharide (LPS) of homologous strains. The toxin neutralizing assay had 46% sensitivity and 94% specificity. It appeared useful in distinguishing patients with acute diarrhoea apparently caused by *Aeromonas* spp.

In a single study of intestinal secretory IgA, response was considered to be an indicator of enteropathogenicity. Eleven of 12 subjects shedding *A. sobria* or *A. hydrophila* had a fourfold or higher titre against the homologous strain. LPS was the predominant antigen. There was no rise in titre evident for two patients who were shedding *A. caviae*[123].

12.3.1 VIRULENCE FEATURES

Several virulence features have been reported for *Aeromonas* spp., with the investigators ascribing the characteristic to a particular species which may or may not have been correctly identified at the time of the study. These features include aerolysin[124]; cytotonic enterotoxin[125,126]; and Chinese hamster ovary (CHO) cell elongating toxin[127-129].

Other features include proteases[130]; endotoxin[131]; enterovasiveness in HEp-2 cells[132-135]; surface adhesions related to outer membrane proteins and fimbriae[136-138].

Extensive genetic analysis of the *pil* gene of *Aeromonas* revealed that it is actually a cluster of genes (*tap*ABCD) that are homologous to *P. aeruginosa* type IV pilus biogenesis genes (*pil*ABCD). It was shown that *tap*D is required for extracellular secretion of aerolysin and protease, indicating that *tap*D may play a central role in the virulence of *A. hydrophila*[138].

12.3.2 BIOTYPING AND SEROTYPING

Biotyping has been attempted on the basis of comparative virulence features, e.g. between either clinical and environmental groups or between species[71,139-148]. Prior to the recent undertaking of the taxonomy of the aeromonads and existing methods to identify all 14 of the hybridization groups of phenospecies, investigators believed that certain biochemical characteristics were indicative of virulence. Most of those findings have now been shown to be erroneous[71,149,150], although one method, pyrazinamidase activity, may serve as a phenotypic and/or virulence marker[151].

Other biotyping methods include phage typing[152]; esterases[153]; adherence patterns[137,154]; ribotyping[155-157]; suicide phenomenon[158]; serotyping[159,160]; S-layer and LPS analysis[131].

12.3.3 SOURCES OF *AEROMONAS* SPP.

A single commonly repeated question relates to the source of *Aeromonas* during outbreaks or even in sporadic cases. The possibility of human to human dissemination was described above in household contacts[62].

Sources from the environment include drinking water, estuaries, aquatic waters and aquaria[33,161–170], seafoods[171,173,174], colonic irrigation machine[175], red meat[172,176], vegetables[177,178], domestic and wild fowl[18,179], and milk and dairy products[180].

12.4 CAUSATION – FACT OR FICTION?

Several excellent reviews or reports have been written which address the issue of causation[68,181–187]. Thus far the opinions have been equivocal or negative pending further investigation and information.

There are particular facts that can be drawn from the data collected thus far. In most regions and studies, *Aeromonas* rates are higher in affected populations, in both adults and children. Virgin populations during early arrival to a region are affected most frequently. Also, paediatric populations are most heavily affected, especially with *A. caviae*.

There are numerous sources of *Aeromonas* spp., which would suggest that there should be more illness than is observed. Perhaps as suggested by Daily and colleagues[140], and others since[94,187–189], there are particular biotypes of *Aeromonas* which cause illness, much as seen with *E. coli*. This hypothesis could explain the conundrum of multiple phenotypes seen at times in single parents. *Aeromonas* appears to have a substantial array of virulence features, certainly enough to cause diarrhoeal disease. While there are numerous biotyping methods available, we still require others. Perhaps we are still not using the correct methodologies to differentiate the enteropathogens from non-pathogens. Certainly, more well designed epidemiological studies are needed, especially in the absence of major outbreaks. However, the individual cases and small outbreaks mentioned above lend greater credence to causality now than was possible previously.

The spectre of the one failed human study[40] and absence of adequate animal models are understandable obstacles, but considering the various criteria of causality which can be applied to the existing data, one could make a case for *Aeromonas* as a probable causal agent based on consistency, temporal measure and strength. Specificity and coherence (biological plausibility) require further evidence.

I would propose another measure – the concept of cruciality, that is, how often cases are seen on a regular basis in locales around the world. Based on the large accumulation of data on humans, it certainly appears that *Aeromonas* is isolated in large numbers of cases of enteric disease and disappears when remission occurs. There appears to be sufficient evidence to consider *Aeromonas* (possibly certain biotypes) as putative enteropathogens, and to recommend routine culture and antibiotic treatment in acute cases, to continue active research in all areas and to recommend preventative measures in food preparation, to treat drinking water and to follow appropriate hygienic procedures in hospitals, nursing homes and day care centres. These views are consistent with the recent publication by the Institute of Medicine, National Academy of Sciences, which cites *Aeromonas* as an emerging microbial threat to health in the USA[190].

NOTE IN PROOF

It has come to my attention that Zengshan *et al.*[191] reported an outbreak in a population of 115 persons attributed to ingestion of pork contaminated with *Aeromonas hydrophila*, which was also isolated from the fresh pork used in the preparation of the meal. The convalescent serum titres were as high as 1:100.

12.5 REFERENCES

1 Kluyver AJ, van Niel CB. Prospects for a natural system of classification of bacteria. *Zbl. Bakteriol. II. Orig* 1936; **94**: 369–403.
2 Joseph SW, Carnahan A. The isolation, identification, and systematics of the motile *Aeromonas* species. *Annu Rev Fish Dis* 1994; **4**: 315–43.
3 Colwell RR, MacDonnell MT, DeLey J. Proposal to recognize the family *Aeromonadaceae* fam. nov. *Int J Syst Bacteriol* 1986; **36**: 473–7.
4 Lee JV, Bryant TN. A numerical taxonomic study of *Aeromonas*. *J Appl Bacteriol* 1984; **57**: 17–18.
5 Arduino MF, Hickman-Brenner FW, Farmer JJ III. Phenotypic analysis of 132 *Aeromonas* strains representing 12 DNA hybridization groups. *J Diarrh Dis Res* 1988; **6**: 137.
6 Altwegg M, Steigerwalt AG, Altwegg-Bissigg R, Luthy-Hottennstein J, Brenner DJ. Biochemical identification of *Aeromonas* genospecies isolated from humans. *J Clin Microbiol* 1990; **28**: 258–64.
7 Kuijper EJ, Steigerwalt AG, Schoenmaker BSCIM, Peters MF, Zanen HC, Brenner DJ. Phenotypic characteristics and DNA relatedness in human fecal isolates of *Aeromonas* spp. *J Clin Microbiol* 1989; **27**: 132–8.
8 Popoff MY, Coynault C, Kiredjian M, Lemelin M. Polynucleotide sequence relatedness among motile *Aeromonas* species. *Curr Microbiol* 1981; **5**: 109–11.
9 Fanning GR, Hickman-Brenner FW, Farmer JJ III, Brenner DJ. DNA relatedness and phenotypic analysis of the genus *Aeromonas*. *Abstr Annu Meet Am Soc Microbiol* 1985; C116.
10 Carnahan AM, Joseph SW. Systematic assessment of geographically and clinically diverse aeromonads. *System Appl Microbiol* 1993; **16**: 72–84.
11 Colwell RR, Kaper J, Joseph SW. *Vibrio cholerae, Vibrio parahaemolyticus* and other vibrios: occurrence and distribution in Chesapeake Bay. *Science* 1977; **198**: 394–6.
12 Colwell RR, Huq A. Vibrios in the environment: viable but non-culturable *Vibrio cholerae* In Wachsmuth IK, Blake PA, Osvik O (eds) Vibrio cholerae *and cholera: molecular to global perspectives 1994*. Washington, DC: American Society for Microbiology, 1994: 117–33.
13 Zimmerman OER. Die bakterien unserer Trink und Nutzewässer. *Ber*

Naturwiss Ges Chemnutz 1890; **11**: 86.

14 Miller RM, Chapman WR. *Epistylis* sp and *Aeromonas hydrophila* infections in fishes from North Carolina reservoirs 1996. *Prog Fish Cult* 38: 165–8.

15 Russell FH. An epidemic: septicemic disease among frogs due to the *Bacillus hydrophilus fuscus*. *JAMA* 1898; 1442–9.

16 Camin JH. Mite transmission of hemorrhagic septicemia in snakes. *J Parasitol* 1948; **34**: 345–54.

17 Pierce RL, Daley CA, Gates CE, Wohlgemuth K. *Aeromonas hydrophila* septicemia in a dog. *J Am Vet Med Assoc* 1973; **162**: 469.

18 Shane SM, Gifford DH. Prevalence and pathogenicity of *Aeromonas hydrophila*. *Avian Dis* 1984; **29**: 681–9.

19 Aiken RS, Barlin B, Miles AA. A case of botulism. *Lancet* 1936; **ii**: 780.

20 Miles AA, Halman ET. A new species of microorganism (*Proteus melanovogenes*) causing black rot in eggs. *J Hyg* 1937; **37**: 79–97.

21 Hill KR, Caselitz F-H, Moody LM. A case of acute metastatic myositis caused by a new organism of the family *Pseudomonadaceae*: a preliminary report. *W Ind Med J* 1954; **3**: 9–11.

22 Caselitz F-H. Ein neues Bakterium der Gattung *Vibrio Müller, Vibrio jamaicensis*. *Z Tropenmed Parasitol* 1955; **6**: 52–3.

23 Lautrop H. *Aeromonas hydrophila* isolated from human faeces and its possible pathological significance. *Acta Pathol Microbiol Scand* 1961; **51**: 299–301.

24 Rosner R. *Aeromonas hydrophila* as the etiologic agent in a case of severe gastroenteritis. *Am J Clin Pathol* 1964; **42**: 402–4.

25 Kok N. *Aeromonas hydrophila* s. liquefaciens isolated from tonsillitis in man. *Acta Pathol Microbiol Scand* 1967; **72**: 599–602.

26 Caselitz FH. Zur Frage von *Pseudomonas aeruginosa* und verwandter Mikroorganismen als Enteritiserreger. *Z Tropinemed Parasitol* 1958; **9**: 269–71.

27 Schubert RHW. Genus II. *Aeromonas* Kluyver and Van Niel 1936, In: Buchanan RE, Gibbons NW (eds) *Bergey's Manual of Determinative Bacteriology* (8th edn), Baltimore, MD: Williams and Wilkins, 1974: 345–8.

28 von Graevenitz A, Mensch AH. The genus *Aeromonas* in human bacteriology. *New Engl J Med* 1968; **278**: 245–9.

29 Slotnick IJ. *Aeromonas* species isolates. *Ann NY Acad Sci* 1970; 174: 503–10.

30 Rautelin H, Hanninen ML, Sivonen A, Turunen U, Valtonen V. Chronic diarrhoea due to a single strain of *Aeromonas caviae*. *Eur J Clin Microbiol Infect Dis* 1995; **14**: 51–3.

31 Rahman AFMS, Willoughby JMT. Dysentery-like syndrome associated with *Aeromonas hydrophila*. *BMJ* 1980; **281**: 976.

32 Champsaur H, Andremont A, Mathieu D, Rottman E, Auzepy P. Cholera-like illness due to *Aeromonas sobria*. *J Infect Dis* 1982; **145**: 248–54.

33 Joseph SW, Daily OP, Hunt WS, Seidler RJ, Allen DA, Colwell RR. *Aeromonas* primary wound infection of a diver in polluted waters. *J Clin Microbiol* 1979; **10**: 46–9.

34 Moyer NP, Larew MS. Recurrent gastroenteritis caused by *Aeromonas* species: a case history. *J Diarrh Dis Res* 1988; **6**: 144.

35 Carnahan A, Chakraborty T, Fanning GR, Verma D, Ali A, Janda JM, Joseph SW. *Aeromonas trota* sp. nov.: an ampicillin susceptible species isolated from clinical specimens. *J Clin Microbiol* 1991; **29**: 1206–10.

36 Bloom HG, Bottone EJ. *Aeromonas hydrophila* diarrhoea in a long-term care setting. *J Am Geriatr Soc* 1990; **38**: 804–6.

37 Bottone E. Personal communication, 1993.

38 Sakazaki R, Tamura K, Kato T, Obara Y, Yamai S, Hobo K. Studies on the enteropathogenic, facultatively halophilic bacteria, *Vibrio parahaemolyticus*. III.

Enteropathogenicity. *Jap J Med Sci Biol* 1968; **21**: 325–31.

39 Sakazaki R, Tamura K, Prescott LM, Dencic Z, Sanyal SC, Sinha R. Bacteriological examination of diarrhoeal stools in Calcutta. *Indian J Med Res* 1971; **59**: 1025–34.

40 Morgan DR, Johnson PC, DuPont HL, Satterwhite TK, Wood LV. Lack of correlation between known virulence properties of *Aeromonas hydrophila* and enteropathogenicity for humans. *Infect Immun* 1985; **50**: 62–5.

41 Gelbart SM, Prabhudesai M, Magee SM. A case report: *Aeromonas sobria* gastroenteritis in an adult. *Am J Clin Pathol* 1985; **83**: 389–91.

42 Dickinson RJ, Wight DGD. Chronic colitis after *Aeromonas* infection. *Gut* 1989; **30**: 1436–7.

43 Pelayo Melero MJ, Segura GA, Cortes LE, Lu Cortez L. Acute gastroenteritis from *Aeromonas sobria* and pancreatis: apropos of a case (letter). *Med Clin (Barc)* 1993; **101**: 238 (In Spanish).

44 delVal A, Molés J-R, Ganigues V. Very prolonged diarrhoea associated with *Aeromonas* diarrhoea. *Am J Gastroenterol* 1990; **85**: 1535.

45 Abgonlahor DE. The role of *Aeromonas* in acute diarrhoeal disease in Nigeria. *Cent Afr J Med* 1983; **29**: 142–5.

46 Block K, Braver JM, Farraye FM. *Aeromonas* infection and intramural intestinal hemmorhage as a cause of a small bowel obstruction. *Am J Gastroenterol* 1994; **89**: 1902–3.

47 Krovacek K, Dumontet S, Ericksson E, Balada SB. Isolation and virulence profiles of *Aeromonas hydrophila* implicated in an outbreak of food poisoning in Sweden. *Microbiol Immunol* 1995; **39**: 655–61.

48 Flynn TJ, Knipp IG. Seafood shucking as an etiology for *Aeromonas hydrophila* infection. *Arch Intern Med* 1987; **147**: 1816–17.

49 Altwegg M, Martinetti Lucchini G, Lüthy-Hottenstein J, Rohrbach M. *Aeromonas*-associated gastroenteritis after consumption of contaminated shrimp. *Eur J Clin Microbiol Infect Dis* 1991; **10**: 44–5.

50 Nathwani D, Laing RB, Harvey G, Smith CC. Treatment of symptomatic enteric *Aeromonas hydrophila* infection with ciprofloxacin. *Scand J Infect Dis* 1991; **23**: 653–4.

51 Glover D, Ross A, Lugsdin J. Gastroenteritis outbreak at an industrial camp – British Columbia. *Can Commun Dis Rep* 1992; **18**: 66–8.

52 Sims RV, Hauser RJ, Adewale AO, Maislin G, Skeie S, Lavizzo-Mourey RJ, Rubin H. Acute gastroenteritis in three community-based nursing homes. *J Gerontol A Biol Sci Med Sci* 1995; **50**: M252–6.

53 Ashdown LR, Koehler JM. The spectrum of *Aeromonas*-associated diarrhoea in tropical Queensland, Australia. *Southeast Asian J Trop Med Publ Health* 1993; **24**: 347–53.

54 Haberberger RL, Mikhail IA, Burans JP, Hyams KC, Glenn JC, Diniega BM, Sorgen S, Mansour N, Blacklow NR, Woody JN. Traveller's diarrhoea among United States military personnel during joint American–Egyptian armed forces exercises in Cairo, Egypt. *Mil Med* 1991; **156**: 27–30.

55 Oyofo A, el-Gendy A, Wasfy MO, el-EtrsH Churilla A, Murphy J. A survey of enteropathogens among United States military personnel during operation Bright Star '94, in Cairo, Egypt. *Mil Med* 1995; **160**: 331–4.

56 Pazzaglia G, Podgore J, Mercado W, Martinez A, Urteaga A, Echevarry E. The etiology of childhood diarrhoea in northern coastal Peru: the 1989 Fuerzas Unidas humanitarian civic action-a model of international and interservice cooperation, community service, and scientific opportunity. *Mil Med* 1991; **156**: 402–5.

57 Pazzaglia G, Escamilla J, Batchelor R. The etiology of diarrhoea among

American adults living in Peru. *Mil Med* 1991; **156**: 484–7.

58   Mikhail IA, Fox E, Haberberger RL, Ahmed MH, Abbate EA. Epidemiology of bacterial pathogens associated with infectious diarrhoea in Djibouti. *J Clin Microbiol* 1990; **28**: 956–61.

59   Casalino M, Yusuf MW, Nicoletti M, Bazzicalupo P, Coppo A, Colonna B, Capelli C, Bianchini C, Falbo V, Ahmed HJ, Omar KH, Maxamuud KB, Maimone F. A two-year study of infections associated with diarrhoeal diseases in children in urban Somalia. *Trans R Soc Trop Med Hyg* 1988; **82**: 637–41.

60   Echeverria P, Blacklow NR, Sanford LB, Cukor GG. Traveller's diarrhoea among American Peace Corps volunteers in rural Thailand. *J Infect Dis* 1981; **143**: 767–71.

61   Pitarangsi C, Echeverria P, Whitmire R, Tirapat C, Formal S, Dammin GJ, Tingtalapong M. Enteropathogenicity of *Aeromonas hydrophila* and *Plesiomonas shigelloides*: Prevalence among individuals with and without diarrhoea in Thailand. *Infect Immun* 1982; **35**: 666–73.

62   Echeverria P, Seriwatana J, Taylor DN, Yanggratoke S, Tirapat C. A comparative study of enterotoxigenic *Escherichia coli, Shigella, Aeromonas* and *Vibrio* as etiologies of diarrhoea in Northeastern Thailand. *Am J Trop Med Hyg* 1985; **34**: 547–54.

63   Taylor DN, Blaser MJ, Blacklow N, Echeverria P, Pitarangsi C, Cross J, Weniger BG, Polymicrobial aetiology of travellers' diarrhoea. *Lancet* 1985; **i**: 381–3.

64   Taylor DN, Echeverria P. Etiology and epidemiology of travellers' diarrhoea in Asia. *Rev Infect Dis* 1986; **8**: S136–41.

65   Reina J, Riera M, Hervas J. Etiology of bacterial gastroenteritis in patients infected by the human immunodeficiency virus. *Rev Clin Esp* 1993; **193**: 428–30. (In Spanish).

66   King GE, Werner SB, Kizer KW. Epidemiology of *Aeromonas* infections in California. *Clin Infect Dis* 1992; **15**: 449–52.

67   Moyer NP. Clinical significance of *Aeromonas* species isolated from patients with diarrhoea. *J Clin Microbiol* 1987; **25**: 2044–48.

68   Holmberg SD, Schell WL, Fanning GR, Wachsmuth IK, Hickman-Brenner FW, Blake PA, Brenner DJ, Farmer III JJ. *Aeromonas* intestinal infections in the United States. *Ann Intern Med* 1986; **105**: 683–9.

69   Hänninen ML, Salmi S, Mattila L, Taipalinen R, Siitonen A. Association of *Aeromonas* spp. with travellers' diarrhoea in Finland. *J Med Microbiol* 1995; **42**: 26–31.

70   Williams TW, Phippin D, Louie TJ. Survey of diarrhoeal illness associated with *Aeromonas hydrophila* in Manitoba. *Abstr Annu Meet Am Soc Microbiol* 1983; C151.

71   Cumberbatch N, Gurwith MJ, Langston C, Sack RB, Brunton JL. Cytotoxic enterotoxin produced by *Aeromonas hydrophila*: relationship of toxigenic isolates to diarrhoeal disease. *Infect Immun* 1979; **23**: 829–37.

72   Burke V, Gracey M, Robinson J, Peck D, Beaman J, Bundell C. The microbiology of childhood gastroenteritis: *Aeromonas* species and other infective agents. *J Infect Dis* 1983; **148**: 68–74.

73   Deodhar LP, Saraswathi K, Varudkar A. *Aeromonas* spp. and their association with human diarrhoeal disease. *J Clin Microbiol* 1991; **29**: 853–6.

74   Jesudason MV, Koshi G. *Aeromonas* species in human septicaemia and diarrhoea. *Indian J Med Res* 1990; **91**: 174–6.

75   Kay BA, Harris JR, Clemens JD, Sack DA, Huq A, Rahman R, Hasan AKJ, Ansaruzzaman M. Comparisons between *Aeromonas hydrophila, Aeromonas sobria* and *Aeromonas caviae* with diarrhoeal patients in Bangladesh. *Abstr*

Annu Meet Am Soc Microbiol 1986; C282: 375.

76 Rahim Z, Kay BA. Incidence of *Aeromonas* spp. and *Plesiomons shigelloides* during a diarrhoeal epidemic in Bangladesh. *J Diarrh Dis Res* 1988; **6**: 144.

77 Kilpatrick ME, Escamilla J, Bourgeois AL, Adkins HJ, Rockhill RC. Overview of four U.S. Navy overseas research studies on *Aeromonas*. *Experientia* 1987; **43**: 365–6.

78 Qadri SMH, Zafar M, Lee GC. Can isolation of *Aeromonas hydrophila* from human feces have any clinical significance? *J Clin Gastroenterol* 1991; **13**: 537–40.

79 Golik A, Modai D, Gluskin I, Schechter I, Cohen N, Eschar J. *Aeromonas* in adult diarrhoea: an enteropathogen or an innocent bystander? *J Clin Gastroenterol* 1990; **12**: 148–52.

80 Gluskin I, Batash D, Shoseyov D, Mor A, Kazak R, Azizi E, Boldur I. A 15 year study of the role of *Aeromonas* spp. in gastroenteritis in hospitalised children. *J Med Microbiol* 1992; **37**: 315–18.

81 Alabi SA, Odugebemi T. Occurrence of *Aeromonas* species and *Plesiomonas shigelloides* in patients with and without diarrhoea in Lagos, Nigeria. *J Med Microbiol* 1990; **21**: 45–8.

82 Nishikawa Y, Kishi T. Isolation and characterization of motile *Aeromonas* from human, food and environmental specimens. *Epidemiol Infect* 1988; **101**: 213–23.

83 Millership SE, Curnow SR, Chattopadhyay B. Faecal carriage rate of *Aeromonas hydrophila*. *J Clin Pathol* 1983; **36**: 920–3.

84 Shread P, Donovan TJ, Lee JV. A survey of the incidence of *Aeromonas* in human feces. *Soc Gen Microbiol Q* 1981; **8**: 184.

85 Cataras M, Buttiaux R. Les *Aeromonas* dans les matierès fecales humanes. *Ann Inst Pasteur Lille* 1965; **16**: 85–9.

86 Figura N, Marri L, Verdiani S, Ceccherini C, Barberi A. Prevalence, species differentiation, and toxigenicity of *Aeromonas* strains in cases of childhood gastroenteritis and in controls. *J Clin Microbiol* 1986; **23**: 595–9.

87 Agger WA, McCormick JD, Gurwith MJ. Clinical and microbiological features of *Aeromonas hydrophila*-associated diarrhoea, *J Clin Microbiol* 1985; **21**: 909–913.

88 San Joaquin VH, Pickett DA. *Aeromonas*-associated gastroenteritis in children. *Ped Infect Dis J* 1988; **7**: 53–7.

89 Kotloff KL, Wasserman SS, Steciak JY, Tall BD, Lasonsky GA, Nair P, Morris JG Jr, Levine MM. Acute diarrhoea in Baltimore children attending an outpatient clinic. *Ped Infect Dis J* 1988; **7**: 753–9.

90 Brake S, Osterhout G. Recovery of *Aeromonas* spp. in the clinical laboratory. *Abst Annu Meet Am Soc Microbiol* 1990; C127: 365.

91 Showalter CA, Quinn PJ, Holton PA, Holcomb LA, Moyer NP. Isolation of *Aeromonas* spp. from human fecal specimens. *2nd International Workshop on* Aeromonas/Plesiomonas 1988; **P16**: 47.

92 Morgan D, Wood L. Is *Aeromonas* sp. a foodborn pathogen? Review of the clinical data. *J Food Saf* 1988; **9**: 59–72.

93 Janda JM, Abbott SL, Morris JG Jr. *Aeromonas, Plesiomonas* and *Edwardsiella*. In Blaser MJ, Smith PD, Ravdin JI, Greenberg HB, Guerrant RL (eds) *Infections of the Gastrointestinal Tract*. Raven Press, New York 1995; 905–17.

94 Janda JM, Reitano M, Bottone EJ. Biotyping of *Aeromonas* isolates as a correlate to delineating species-associated disease spectrum. *J Clin Microbiol* 1984; **19**: 44–7.

95 Carnahan AM, Watsky D, Peeler R. A 3-year retrospective study of clinical *Aeromonas* isolates using Aerokey II. *Abstr Annu Meet Am Soc Microbiol* 1992; **C295**.

96 Phillips DL, Pien FD, Leong TA. Clinical features of *Aeromonas* enteritis in Hawaii. *Trans R Soc Trop Med Hyg* 1990; **84**: 126–8.

97 Pazzaglia G, Escalante JR, Sack RB, Rocca C, Benavides V. Transient intestinal colonization by multiple phenotypes of *Aeromonas* species during the first week of life. *J Clin Microbiol* 1990; **28**: 1842–6.

98 Namdari H, Bottone E. Microbiologic and clinical evidence supporting the role of *Aeromonas caviae* as a pediatric enteric pathogen. *J Clin Microbiol* 1990; **28**: 837–40.

99 Karunakaran T, Sur BG. Characterization of hemolytic activity from *Aeromonas caviae*. *Epidemiol Infect* 1994; **112**: 291–8.

100 Cohen MB. Etiology and mechanisms of acute infectious diarrhoea in infants in the United States. *J Ped* 1991; **4**: 534–9.

101 Bechtel JL, Campos JM. Recovery of *Aeromonas hydrophila* and other enteric bacterial pathogens from pediatric outpatients with diarrhoea. *Abstr Annu Meet Am Soc Microbiol* 1984; C-80.

102 Williams D, Bochini JA Jr, Rambin ED, Black-Payne C. The prevalence of *Aeromonas* species in acute enteritis in children in Northwest Louisiana. *Abstr Annu Meet Am Soc Microbiol* 1988; C119.

103 Baldy LM, Campos M, Combee C. The incidence of enteric pathogens in the Florida panhandle over the past five years. *Abstr Annu Meet Am Soc Microbiol* 1986; C-285.

104 Leggiadro RJ, Szabo K, Epidemiology of *Aeromonas hydrophila* stool isolates in a pediatric population. *Ped Infect Dis* 1986; **5**: 495.

105 De la Morena ML, Van R, Singh K, Brian M, Murray BE, Pickering LK. Diarrhoea associated with *Aeromonas* species in children in day care centers. *J Infect Dis* 1993; **163**: 215–8.

106 Hunt GH, Price EH, Patel U, Messenger L, Stow P, Salter P. Isolation of *Aeromonas* sp from faecal specimens. *J Clin Pathol* 1987; **40**: 1382–84.

107 Wilcox MH, Cook AM, Eley A, Spencer RC. *Aeromonas* spp as a potential cause of diarrhoea in children. *J Clin Pathol* 1992; **45**: 959–63.

108 Geiss HK, Fogel W, Sonntag H-G. Isolation rates of *Aeromonas* species in stool specimens of healthy and diarrhoeic patients. *Immun Infekt* 1988; **16**: 115–17.

109 Mégraud F. Incidence and virulence of *Aeromonas* species in feces of children with diarrhoea. *Eur J Clin Microbiol* 1986; **5**: 311–16.

110 Reina J, Hervas J, Serra A, Bonell N. Clinical and microbiologic characteristics of 282 patients with mesophilic *Aeromonas* isolated from feces. *Enferm Infec Microbiol Clin* 1993; **11**: 366–72.

111 Pazzaglia G, Salazar E, Yi A, Chea E. Hospital case-control study of *Aeromonas*-associated diarrhoea in Peruvian infants. *Abstr Annu Meet Am Soc Microbiol* 1989; C-252.

112 Figueroa G, Galeno H, Soto V, Troncoso M, Hinrichsen V, Yudelevich A. Enteropathogenicity of *Aeromonas* species isolated from infants: a cohort study. *J Infect* 1988; **17**: 205–13.

113 Ghanem EH, Mussa ME, Eraki HM. *Aeromonas*-associated gastroenteritis in Egypt. *Zentralbl Mikrobiol* 1993; **148**: 441–7.

114 Gracey M, Burke V, Rockhill RC, Suharyono, Sunoto. *Aeromonas* species as enteric pathogens. *Lancet* 1982; **i**: 223–4.

115 Santoso H, Agung IGN, Robinson J, Gracey M. Faecal *Aeromonas* spp. in Balinese children. *J Gastroenterol Hepatol* 1986; **1**: 115–18.

116 Bhatia BD, Agarwal DK, Singla PN, Sanyal SC. Environmental factors and microbial flora in hospitalized children with diarrhoea. *J Ind Acad Ped* 1980; **17**: 354–60.

117 Kim KH, Suh IS, Kim JM, Kim CW, Cho YJ. Etiology of childhood diarrhoea in

Korea. *J Clin Microbiol* 1989; **27**: 1192–6.
118 Palfreeman SJ, Waters LK, Norris M. *Aeromonas hydrophila* gastroenteritis. *Aust NZ J Med* 1983; **13**: 524–5.
119 Guglielmetti P, Figura N, Zancki A. *Aeromonas* enteritis: a serological study. *Microecol Ther* 1989; **18**: 263–6.
120 Gosling PJ, Turnbull PCB, Lightfoot NF. Clinical and serological aspects of *Aeromonas* associated gastroenteritis. *Abst Annu Meet Am Soc Microbiol* 1988; C-114.
121 Khaitovich AB, Galtseva GV. Serological diagnosis of diarrhoea caused by some representatives of the genera *Vibrio* and *Aeromonas*. *J Hyg Epidemiol Microbiol Immunol* 1983; **27**: 211–18.
122 Kuijper EJ, van Alphen L, Peeters MF, Brenner DJ. Human serum antibody response to the presence of *Aeromonas* spp in the intestinal tract. *J Clin Microbiol* 1990; **28**: 584–90.
123 Jiang ZD, Nelson AC, Mathewson JJ, Ericsson CD, DuPont HL. Intestinal secretory immune response to infection with *Aeromonas* species and *Plesiomonas shigelloides* among students from the United States in Mexico. *J Infect Dis* 1991; **164**: 979–82.
124 Chakraborty TM, Montenegro MA, Sanyal SC, Helmuth R, Bulling E, Timmis KN. Cloning of enterotoxin gene from *Aeromonas hdyrophila* provides conclusive evidence of production of a cytotoxic enterotoxin. *Infect Immun* 1984; **46**: 435–41.
125 Chopra AK, Houston CW. Purification and partial characterization of a cytotoxic enterotoxin produced by *Aeromonas hydrophila*. *Can J Microbiol* 1989; **35**: 719–27.
126 Gosling PJ, Turnbull PCB, Lightfoot NF, Pether JVS, Lewis RJ. Isolation and purification of *Aeromonas sobria* cytotonic enterotoxin and β-haemolysin. *J Med Microbiol* 1993; **38**: 227–34.
127 McCardell BA, Madden JM, Kotharg MH, Sathyanoorthy V. Purification and characterization of a CHO cell elongating toxin from *Aeromonas*. *Abst Annu Meet Am Soc Microbiol* 1994; 3–83.
128 Chopra AK, Pham R, Houston CW. Cloning and expression of putative cytotoxic enterotoxin-encoding genes from *Aeromonas hydrophila*. *Gene* 1994; **137**: 87–91.
129 Burke V, Robinson J, Atkinson HM, Dibley M, Berry RJ, Gracey M. Exotoxins of *Aeromonas hydrophila*. *Austr J Exper Biol Med Sci* 1981; **59**: 753–61.
130 Ljungh Å. *Aeromonas* – Toxins and other virulence factors. *Experientia* 1987; **43**: 367–8.
131 Kokka RP, Janda JM, Oshiro L, Altwegg M, Shimada T, Sakazaki R and Brenner DJ. Biochemical and genetic characterization of autoagglutinating phenotypes of *Aeromonas* associated with invasive and non-invasive disease. *J Infect Dis* 1991; **161**: 890–4.
132 Lawson MA, Burke V, Chang BJ. Invasion of Hep-2 cells by fecal isolates of *Aeromonas hydrophila*. *Infect Immun* 1985; **47**: 680–3.
133 Nishikawa Y, Hase A, Ogawasara J, Scotland SM, Smith HR, Kimura T. Adhesion to and invasion of human colon carcinoma Caco-2 cells by *Aeromonas* strains. *J Med Microbiol* 1994; **40**: 55–61.
134 Neves MS, Neves MP, Milhomen AM. *Aeromonas* species exhibit aggregative adherence to HEp-2 cells. *J Clin Microbiol* 1994; **32**: 1130–1.
135 Thornley JP, Eley A, Shaw JG. *Aeromonas caviae* exhibits aggregative adherence to HEp-2 cells. *J Clin Microbiol* 1994; **32**: 2631–2.
136 Atkinson HM, Adams D, Sawas RS, Trust TJ. *Aeromonas* adhesion antigens. *Experientia* 1987; **43**: 372–9.

137 Burke V, Cooper M, Robinson J, Gracey M, Lesmana M, Echeverria P, Janda JM. Hemagglutination patterns of *Aeromonas* spp. in relation to biotype and source. *J Clin Microbiol* 1984; **19**: 39–43.

138 Pepe CM, Eklund MW, Strom MS. Cloning of an *Aeromonas hydrophila* type IV pilus biogenesis gene cluster: complementation of pilus assembly functions and characterization of a type IV leader peptidase/*N*-methyltransferase required for extracellular protein secretion. *Mol Microbiol* 1996; **19**: 857–69.

139 Burke V, Robinson J, Beaman J, Gracey M, Lesmana M, Rockhill R, Echeverria P, Janda JM. Correlation of enterotoxicity with biotype in *Aeromonas* spp. *J Clin Microbiol* 1983; **18**: 1196–1200.

140 Daily OP, Joseph SW, Coolbaugh JC, Walker RI, Merrell BR, Rollins DM, Seidler RJ, Colwell RR, Lissner CR. Association of *Aeromonas sobria* with human infection. *J Clin Microbiol* 1981; **13**: 769–77.

141 Barer MR, Millership SE, Tabaqchali S. Relationship of toxin production to species in the genus *Aeromonas*. *J Med Microbiol* 1986; **22**: 303–9.

142 Turnbull KB, Lee JV, Miliotis MD, van de Walle S, Koornhof HJ, Jeffery L, Bryant TN. Toxin production in relation to taxonomic grouping and source of isolation of *Aeromonas* species. *J Clin Microbiol* 1984; **19**: 175–80.

143 Kindschuh M, Pickering LK, Clearly TG, Ruiz-Palacois G. Clinical and biochemical significance of toxin production by *Aeromonas hydrophila*. *J Clin Microbiol* 1987; **25**: 48–51.

144 Watson IM, Robinson JO, Burke V, Gracey M. Invasiveness of *Aeromonas* spp. in relation to biotype, virulence factors, and clinical features. *J Clin Microbiol* 1985; **22**: 48–51.

145 Ljung Å, Popoff M, Wadström T. *Aeromonas hydrophila* in acute diarrhoeal disease: detection of enterotoxin and biotyping of strains. *J Clin Microbiol* 1977; **6**: 96–100.

146 Millership SE, Barer MR, Tabaqchali S. Toxin production by *Aeromonas* spp. from different sources. *J Med Microbiol* 1986; **22**: 311–14.

147 Pin C, Marin ML, Selgas D, Garcia ML, Tormo J, Casas C. Differences in production of several extracellular virulence factors in clinical and food *Aeromonas* spp. strains. *J Appl Bacteriol* 1995; **78**: 175–9.

148 Krovacek K, Paquale V, Baloda SB, Soprano V, Conte M, Dumontet S. Comparison of putative virulence factors in *Aeromonas hydrophila* strains isolated from the marine environment and human diarrhoeal cases in Southern Italy. *Appl Environ Microbiol* 1994; **60**: 1379–82.

149 Burke V, Robinson J, Atkinson HM, Gracey M. Biochemical characteristics of enterotoxigenic *Aeromonas* spp. *J Clin Microbiol* 1982; **15**: 48–52.

150 Gosling PJ. Biochemical characteristics, enterotoxigenicity and susceptibility to antimicrobial agents of clinical isolates of *Aeromonas* species encountered in the western region of Saudi Arabia. *J Med Microbiol* 1986; **22**: 51–5.

151 Carnahan A, Hammontree L, Bourgeois L, Joseph SW. Pyrazinamidase as a phenotypic marker for several *Aeromonas* spp. isolated from clinical specimens. *J Clin Microbiol* 1990; **28**: 391–2.

152 Merino S, Benedi V-J, Tomás JM. *Aeromonas hydrophila* strains with moderate virulence. *Microbios* 1989; **59**: 165–73.

153 Picard B, Goullet P. Epidemiological complexity of hospital *Aeromonas* infections revealed by electrophoretic typing of esterases. *Epidemiol Inf* 1987; **98**: 5–14.

154 Bartkova G, Ciznar I. Adherence pattern of non-piliated *Aeromonas hydrophila* strains to tissue cultures. *Microbios* 1994; **77**: 47–55.

155 Moyer NP, Martinetti G, Lüthy-Hottenstein J, Altwegg M. Value of rRNA gene restriction patterns of *Aeromonas* spp. for epidemiological investigations. *Curr*

Microbiol 1992; **24**: 15–21.

156 Moyer NP, Martinetti Luccini G, Holcomb LA, Hall NH, Altwegg M. Application of ribotyping for differentiating aeromonads isolated from clinical and environmental sources. *Appl Environ Microbiol* 1992; **58**: 1940–4.

157 Carey PE, Eley A, Wilcox MH. Assignment of a chemiluminescent universal probe for taxonomical and epidemiological investigations of *Aeromonas* spp. isolates *J Clin Pathol* 1994; **47**: 642–6.

158 Namdari H, Bottone E. Correlation of the suicide phenomenon in *Aeromonas* species with virulence and enteropathogenicity. *J Clin Microbiol* 1988; **26**: 2615–19.

159 Janda JM, Guthertz LS, Kokka RP, Shimada T. *Aeromonas* species in septicemia: laboratory characteristics and clinical observations. *Clin Infect Dis* 1994; **19**: 77–83.

160 Merino S, Camprubi S, Tomas JM. Detection of *Aeromonas hydrophila* serogroup 0:34 in faeces using an enzyme-linked immunosorbent assay. *J Diarrh Dis Res* 1993; **11**: 30–4.

161 Schubert RH. Aeromonads and their significance as potential pathogens in water. *Soc Appl Bacteriol Symp Ser* 1991; **29**: 1313–55.

162 Kirov SM. The public health significance of *Aeromonas* spp. in foods. *Int J Food Microbiol* 1993; **20**: 179–98.

163 Knøchel S, Jeppesen C. Distribution and characteristics of *Aeromonas* in food and drinking water in Denmark. *Int J Food Microbiol* 1990; **10**: 317–22.

164 Picard B, Goullet P. Seasonal prevalence of nosocomial *Aeromonas hydrophila* infection related to *Aeromonas* in hospital water. *J Hosp Infect* 1987; **10**: 152–5.

165 Millership SE, Stephenson JR, Tabaqchali S. Epidemiology of *Aeromonas* species in a hospital. *J Hosp Infect* 1988; **11**: 169–75.

166 San Joaquin VH, Pickett DA, Welch DF, Finkhouse BD. *Aeromonas* species in aquaria: a reservoir of gastrointestinal infections? *J Hosp Infect* 1989; **13**: 173–7.

167 Schets FM, Medema GJ. Prevention of toxicity of metal ions to *Aeromonas* and other bacteria in drinking-water samples using nitrilotriacetic acid (NTA) instead of ethylenediaminetetraacetic acid (EDTA). *Lett Appl Microbiol* 1993; **16**: 75–6.

168 Coolbaugh JC, Daily OP, Joseph SW, Colwell RR. Bacterial contamination of divers during training exercises in coastal waters. *Marine Technol Soc J* 1981; **15**: 15–21.

169 Seidler RJ, Allen DA, Lockman H, Colwell RR, Joseph SW, Daily OP. Isolation, enumeration, and characterization of *Aeromonas* from polluted waters encountered in diving operations. *Appl Environ Microbiol* 1980; **39**: 1010–18.

170 LeChevallier MW, Evans TM, Seidler RJ, Daily OP, Merrell BR, Rollins DM, Joseph SW. *Aeromonas sobria* in chlorinated drinking water supplies. *Microb Ecol* 1982; **8**: 325–33.

171 Rahim Z, Sanyal SC, Aziz KMS, Huq MI, Chowdhury AA. Isolation of enterotoxigenic, hemolytic and antibiotic-resistant *Aeromonas hydrophila* strains from infected fish in Bangladesh. *Appl Environ Microbiol* 1984; **48**: 865–7.

172 Fang G, Araujo V, Guerrant RL. Enteric infections associated with exposure to animals or animal products. *Infect Dis Clin North Am* 1991; **5**: 681–701.

173 Hudson JA, Mott SJ. Growth of *Listeria monocytogenes, Aeromonas hydrophila* and *Yersinia enterocolitica* on cold-smoked salmon under refrigeration and mild temperature abuse. *Food Microbiol* 1990.

174 Abeyta C Jr, Wekell MM. Potential sources of *Aeromonas hydrophila*. *J Food Saf* 1988; **9**: 11–22.

175 Sorvillo F, Mascola L, Kilman L. Bondage, dominance, irrigation, and *Aeromonas hydrophila:* California dreamin'. *JAMA* 1989; **262**: 697–8.

176 Majeed KV, MacRae IC. Experimental evidence for toxin production by *Aeromonas hydrophila* and *Aeromonas sobria* in a meat extract at low temperatures. *Int J Food Microbiol* 1991; **12**: 181–8.

177 Beuchat LR. Behaviour of *Aeromonas* species at refrigeration temperatures. *Int J Food Microbiol* 1991; **13**: 217–24.

178 Callister S, Agger W. Enumeration and characterization of *Aeromonas hydrophila* and *Aeromonas caviae* isolated from grocery store produce. *Appl Environ Microbiol* 1987; **53**: 249–53.

179 Gluender G. Occurrence of *Aeromonas hydrophila* in birds. *Zentral Veterinaermed Reihe* 1988; **35**: 331–7.

180 Freitas AC, Nunes MP, Milhomem AM, Ricciardi ID. Occurrence and characterization of *Aeromonas* species in pasteurized milk and white cheese in Rio de Janeiro, Brazil. *J Food Prot* 1992; **56**: 62–5.

181 Buchanan RL, Palumbo SA. *Aeromonas hydrophila* and *Aeromonas sobria* as potential food poisoning species: a review. *J Food Saf* 1985; **7**: 15–29.

182 Burke V, Gracey M. *Aeromonas* species in human diarrhoeal disease. *J Gastroenterol Hepatol* 1986; **1**: 237–49.

183 Holmberg SD, Farmer JJ III. *Aeromonas hydrophila* and *Plesiomonas shigelloides* as causes of intestinal infections. *Rev Infect Dis* 1987; **6**: 635–9.

184 Sack DA. Intestinal infection with *Aeromonas* and *Plesiomonas* in humans. *2nd International Workshop on* Aeromonas *and* Plesiomonas 1988; 20–1.

185 Gross RJ. *Aeromonas* in enteric infections: Introductory comments. *Experientia* 1987; **43**: 363.

186 Gray J. *Aeromonas* in the gut. *J Infect* 1987; **15**: 197–200.

187 Altwegg M, Geiss HK. *Aeromonas* as a human pathogen. *CRC Crit Rev Microbiol* 1989; **16**: 253–86.

188 Janda JM, Bottone EJ, Reitano M. *Aeromonas* species in clinical microbiology: significance, epidemiology, and speciation. *Diagn Microbiol Infect Dis* 1983; **1**: 221–8.

189 Janda JM, Duffey PS. Mesophilic aeromonads in human disease: current taxonomy, laboratory identification, and infectious disease spectrum. *Rev Infect Dis* 1988; **10**: 980–97.

190 Lederberg J, Shope RE, Oaks SC Jr (eds). *Emerging Infections – Microbial Threats to Health in the United States.* Washington, DC: National Academy Press, 1992; 65.

191 Zeng-shan L, Guilian C, Shumei F. An epidemic of food poisoning by *Aeromonas hydrophila. Chinese J Prev Med* 1988; **22**: 333–4.

13 Outlook

BRIAN AUSTIN
Heriot-Watt University, Edinburgh, Scotland

MARTIN ALTWEGG
University of Zürich, Zürich, Switzerland

PETER J. GOSLING
Department of Health, London, England

Predicting the future is notoriously unreliable. However, it seems reasonable to extrapolate current trends to consider likely developments in the foreseeable future. The emergence of aeromonads as human pathogens has inevitably led to increased research activity. Yet clarification is still needed about the precise pathogenic mechanisms of aermonads in human disease, and especially their role in diarrhoea. Nevertheless, one result of the enhanced research activity has been the description of many new species. However, many workers fail to equate isolates to hybridization groups (HG), and this shortcoming must be rectified in future publications. Notwithstanding, it is predicted that other new taxa remain to be described, particularly from the natural environment. In addition, the taxonomic status of so-called 'atypical' isolates, for example of *Aeromonas salmonicida*, needs to be clarified. Perhaps new subspecies will be proposed to accommodate these unusual or 'atypical' cultures. While on this theme, perhaps the taxonomic status of *Haemophilus piscium* will finally be resolved. It remains for future research to work towards congruence in taxonomic methods. The incongruence in data from phenotypic and genotypic methods is unhelpful. There must thus be moves towards a unified taxonomy.

Improvements in taxonomy will inevitably aid the identification processes. However, it should be emphasized that there is often a reticence to determine the true taxonomic status of isolates. This current trend of scientists to present the results of a comparatively few biochemical tests must cease. After all, with all the developments in the taxonomy of *Aeromonas*, there may be justifiable confusion in just what is meant by *A. hydrophila*. Do authors mean the entire genus, HG 1, or mesophilic strains of HG 1 to HG 3? It is certainly argued that reliable and rapid techniques are sorely needed to ensure the effective identification of aeromonads. The roles of serology and molecular biology are particularly important. Since the advent of monoclonal antibodies and increasingly sensitive assays, serological techniques have increased in value. The development of enzyme-linked immunosorbent

The Genus Aeromonas. Edited by B. Austin, M. Altwegg, P.J. Gosling and S. Joseph
© 1996 John Wiley & Sons Ltd

assays, imunohistochemistry and dot blot techniques has been invaluable. Now, in the age of molecular biology, gene probe technology must provide the next significant breakthrough in identification and diagnosis, particularly with regard to simplified methodologies, suitable for use in routine microbiological laboratories.

The current momentum of ecology-based research seems likely to be maintained. However, the precise role of aeromonads in the aquatic environment remains to be established. Despite conflicting results, it is unclear whether or not *A. salmonicida* can survive in a pathogenic form in the aquatic environment. Perhaps effective selective media for the pathogen would clarify aspects of ecology and also the role of so-called carrier fish, that is those that contain the pathogen but do not express the disease.

What is the significance (if any) of the increasing recognition of aeromonads in food and drinking water? Do the organisms from these sources pose any threat to human health? Answers to these questions are certainly important.

Clarification is needed about the nature of the pathogenic mechanism of aeromonads in the many syndromes with which they have been associated. Contradictory evidence exists, which may reflect differences between methodologies and the inherent distinction between strains (of the same taxon). In this respect, the value of using single isolates in pathogenicity studies is questioned. Certainly, there is confusion over the precise role of toxins, such as enterotoxins (do they actually occur in aeromonads?) in disease. Undoubtedly, future research will clarify the role of these toxins.

Finally, disease control should be mentioned. Advances in vaccine development for fish against *A. salmonicida* have led to seemingly useful products. What about other aeromonad taxa encountered by fish, humans and other animals – will these ever warrant detailed vaccine development programmes or will there be an ongoing reliance on antimicrobial compounds?

Future research may indeed offer an interesting insight into these fascinating bacteria.

Index